Operations Research and Artificial Intelligence: The Integration of Problem-Solving Strategies

Operations Research and Artificial Intelligence: The Integration of Problem-Solving Strategies

edited by
Donald E. Brown
Chelsea C. White, III
University of Virginia

Kluwer Academic Publishers
Boston/Dordrecht/London

Distributors for North America:
Kluwer Academic Publishers
101 Philip Drive
Assinippi Park
Norwell, Massachusetts 02061 USA

Distributors for all other countries:
Kluwer Academic Publishers Group
Distribution Centre
Post Office Box 322
3300 AH Dordrecht, THE NETHERLANDS

Library of Congress Cataloging-in-Publication Data
Operations research and artificial intelligence : the integration of
  problem-solving strategies / edited by Donald E. Brown, Chelsea C.
  White, III
      p. cm.
    Includes bibliographical references and index.
    ISBN 0-7923-9106-3
    1. Artificial intelligence.   2. Operations research.   3. Decision
  -making.   I. Brown, Donald E.   II. White, Chelsea C., 1945–
  Q335.064   1990
  006.3—dc20   90-39443
                                                                      CIP

Copyright © 1990 by Kluwer Academic Publishers

All rights reserved. No part of this publication may be reproduced,
stored in a retrieval system or transmitted in any form or by any
means, mechanical, photocopying, recording, or otherwise, without
the prior written permission of the publisher, Kluwer Academic
Publishers, 101 Philip Drive, Assinippi Park, Norwell, Massachusetts
02061.

Printed in the United States of America

Contents

Contributors		vii
Preface		ix
Introduction D.E. Brown and C.C. White, III		1
I.	Search	7
	Toward the Modeling, Evaluation and Optimization of Search Algorithms O. Hansson, G. Holt and A. Mayer	11
	Genetic Algorithms Applications to Set Covering and Traveling Salesman Problems G.E. Liepins, M.R. Hilliard, J. Richardson and M. Palmer	29
	Discovering and Refining Algorithms Through Machine Learning M.R. Hilliard, G.E. Liepins and M. Palmer	59
II.	Uncertainty Management	79
	Using Probabilities as Control Knowledge to Search for Relevant Problem Models in Automated Reasoning R.K. Bhatnagar and L.N. Kanal	83
	On the Marshalling of Evidence and the Structuring of Argument D.A. Schum	105
	Hybrid Systems for Failure Diagnosis E. Paté-Cornell and H. Lee	141
III.	Imprecise Reasoning	167
	Default Reasoning Through Integer Linear Programming S.D. Post and C.E. Bell	171
	The Problem of Determining Membership Values in Fuzzy Sets in Real World Situations E. Triantaphyllou, P.M. Pardalos and S.H. Mann	197

IV.	Decision Analysis and Decision Support	215
	Applications of Utility Theory in Artificial Intelligence Research P.H. Farquhar	219
	A Multicriteria Stratification Framework for Uncertainty and Risk Analysis J. Barlow and F. Glover	237
	Dispute Mediation: A Computer Model K. Sycara	249
V.	Mathematical Programming and AI	275
	Eliciting Knowledge Representation Schema for Linear Programming Formulation M.M. Sklar, R.A. Pick, G.B. Vesprani and J.R. Evans	279
	A Knowledge Base for Integer Programming—A Meta-OR Approach F. Zahedi	317
VI.	Performance Analysis and Complexity Management of Expert Systems	369
	Validator, A Tool for Verifying and Validating Personal Computer Based Expert Systems M. Jafar and A.T. Bahill	373
	Measuring and Managing Complexity in Knowledge-Based Systems: A Network and Mathematical Programming Approach D.E. O'Leary	387
	Pragmatic Information-Seeking Strategies for Expert Classification Systems L.A. Cox, Jr.	427
VII.	Applications	449
	A Knowledge- and Optimization-Based Approach to Scheduling in Automated Manufacturing Systems A. Kusiak	453
	An Integrated Management Information System for Wastewater Treatment Plants W. Lai, P.M. Berthouex and D. Hindle	481
About the Authors		497
Index		507

Contributors

Professor A. Terry Bahill, Systems and Industrial Engineering, University of Arizona, Tucson, AZ 85721

Professor Colin E. Bell, College of Business Administration, Department of Management Sciences, The University of Iowa, Iowa City, IA 52242

Professor P.M. Berthouex, Department of Civil & Environmental Engineering, University of Wisconsin-Madison, Madison, WI 53706

Mr. Tony Cox, US WEST Advanced Technologies, 6200 South Quebec Street, Englewood, CO 80111

Professor Peter H. Farquhar, Graduate School of Industrial Administration, Carnegie-Mellon University, Pittsburgh, PA 15213-3890

Professor Fred Glover, Graduate School of Business, University of Colorado, Boulder, CO 80309-0419

Dr. Michael R. Hilliard, Research Associate, Martin Marietta Energy Systems, Inc., P.O. Box 2008, Oak Ridge, TN 37831-6366

Professor Laveen N. Kanal, Department of Computer Science, The University of Maryland, College Park, MD 20742

Professor Andrew Kusiak, Industrial & Management Engineering, The University of Iowa, Iowa City, IA 52242

Dr. Gunar Liepins, Oak Ridge National Laboratory, P.O. Box 2008, Oak Ridge, TN 37831

Professor Andrew Mayer, Computer Science Division, University of California at Berkeley, Berkeley, CA 94720

Professor Daniel E. O'Leary, Graduate School of Business, University of Southern California, Los Angeles, CA 90089-1421

Professor Panos M. Pardalos, Department of Computer Science, The Pennsylvania State University, University Park, PA 16802

Professor M.E. Pate-Cornell, Industrial Engineering & Engineering Management, Stanford University, Stanford, CA 94305

Professor David A. Schum, Operations Research & Applied Statistics, George Mason University, Fairfax, VA 22030

Professor Margaret M. Sklar, Department of Management, Marketing, and CIS, School of Business, Northern Michigan University, Marquette, MI 49855

Professor Katia Sycara, The Robotics Institute, Carnegie-Mellon University, Pittsburgh, PA 15213-3890

Professor Fatemeh Zahedi, Management Sciences Department, University of Massachusetts - Boston, Harbor Campus, Boston, MA 02125-3393

Preface

The purpose of this book is to introduce and explain research at the boundary between two fields that view problem solving from different perspectives. Researchers in operations research and artificial intelligence have traditionally remained separate in their activities. Recently, there has been an explosion of work at the border of the two fields, as members of both communities seek to leverage their activities and resolve problems that remain intractable to pure operations research or artificial intelligence techniques. This book presents representative results from this current flurry of activity and provides insights into promising directions for continued exploration.

This book should be of special interest to researchers in artificial intelligence and operations research because it exposes a number of applications and techniques, which have benefited from the integration of problem solving strategies. Even researchers working on different applications or with different techniques can benefit from the descriptions contained here, because they provide insight into effective methods for combining approaches from the two fields. Additionally, researchers in both communities will find a wealth of pointers to challenging new problems and potential opportunities that exist at the interface between operations research and artificial intelligence.

In addition to the obvious interest the book should have for members of the operations research and artificial intelligence communities, the papers here are also relevant to members of other research communities and development activities that can benefit from improvements to fundamental problem solving approaches. Included in this category are engineers and physical and social scientists, who require improved decision making techniques or greater understanding of processes involved in problem solving in complex domains.

Most of the papers in this book were presented at the Joint National Meetings of the Operations Research Society of America and The Institute for Management Science. Over the past three years there were roughly 400 papers presented at these meetings that incorporated results from artificial intelligence. Officers and council members of the Artificial Intelligence Technical Section of the Operations Research Society of America decided to organize and present significant results from among these papers. It was decided early in this process, that rather than simply collect papers and bind them, a formal review process should be instituted. Hence, the papers collected here represent the results of a two tiered review process, designed to distill and present the more significant results from these meetings.

We acknowledge the support received in the preparation of this work from the members of the Artificial Intelligence Technical Section of the Operations Research Society of America. The project was particularly encouraged by the first chairperson of the Technical Section (at the time it was a Special Interest Group), Frank Morisano, and received the complete support of his successors, Jerry May and Gunar Liepens. We also appreciate the support of the referees, who assisted us in reviewing the submitted papers. We experienced a very high return rate on review requests for this volume, which made it much easier to compile the papers. Finally, we owe special thanks to Annelise Tew, who assisted us throughout the preparation of this book: maintaining files, calling referees, calling authors, reviewing formats, and generally ensuring our plans were well executed.

Donald E. Brown
Chelsea C. White, III

Charlottesville, Virginia

Operations Research and Artificial Intelligence: The Integration of Problem-Solving Strategies

Introduction

Donald E. Brown and Chelsea C. White, III
Department of Systems Engineering
University of Virginia
Charlottesville, Va. 22901

This book contains papers that demonstrate some of the important results from integrating problem solving techniques typically associated with operations research (OR) with those typically associated with artificial intelligence (AI). The papers presented here exemplify what we believe is a stage in the natural evolution of both fields toward more powerful strategies of problem solving. These strategies will find usefulness for both decision aiding, a goal of OR, and automatic decision making, which is the pursuit of AI.

Historically, the OR and AI research communities worked in relative isolation from one another. On the one hand this separation is remarkable, because both disciplines are deeply concerned with questions of human problem solving and decision making, both are highly computer dependent, and both share some common conceptual frameworks (e.g graphs, probability theory, and heuristics). On the other hand, the separation is understandable in that OR has sought optimal methods in decision making through formal mathematical structures. AI has emphasized goal seeking and the use of workable, although suboptimal, strategies more closely associated with human performance. While the fields do share some common conceptual frameworks, there are many more that are distinct to each field. For example, AI has a strong foundation in logic with work that emphasizes automatic theorem proving, while OR has instead emphasized the mathematics of optimization and the quantification of preference through utility and value functions.

These differences aside, the complexity of many, if not most real-world decision problems has exposed the limitations of OR and AI tools, and has caused the two communities to seek solution approaches that integrate these tools. Perhaps the most significant call for integrative approaches to complexity came from Simon (1987), who stressed the common problem solving foundations of the fields. From the OR perspective a more formal organizational statement supporting the general contention that significant advances in problem solving strategies are attainable through the synthesis of tools and techniques from AI and OR is in the CONDOR (1988) report.

This book provides evidence of both the trend in each community toward integrative approaches, and the advantages from pursuing these approaches. The book is organized around the major areas of contemporary research in what we call the OR/AI interface - the topics of mutual interest to both communities. The specific areas reported in this book, which constitute its sections, are search, uncertainty management, imprecise reasoning, decision analysis and decision support, mathematical programming and AI, performance analysis and complexity management of expert systems, and applications. The first five are established areas with considerable supporting literature in their respective fields and more recently in the OR/AI interface. The sixth area, performance analysis and complexity management of expert systems, is of more recent origin reflecting concerns that have become evident with the growth and expanding applications base for expert systems. The applications section provides evidence that work in synthesizing OR and AI tools is of significance to the resolution of important contemporary problems.

In the remainder of this section we place the papers in this book in the context of exemplary work within the OR/AI interface. The purpose is to both introduce the contributions and also to provide a context for the issues and approaches they discuss. Perforce the discussion is brief and the references to relevant literature severely constrained, but this is done with a view toward allowing the reader to quickly locate areas of strong interest. Additional details regarding each of the papers is contained in introductory material at the beginning of each section. The discussion here is organized around the six major sections of the book beginning with search.

Search is the systematic exploration of a state space for recognizable goal states. While search is a traditional concern of AI, it is closely related to optimal seeking and heuristic methods in OR. Glover(1985) explored various AI-based search procedures in the context of optimization and integer programming. One particularly important set of AI techniques, genetic algorithms (Holland 1975, and Davis 1987), uses operators which mimic biological reproduction and mutation to perform global optimization. Genetic algorithms are now in active use by members of both fields. Two of the papers, Liepens, et al., and Hilliard, et al., in the search section of this book explore the use of genetic algorithms for set covering, the traveling salesman problem, and scheduling. In the latter paper, Hilliard et al. also demonstrate the important connection between search and machine learning through the use of classifier systems based on genetic algorithms. The other paper in the search section of this book by Hansson, et al., examines the use of decision analysis techniques to improve the performance of search algorithms. This is an example of the use of OR methods to improve the effectiveness of traditional AI approaches.

Both AI and OR require methods that effectively cope with uncertainty and imprecision. The importance of this topic in both communities is evident in the large number of workshops and special sessions devoted to the area, and in the resulting extensive literature. Examples of the breadth and depth of the work in this area are Kanal and Lemmer (1986), Schacter (1986), Dalkey (1986), Spiegelhalter (1986), and Pearl (1988). This book contains five contributions to uncertain and imprecise reasoning, which explore key issues of interest to both OR and AI researchers in the area. The contributions are divided into two sections: Uncertainty Management and Imprecise Reasoning. The first paper in the uncertainty management section by Bhatnagar and Kanal adopts the unique and important perspective of examining causal models that support interesting inferences rather than the more typical perspective of building inference structures based on preestablished causal models. The Schum paper which follows explores the issues of combining evidence to support reasoning tasks. The final paper of the section by Pate-Cornell demonstrates that a hybrid system using probabilistic methods for uncertainty management, optimization techniques, and heuristic procedures can provide effective decision support.

In addition to uncertainty, most real-world problems are plagued by imprecision. One aspect of this imprecision involves the lack complete information required for decision making. Some form of default reasoning is typically necessary. The paper by Post and Bell in the section on imprecise reasoning applies integer programming to the problem and the result is the least exception logic model for default reasoning. This approach specifically aids in the modeling of default rules. Fuzzy sets represent an approach to other aspects of imprecision. The second paper in the imprecise reasoning section examines a problem at the foundation of any application of fuzzy sets: the determination of the membership functions. This paper specifically investigates OR approaches to the problem and points to directions for additional work.

Decision analysis is a normative OR approach to alternative selection under conditions of risk and uncertainty. It is an area with numerous successes, but there are also numerous opportunities for improvement through the effective integration of AI tools. For example, Kowalik and Keeney (1986) report several projects that merge expert systems and mathematically-based models. Horvitz, et al. (1988) examines the integration of decision theory and AI in terms of representation, inference, knowledge engineering, and explanation. This book contains three papers on the integration of AI with decision analysis and decision support. The paper by Farquhar shows the applicability of utility theory to intelligent systems. Barlow and Glover in the same section provide the foundation for a decision aid employing multicriteria optimization and user feedback to obtain successively improving solutions. The Sycara paper, which concludes the section, deals with conflict resolution in negotiation/mediation and is important to the rapidly growing field of group decision support.

Mathematical programming is in some sense a cornerstone of OR and is arguably the most widely used and visible extension of the field. Yet, there is a common concern within the OR community that the applications base can be expanded still further with improvements to modeling and aids to model understanding. Geoffrion's (1987) structured modeling, Murphy and Stohr's (1985) use of expert systems, and Greenberg's (1989) neural network approach all exemplify OR/AI responses to this problem. Two papers in this volume examine important issues in this area. The Sklar, et al. paper contributes to model formulation by examining the classification and knowledge representation schemes used by experts in linear programming. The Zahedi paper uses AI techniques to capture lessons learned from applications of integer programming in order to provide guidance for future applications.

The section on performance analysis and complexity management of expert systems represents a relatively new direction within the OR/AI interface. The first two papers in this section are important to developers of expert systems. The Jafar and Bahill paper provides an approach to validation of expert systems, which is of growing concern as the number of fielded systems increases. The O'Leary paper which follows develops a network representation of a rule base as a vehicle for complexity management in the development of expert systems. The final paper in this section concerns the development of classifier systems. In this paper, Cox presents an approach to reducing the cost of classification in real-world situations where the information needed for classification can be costly to acquire.

The final section in this book describes important applications that illustrate the effective integration of techniques from AI and OR. The paper by Kusiak shows the effective combination of a heuristic procedure and a knowledge-based system in scheduling problems. The Lai, et al. paper describes a hybrid approach to aid in the operation and control of wastewater treatment facilities.

All of this recent work, as reported in this book and elsewhere, strongly suggests that many in the OR community now recognize that formal models are not completely capturing the qualitative and knowledge laden components of unstructured decision problems. At the same time this work also shows that there are members of the AI community who appreciate the generality and extensibility of prescriptive approaches to decision making. As a result, OR and AI practitioners and researchers are looking at the opposite community for concepts and techniques that will enhance the strengths and reduce the limitations inherent in the problem solving tools of their own discipline.

References

CONDOR (1988), "Operation research: the next decade," *Operations Research*, 36, 619-637.

Dalkey, N.C. (1986), "Models vs. inductive inference for dealing with probabilistic knowledge," Proceedings of the Workshop on Uncertainty in Artificial Intelligence, University of Pennsylvania, August 8-10, 63-70.

Davis, L., ed. (1987), **Genetic Algorithms and Simulated Annealing**, Morgan Kaufmann, Los Angeles

Geoffrion, A.M. (1987), "An introduction to structured modeling," *Management Science*, 33, 547-548.

Glover, F. (1985), "Future paths for integer programming and links to artificial intelligence," CAAI Report 85-8, Graduate School of Business, University of Colorado, Boulder, Co.

Greenberg, H.J. (1989), "Neural networks for an intelligent mathematical programming system," in **Impacts of Recent Computer Advances on Operations Research**, R. Sharada, B.L. Golden, E. Wasil, O. Balci, and W. Stewart, eds., North-Holland, New York, 313-320.

Holland, J.H. (1975), **Adaptation in Natural and Artificial Systems**, The University of Michigan Press, Ann Arbor, Michigan.

Horvitz, E.J., J.S. Breese, and M. Henrion (1988), "Decision theory in expert systems and artificial intelligence," Int. J. Approximate Reasoning, forthcoming.

Kowalik, J.S., ed. (1986), **Coupling Symbolic and Numerical Computing in Expert Systems**, North Holland.

Kanal, L.N. and J.F. Lemmer, eds. (1986), Uncertainty in Artificial Intelligence, North Holland, Amsterdam.

Murphy, F.H. and E.A. Stohr (1985), "An intelligent system for formulating linear programs," *Decision Support Systems*, 2, 39-47.

Pearl, J. (1988), **Probabilistic Reasoning in Intelligent Systems**, Morgan Kaufmann, San Mateo, Ca.

Schacter, R.D. (1986), "DAVID: influence diagram processing system for the Macintosh," Proceedings of the Workshop on Uncertainty in Artificial Intelligence, University of Pennsylvania, August 8-10, 243-248.

Simon, H.A. (1987), "Two heads are better than one: the collaboration between AI and OR," *Interfaces*, 17, 8-15.

Spiegelhalter, D.J. (1986), "A statistical view of uncertainty in expert systems," in **Artificial Intelligence and Statistics**, W.A. Gale, ed., Addison-Wesley, Reading, Ma.

ました # I. SEARCH

The first of the three papers that comprise this section ("Toward the Modeling, Evaluation and Optimization of Search Algorithms" by Hansson et al.) applies two techniques familiar to Operations Research, Markov modeling and multiattribute utility theory, to the optimization of heuristic search, the fundamental technique of Artificial Intelligence. AI systems, even simple heuristic search algorithms, can benefit greatly from mathematical models of their own behavior, and from decision-theoretic control of their actions and computations. For example, the authors show that it is possible to automatically select among search strategies by accurately modeling their performance with a simple Markov model, and capturing the preferences of a user with a multiattribute utility theory function. The selection of strategies is thus automatically tailored to both the user and the problem instance at hand. The authors see tremendous potential for the integration of quantitative techniques in the development of flexible, reflective AI systems, which can allocate scarce resources to solve problems even under great uncertainty.

The research reported in the paper by Liepins et al. ("Genetic Algorithms Applications to set Covering and Traveling Salesman Problems") is motivated by the importance of the set covering and traveling salesman problems. For example, these problems are often encountered in routing and diagnosis problems. Any algorithm that could improve on current solution techniques is important. At the same time, it is important to establish, understand, and extend the proper domain of application of genetic algorithms.

This paper reports results of a controlled study of selected variants of genetic algorithms to these problems. Two general conclusions seem to emerge. First, the genetic algorithm variants studied in this paper for the traveling salesman problem were less than distinguished performers. Apparently, because their formulation did not adequately preserve the city to city link information, "basic building blocks" were able to find continually improving solutions. Second, for set covering problems, a penalty function genetic algorithm formulation produced good results, especially when the penalty function provided an adequate estimate of the "cost to completion" for infeasible solutions. This corresponds to observations in TABU search that an oscillating strategy appears useful. Solutions should be approached both from outside as well as within the feasibility region.

These two findings are expected to be the focus of continuing algorithm research. For problems that do not naturally map into a binary chromosome representation with standard crossover, representations and crossover operators that preserve basic building blocks need to be developed.

Whether a "meta-theory" will emerge to help guide the choice of representation remains uncertain. The specification of tight, inexpensive, easily estimated penalty functions will almost certainly receive continued attention.

In the third paper ("Discovering and Refining Algorithms Through Machine Learning" by Hilliard et al.) several questions provide the impetus for a discussion of some initial experiments using genetic algorithms and classifier systems to learn heuristics for scheduling problems. What role can a computer play in developing algorithms? Can a machine, through experimentation with numerous examples develop a method for solving a class of problems? How much information and structure do we need to provide to an automated learning system to assist it in finding good algorithms? Can a machine learning approach help to refine heuristics currently implemented? While people will probably provide the insight and large scale concepts for algorithm development, there is some evidence that useful tools can be developed which may help to design and even to discover new algorithms. In some cases the efforts produce rule sets, in others they supplement an existing heuristic algorithm by refining a portion of the heuristic choices. While the results are preliminary, the concepts provide a new area for research and development which may produce systems in the future which can develop general methods for solving classes of problems.

Toward the Modeling, Evaluation and Optimization of Search Algorithms*

Othar Hansson	Gerhard Holt	Andrew Mayer
C.S. Division	Bear, Stearns Inc.	C.S. Division
Univ. of California	245 Park Avenue	Univ. of California
Berkeley, CA 94720	New York, NY 10067	Berkeley, CA 94720

Abstract

A real-world problem-solver must balance conflicting objectives, and therefore measures the quality of a solution to his problem along multiple axes of value. However, algorithms for finding optimal multi-objective solutions usually require exponential computation time. Commonly, problem-solvers prefer computationally inexpensive methods which find non-optimal solutions [18]. Thus, there exists a clear trade-off between the desirability of an algorithm's output, and the costs of computing that output. In fact, the choice facing a problem-solver is not among different *solutions*, but rather among the different *algorithms* which find these solutions. To make a wise choice, each available algorithm should be evaluated based on both its output and its processing costs[6].

This paper offers directions for automating the process of algorithm selection in state-space problem-solving. We describe a method consisting of two components: (1) a mathematical tool for modeling the performance of search algorithms and (2) a utility function for encoding the preferences of a problem-solver. Together they enable the selection of the particular search algorithm which will maximize a problem-solver's expected utility. We empirically demonstrate the success of this optimization technique on a sample problem. Finally, we discuss the generalization of this work to search strategies which are themselves utility-directed.

---

*This research was initiated while the authors were students at Columbia University. This paper was prepared with the support of Heuristicrats, the National Aeronautics and Space Administration, and the Rand Corporation.

1 State-Space Problem-Solving

Problem Representation

The state-space approach to problem-solving [14] considers a problem P as a quadruple, $(S = \{S_0, S_1, \ldots\}, O \subset S \times S, I \in S, G \subset S)$. S is the set of possible *states* in the problem, O is the set of *operators* or transitions between states, I is the one *initial state*, and G is the set of *goal states*. In theory, any problem can be represented as a state-space graph, where the states are nodes, and the operators are directed, weighted arcs between nodes. The weight associated with each operator, O_i, is the cost of applying it, $C(O_i)$.

The problem is said to be solved when a sequence of operators has been found which forms a path between I and a state in G. A minimal cost sequence of operators will be called a *shortest-path* solution, whose cost will be denoted by $C_{opt}(I)$, and referred to as the *distance* of I from the nearest goal state. Such a path is found by performing a *search* through the state-space. The primitive action performed by search algorithms is the *state expansion*. This consists of *generating* (enumerating) the successors of the state, which then become available for subsequent expansion. Continuing this process results in a tree of states rooted at the initial state, or a *search-tree*.

Optimal Search Algorithms

The simplest class of problem-solving techniques are the brute-force search algorithms, which are designed to guarantee shortest-path solutions. An example of such an algorithm is uniform-cost search, in which a problem-solver expands states in order of their proximity to the initial state. As this process never expands a state with cost x until all states of cost $y < x$ have been expanded, the first state in G which is found will be the closest one to I. Hence, the path between them will have minimal cost. Unfortunately, the number of states in the search-tree grows exponentially with its depth. Thus, brute-force search requires exponential computation time.

More sophisticated methods of guiding a search have been devised in hopes of finding optimal solutions faster. Typically, such methods take the form of branch-and-bound, wherein partial solutions (equivalently, classes of solutions) are enumerated ("branch"), and possibly eliminated from future consideration by an estimate of solution cost ("bound"). One such method is the A^* algorithm[11], in which the bounding estimates are provided by admissible heuristic evaluation functions. A powerful heuristic will offer a tight bound, and will sharply focus a search. Although a major improvement over brute force techniques, most heuristic search algorithms still require exponential computation time.

To combat exponential time complexity, many approximation, or ϵ-optimal algorithms have been developed. Such algorithms can dramatically reduce computation time by only guaranteeing solutions whose cost is within some factor (ϵ) of optimal. Several ϵ-optimal variants of A^*, for example, are discussed in [16]. Unfortunately, there are some problems for which all known approximation algorithms still require exponential computation time (e.g., some versions of the Traveling-Salesman Problem)[13,15].

Probabilistic Search Algorithms

However, a far richer class of search algorithms results from an explicit distinction between the two actions a problem-solver engages in during search. Acts of *execution* are those in which a problem-solver *physically* applies an operator (moves) in order to change his state in the real world. This activity is essential to problem-solving because a path from I to G must eventually be traversed for a goal to be attained. Acts of *simulation* are those in which a problem-solver *conceptually* applies an operator in order to reason about the consequences of possible actions. Although not formally required for problem-solving, simulation is omnipresent as it yields information which will guide the problem-solver toward superior moves.

Characterizing search algorithms as sequences of these two types of actions admits of dynamic processes which interleave simulation and execution. We will refer to these as *on-line* algorithms. On-line algorithms "loop" through the following steps, until a goal state is reached:

Simulation: Perform 0 or more state expansions
Execution: Apply 1 or more operators

Probabilistic in nature, an on-line algorithm provides no strict guarantees about the solutions it will produce. Instead, we may think of it as drawing a solution from a distribution over possible solutions. When viewed over many problem instances, the characteristics of this distribution will reveal the quality of the algorithm. A method for estimating the distribution associated with a given algorithm is the subject of the following section.

Note that the optimal and ϵ-optimal algorithms discussed earlier are proper subsets of this broad class of on-line algorithms. Such *off-line* algorithms are static design processeses in which simulation wholly precedes execution. Simulation proceeds until the problem-solver has proven the (ϵ-)optimality of a sequence of operators, at which point that sequence is executed.

Strategy Selection

It is convenient to think of search algorithms as being parameterized, in that different values of key control parameters can tailor an algorithm's performance. Familiar parameters include the optimality bound ϵ, and the heuristic function h. For clarity, we say that each such algorithm, together with settings for each of its parameters, constitutes a *problem-solving strategy*. To a human problem-solver faced with a problem instance, setting parameters and running an algorithm is indeed a strategy for finding a solution. The human's only decision is the choice among available strategies, i.e., selecting the most appropriate algorithm and setting its parameters.

Making this choice wisely requires that the problem-solver estimate and evaluate the eventual consequence, or *outcome*, of using each candidate strategy. *Attributes* of these outcomes will include the computational resources (e.g., time, memory) consumed during simulation, the physical resources (e.g., money, fuel) consumed by execution, and the rewards achieved by attaining a goal.

The next section discusses how modeling a search strategy as a Markov process can yield accurate estimates of the probability distribution over the strategy's possible outcomes. The subsequent section discusses how utility theory (specifically MAUT) can be used to assess a problem-solver's subjective preferences among the multiattribute outcomes. Together, these tools enable the automation of strategy selection to suit a given problem-solver and problem instance.

2 Modeling of On-line Strategies

As a search strategy operates, seeking an ultimate goal, it must decide repeatedly which local steps to take. These local steps amount to individual decisions concerning which operator to select at any point in time. The overall performance of a search strategy in solving a given problem emerges as an epiphenomemon of its ability to make these individual decisions judiciously. In this section, an estimate of a strategy's outcome is derived from a decision quality metric, using a Markov chain. This model allows us to compute a probability distribution over the possible outcomes of using a strategy σ on a problem instance P.

A Simple Markov Model

We begin with a simple path-planning example where the problem-solver is concerned only with a single attribute – the *length* of the solution-path – and where the operators O all have unit cost (the model will subsequently be generalized). Consider the first move made by a search strategy, to apply an operator $O_{0,1}$ to a state S_0 yielding a state S_1. If the shortest-path from S_0 to a state in G has cost $C_{opt}(S_0)$, then the move from S_0 to S_1 was correct *iff* $C_{opt}(S_0) = C_{opt}(S_1) + C(O_{0,1})$.

The probability that a correct move is made from a state with solution cost C_x (for varying C_x) is called the *decision quality* of the strategy. To estimate outcomes (solution quality) from decision quality, we model the individual moves made by a search strategy as transitions in a Markov chain, pictured in Figure 1. Each state S_k^m in the Markov chain corresponds to a subset of S (the states of the problem P) – in this case, let S_k^m be the set of all states $S_i \in S$ for which $C_{opt}(S_i) = k$. Figure 1 shows these two levels, with the states in the Markov model (M) above the states in the state-space of the problem P. Each of the shaded states in M contains the states of the corresponding shade in S.

The goal states G are contained in the absorbing state of the Markov chain $G^m = S_0^m$ (not shown). At any given point, the strategy is at a state S_a in the problem state-space, corresponding to a Markov state $S_i^m \ni S_a$, from which it is to move. With probability $P(O_{i,j}^m \mid \sigma)$, it moves to another state $S_b \in S_j^m$. Upon reaching a goal state (S_0^m) it is absorbed. In this model, the essential difference among search strategies is completely described by the transition probabilities – the decision quality estimates discussed above.

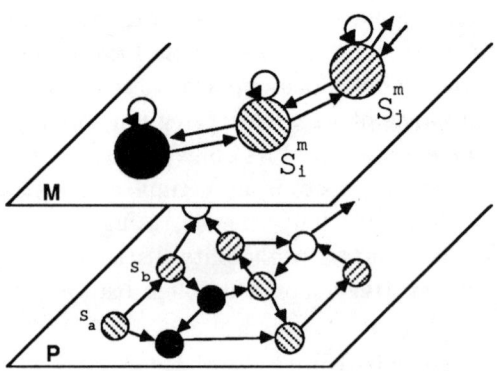

Figure 1: Simple Markov Model

From the $P(O^m_{i,j} \mid \sigma)$, we can easily compute $P(C(I) = \ell \mid \sigma)$, the Bayesian probability that strategy σ will find a solution of length l from an initial state I.

$$P(C(I) = \ell \mid \sigma) = \sum_i P(C(S^m_i) = \ell \mid \sigma) P(I \in S^m_i)$$

$$P(C(S^m_i) = 0 \mid \sigma) = 1 \text{ iff } S^m_i = G^m$$

$$\begin{aligned} P(C(S^m_i) = \ell \mid \sigma) = {} & P(O^m_{i,i-1} \mid \sigma) P(C(S^m_{i-1}) = \ell - 1 \mid \sigma) + \\ & P(O^m_{i,i+1} \mid \sigma) P(C(S^m_{i+1}) = \ell - 1 \mid \sigma) + \\ & P(O^m_{i,i} \mid \sigma) P(C(S^m_i) = \ell - 1 \mid \sigma) \end{aligned}$$

Thus, one can use the Markov model to predict the solution lengths that will be achieved by strategy σ, provided that we can estimate $P(I \in S^m_i)$ and $P(O^m_{i,j} \mid \sigma)$. The former, $P(I \in S^m_i)$, can be provided by a *probabilistic heuristic estimate* [7], composed of a heuristic function h, coupled with an association $P(S_i \in S^m_j \mid h(S_i))$. The association can be learned by applying the heuristic function to a series of sample problems S_i for which $S^m_j \ni S_i$ is known, or by updating based on experience as problems are solved. Details on the uses of such probabilistic heuristic estimates, as well as other methods for learning and expressing the association, can be found in [7] and [10].

The latter, $P(O_{i,j}^m \mid \sigma)$, can be obtained from decision quality experiments, in which one observes the decisions made by a strategy σ on a number of sample problems (such experiments are commonly used in the heuristic search literature to compare algorithms and heuristic functions[1]). For each test problem S_a, we must know $S_i^m \ni S_a$, as well as $S_j^m \ni S_b$ for each state S_b which is adjacent to S_a. The savings of using sample problems is that one need not solve entire problems using each σ, but instead, merely observe many individual moves – each test problem need be solved only once, and the cost of finding a solution can be amortized over arbitrarily many algorithms which are tested. In addition, if one does choose to use a particular σ to solve a problem P, $P(O_{i,j}^m \mid \sigma)$ can be updated based on the solution found.

The General Markov Model

Generalizing, a Markov model of a search strategy is a quadruple ($P = (S, O, I, G)$, $S^m = \{S_0^m, S_1^m, \ldots, S_n^m\}, O^m \subseteq S^m \times S^m, I^m \in S^m, G^m \subseteq S^m$). P is the problem on which we model σ's behavior. S^m is the set of states composing the Markov chain, such that S^m represents a partition of S. O^m is the set of transitions in the Markov chain, induced from O by the partition S^m on S (each $O_{i,j}^m \in O^m$ is associated with $P(O_{i,j}^m \mid \sigma)$, the probability that σ makes a transition from S_i^m to S_j^m). I^m is the unknown starting state in the Markov chain. G^m is the set of absorbing states of the Markov chain, a partition of G.

The equations above generalize into:

$$P(A(I) = \alpha \mid \sigma) = \sum_i P(A(S_i^m) = \alpha \mid \sigma) P(I^m = S_i^m)$$

$$P(A(S_i^m) = \alpha \mid \sigma) = 1 \text{ iff } \left[\left(S_i^m = G_j^m\right) \wedge \left(A(G_j^m) = \alpha\right)\right]$$

$$P(A(S_i^m) = \alpha \mid \sigma) = \sum_j P(O_{i,j}^m \mid \sigma) P(f(A(S_j^m), A(O_{i,j}^m)) = \alpha)$$

where $A(I)$ is the vector of attribute values describing the outcome of applying the strategy σ to problem P, and $P(A(I) \mid \sigma)$ the corresponding probability distribution over such vectors, and where

$$P(I^m = S_i^m) = P(I \in S_i^m)$$

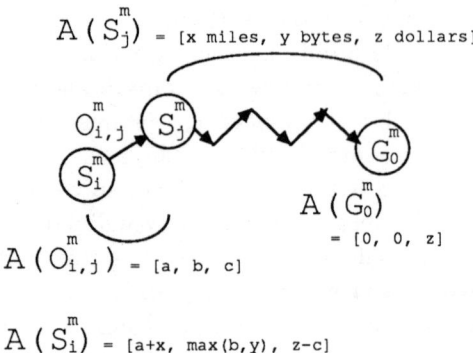

Figure 2: Composition of Solutions

$$P(f(A(S_j^m), A(O_{i,j}^m)) = \alpha) = \sum_{f(\beta,\gamma)=\alpha} P(A(S_j^m) = \beta) P(A(O_{i,j}^m) = \gamma)$$

As illustrated in Figure 2, f is a function on pairs of attribute vectors, combining the attributes of an operator and the state which results from the application of that operator. More precisely, given the attribute values of a particular state, and those of an operator which moves to that state from another, the function gives a set of attribute values for the preceding state, assuming that the operator will be applied when that state is reached. Typically, f will prescribe different types of combinations for different components of the two vectors (e.g., minimization, addition, multiplication, etc.) In Figure 2, for example, the attributes are distance, memory usage and a monetary reward received upon reaching the goal. In this case, f will add the distances, maximize the memory usage, and subtract costs from monetary rewards.

3 Evaluation of On-line Strategies

Using the Markov model to generate a probability density function over the space of possible multiattribute outcomes, the problem-solver need only determine his preference structure among these outcomes in order to choose among the available strategies.

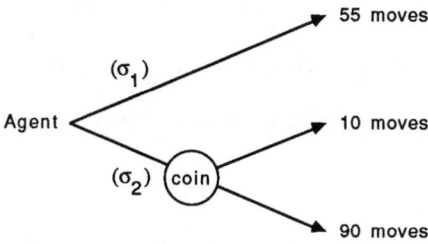

Figure 3: Evaluating strategies

Expected Utility

The theory of expected utility [19] has come to be the central tool in modern decision analysis. Utility is the subjective assignment of value to potential outcomes, when the exact outcome is uncertain. The theory claims that rational decision-makers attach utilities to all possible outcomes, and when faced with a decision under uncertainty, select that outcome with maximum expected utility.

Formally, the choice from a set of strategies $\{\sigma_1 \ldots \sigma_n\}$ each yielding possible outcome vectors $\alpha^1 \ldots \alpha^m$, will be that strategy σ_i which has maximum expected utility, $EU(\sigma_i)$, where

$$EU(\sigma_i) = \sum_{\alpha^j} U(\alpha^j) P(\alpha^j \mid \sigma_i)$$

It is important to stress the inherent subjectivity of the utility assignments – different decision-makers may well have different preference orderings among the outcomes of different strategies. Fortunately, a robust method for determining a decision-maker's preferences exists.

Consider, for example, the choice among strategies depicted in Figure 3. The outcome of σ_1 is a 55-move solution which would be attained with absolute certainty were this strategy selected. However, σ_2 is probabilistic – depending on the outcome of a coin toss one attains either a 10-move solution or a 90-move solution. One problem-solver might assign the following utilities to the different outcomes:

$$u(10 \text{ moves}) = 1.0$$
$$u(55 \text{ moves}) = 0.6$$
$$u(90 \text{ moves}) = 0.0$$

Therefore the expected utilities of the strategies would be:

σ_1 $u(55 \text{ moves}) \cdot p(55 \text{ moves}) = $ $0.6 \cdot 1.0 = 0.6$

σ_2 $(u(10 \text{ moves}) \cdot p(10 \text{ moves}))$
$+(u(90 \text{ moves}) \cdot p(90 \text{ moves})) = $ $(1.0 \cdot 0.5) + (0.0 \cdot 0.5) = 0.5$

Clearly, this problem-solver desires the 10-move solution, but would prefer to avoid the possibility of receiving the 90-move solution. The 55-move solution will satisfice. Based on his expected utility, this *risk averse* problem-solver chooses σ_1 and avoids the possibility of disaster.

A *risk prone* decision-maker might be trying desperately to obtain the shortest solution. He highly values the 10-move solution and sees little difference between the 55- and 90-move solutions, assigning $u(55 \text{ moves}) = 0.1$. The expected utilities for this decision-maker would be:

σ_1 $u(55 \text{ moves}) \cdot p(55 \text{ moves}) = $ $0.1 \cdot 1.0 = 0.1$

σ_2 $(u(10 \text{ moves}) \cdot p(10 \text{ moves}))$
$+(u(90 \text{ moves}) \cdot p(90 \text{ moves})) = $ $(1.0 \cdot 0.5) + (0.0 \cdot 0.5) = 0.5$

To maximize his expected utility, this problem-solver should choose the gamble, σ_2.

A decision-maker may be unable to produce such precise utility assignments on demand. However, by observing his decision-making on sample problems one can infer the utilities that he assigns to certain outcomes and interpolate to produce his overall utility function [17]. For example, if the risk-prone decision-maker prefers a new strategy, σ_3, to either σ_1 or σ_2, one can infer that $EU(\sigma_3) \geq 0.5$. In this manner, observed decisions form the basis for the utility elicitation techniques of decision analysis.

Multiattribute Utility Theory

An extension of utility theory, which describes the behavior of a decisonmaker faced with multiple, and possibly conflicting objectives, is multiattribute utility theory (MAUT). The extension is in allowing one to combine utility functions of individual attributes into a joint utility function. A rigorous presentation may be found in [17].

In brief, the techniques involve assessing the decision-maker's marginal utility of improving each of the attributes. In addition, the utility independence relations of the attributes must be modelled, e.g., by assessing whether the utility function for each attribute is independent of the values of the others. Given this information, one can determine what form the function should take [17,20]. Once constructed, the multiattribute utility function can be evaluated for all potential outcomes, which are specified as vectors of individual attribute values.

Multiattribute utility theory offers a means for coping with uncertainty, subjectivity, and conflicting objectives. Given a probability distribution over multiattribute outcomes $P(A(I) = \alpha \mid \sigma)$, the strategy choice is determined by a user-defined utility function $U(\alpha)$, which encapsulates the preferences of a problem-solver for different possible outcomes.

4 Testing the Model

To demonstrate the applicability and usefulness of the Markov modeling technique, we contrived a simple example which nevertheless reflects the sorts of trade-offs that problem solvers must cope with in the real-world.

A Sample Domain

The Eight Puzzle (Figure 4) is the classic example of a small, well defined, and conceptually simple problem which is sufficiently complex to exhibit interesting phenomena – therefore, it has served as a popular testing ground for problem-solving methods for over twenty years [2,3,16].

The Eight Puzzle consists of a 3x3 frame containing 8 numbered, sliding tiles. One of the positions in the frame does not contain a tile – this space is called the "blank." There is only one legal operator in this state-space – sliding any one of the tiles which are horizontally or vertically adjacent to the blank into the blank's position. A solution to

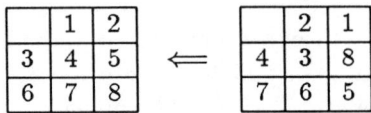

Figure 4: Eight Puzzle goal state and sample initial state

a problem instance is a sequence of operators which transforms a given initial state into a particular goal state (Figure 4).

In an attempt to make this problem more interesting from a multi-attribute perspective, consider the following extension. A terrorist has activated an explosive device which uses an Eight Puzzle as a control mechanism. A robot, with ten megabytes of on-board memory, is sent to defuse the bomb. If it fails to solve the problem within, say, ten minutes, the device will detonate.

Certainly, the robot must try to solve the problem quickly (without excessive computation), but it also ought to minimize the overall solution length as each new move entails the possibility of accidental detonation. Additionally, the robot will fail if its problem-solving strategy requires more memory than its internal capacity.

Modeling Strategies

The family of strategies made available to the robot were derived from Minimin, the simple on-line search algorithm which follows:

> While not at goal state
>> Search forward until depth θ is reached
>> Evaluate frontier nodes with heuristic function
>> Make one move toward minimum-valued frontier node

Fixing the heuristic function to be the ubiquitous Manhattan Distance function simplifies the robot's choice to one of θ different strategies, which are distinguished by entire levels of lookahead in the search-tree, rather than by individual state expansions. Nonetheless, this offers a coarse spectrum between purely random movement ($\theta = 0$) and off-line (boundless-resource) enumeration of solutions ($\theta = C_{opt}(I)$).

To model the performance of Minimin, we used a Markov model where each state S_d^m contained all the states S which are at a distance of d moves (or *depth d*) from the nearest goal state. We estimated $P(O_{i,j}^m \mid \sigma)$ by observing decisions of Minimin (with varying θ) on 10000 sample problems. For each decision, we recorded both the transition in the Markov chain, and the attributes of the operator $A(O_{i,j}^m)$ (to estimate $P(A(O_{i,j}^m) = \gamma)$). For each test problem, the actual distance was determined by performing an A^* search.

Eliciting Utilities

Decision analysis provides a number of protocols for eliciting utility judgements from humans. For the purposes of this paper, we used the "midvalue splitting" technique [17] to assess the utility function for each of the attributes: solution length, memory usage, and simulation time. A multiplicative utility function [17] was deemed appropriate, and the scaling constants for this function were elicited.

The multiplicative utility function has the general form:

$$1 + k \cdot u(\alpha) = \prod_{i=1}^{n}[1 + k \cdot k_i \cdot u_i(\alpha_i)]$$

(where α_i indicates the ith component of vector α), and in our case:

$$u(\alpha) \propto_n (1 + 3.4u_1(\alpha_1)) \cdot (1 + 1.1u_2(\alpha_2)) \cdot (1 + 11u_3(\alpha_3))$$

where

Solution Length (moves)	$u_1(\alpha_1) =$	$1.8 \; (e^{-0.08\alpha_1} - 1)$
Memory Usage (megabytes)	$u_2(\alpha_2) =$	$0.001 \; (1 - e^{0.7\alpha_2})$
Simulation Time (minutes)	$u_3(\alpha_3) =$	$0.1 \; (1 - e^{0.5\alpha_3})$

The above utility function, $u(x)$, can perhaps best be understood in the context of three solutions which it judged to be relatively equivalent. The space attribute provided only miniscule differentiation among utilities of solutions (space posed no problem for the computer in this particular problem so long as it did not exceed the resource bounds):

Solution Length	Memory Usage	Simulation Time
20 moves	≤ 9 mbytes	8 minutes
68 moves	≤ 9 mbytes	6 minutes
93 moves	≤ 9 mbytes	4 minutes

Accuracy of Markov Modeling

Using the Markov model, we were able to predict the performance of the different Minimin strategies, and select the θ which the robot should use to solve a particular problem instance, given its utility function.

As described in Section 4.2, we calculated $P(A(I) \mid \sigma)$. These estimates (of solution length, memory usage, and simulation time) were then compared to actual attribute values produced by running Minimin, with varying θ, on another 5000 random Eight Puzzle instances. On average, the estimates provided by the Markov Model proved to be accurate predictions of the empirical test runs.

Accuracy of Strategy Utility predictions

Experiments on sample puzzles resulted in close correlation between the predicted $EU(\sigma)$ and the actual utility seen in the performance of the strategy. For example, for puzzles of depth $d = 19$, $\theta = 16$ was selected as the optimal lookahead level, and in solving 1000 random puzzles with Minimin, $\theta = 16$ yielded the highest average case utility. Over all d $(0\ldots 31)$ the selection of θ exactly matched the empirically optimal choice 88.3% of the time and came within a single level of optimal 95.4% of the time. In the poorest estimate, θ was misjudged by three levels. It is notable, however, that in these cases where non-optimal strategies were selected, the chosen strategy and the empirically optimal strategy had nearly equivalent empirical utilities (to at least three significant figures).

Of course, the selections of our model are also dependent on the ability of our heuristic estimate to provide $P(I \in S_i^m \mid h(I))$. But the accuracy of the expected utility estimates and strategy selection for states in each S_i^m implies that while the utility of using the selected strategy will suffer if there is great uncertainty in the heuristic, the selected strategy should still outperform all others on average.

5 Further Work

The most obvious extension of this approach is to allow parameterized strategies to have access to the problem-solver's utility function and to a Markov model. The strategies can then *themselves* choose the optimal set of parameters to apply to each problem instance. In this sense, the strategy acts as the *agent* of the problem-solver, attempting to act in his best interests, and to make decisions as he would. This seems desirable, as the problem-solver need only specify his utility function, and need not make any decisions which explicitly control the search.

Of course, the choice of any fixed strategy to apply to the solution of a problem instance has inherent limitations. In general, it may be advantageous to choose a different strategy to apply at each step of the problem's solution. Such decisions, too, could be made internally by a parameterized strategy, which would update the parameter settings to optimize for the current state of the problem.

In the limit, this paradigm may lead to algorithms in which every search action, down to the level of individual state expansions and heuristic evaluations (and possibly further), is chosen so as to maximize the problem-solver's expected utility [4,5,12]. The problem-solver simply specifies his preferences to a utility-directed search algorithm which serves as his agent, and which makes all necessary decisions with his preferences in mind. First steps toward such a rational search algorithm, using explicit representation of utilities and probabilities, are described in [8] and [9].

The concept of pure, utility-directed control of such a system raises new questions about the relationship between algorithms and decision processes. Typically, algorithms are designed by abstracting individual decisions into fixed sequences, effectively "compiling" them with respect to a generic utility function. Probing the "decision-level" which underlies algorithms may lead to a better understanding of algorithm design and to the development of practical techniques for designing efficient algorithms. Ideally, understanding the relation of decisions to algorithms will lead to compilers which take problems and utilities as input, and yield efficient algorithms as output.

Conclusion

This work demonstrates one method for modeling and evaluating the performance of different search algorithms and heuristics. Specifically, performance predictions are used to select a strategy for solving a particular problem instance, incorporating the preferences of a particular human problem-solver. The performance predictions are given by a Markov chain model of the strategies, and the preferences captured by a multiattribute utility function.

A major bottleneck in AI systems has been their inability to effectively allocate scarce computational resources in the face of uncertainty and conflicting objectives – in heuristic search, for example, the problem is to focus effort on those portions of the search-tree relevant to the current decision. Our formulation and early results suggest the power of enlisting precise and proven techniques from other fields, such as Markov modeling and multiattribute utility theory, in the attack on this central problem in AI.

References

[1] Abramson, B., Control strategies for two-player games. *Computing Surveys*, June, 1989.

[2] Doran, J., and D. Michie, Experiments with the graph traverser program. In *Proceedings of the Royal Society, 294 (A)*, pages 235–259, 1966.

[3] Gaschnig, J., A problem similarity approach to devising heuristics: first results. In *Proceedings of the International Joint Conference on Artificial Intelligence*, pages 301–307, Tokyo, 1979.

[4] Good, I. J., A five year plan for automatic chess. *Machine Intelligence*, 2, 1968.

[5] Hansson, O., G. Holt, and A. Mayer, The comparison and optimization of search algorithm performance. Manuscript, December, 1986.

[6] Hansson, O., and A. Mayer, The optimality of satisficing solutions. In *Proceedings of the Fourth Workshop on Uncertainty in AI*, Minneapolis, August, 1988.

[7] Hansson, O., and A. Mayer, Probabilistic heuristic estimates. In *Proceedings of the Second International Workshop on AI and Statistics*, Fort Lauderdale, January, 1989.

[8] Hansson, O., and A. Mayer, Decision-theoretic control of search in BPS. In *Proceedings of the AAAI Spring Symposium on Limited Rationality*, Palo Alto, March, 1989.

[9] Hansson, O., and A. Mayer. Heuristic search as evidential reasoning. In *Proceedings of the Fifth AAAI Workshop on Uncertainty in AI*, Windsor, Ontario, August, 1989.

[10] Hansson, O., and A. Mayer, Probabilistic heuristic estimates. Annals of Mathematics and Artificial Intelligence, to appear.

[11] Hart, P., N. Nilsson, and B. Raphael, A formal basis for the heuristic determination of minimum cost paths. *IEEE Transactions on Systems Science and Cybernetics*, 2:100–107, 1968.

[12] Horvitz, E. J., Reasoning under varying and uncertain resource constraints. In *Proceedings of the National Conference on Artificial Intelligence*, Minneapolis, 1988.

[13] Lawler, E. L., J. K. Lenstra, A. H. G. Rinnooy Kan, and D. B. Shmoys, editors, *The Traveling Salesman Problem: a Guided Tour of Combinatorial Optimization*, John Wiley, New York, NY, 1985.

[14] Newell, A., and H. A. Simon, *Human Problem Solving*. Prentice-Hall, Englewood Cliffs, NJ, 1972.

[15] Papadimitriou C. H. and K. Steiglitz, *Combinatorial Optimization: Algorithms and Complexity*, Prentice-Hall, Englewood Cliffs, NJ, 1982.

[16] Pearl, J., *Heuristics*, Addison-Wesley, Reading, Massachusetts, 1984.

[17] Raiffa, H., and R. L. Keeney, *Decisions with Multiple Objectives: Preferences and Value Tradeoffs*, John Wiley, New York, 1976.

[18] Simon, H.A., *The Sciences of the Artificial*, 2nd edition, M.I.T. Press, Cambridge, 1981.

[19] Von Neumann, J., and O. Morgenstern, *Theory of Games and Economic Behavior*, Princeton University Press, 1944.

[20] von Winterfeldt, D., and W. Edwards, *Decision Analysis and Behavioral Research*, Cambridge University Press, 1986.

GENETIC ALGORITHMS APPLICATIONS TO SET COVERING AND TRAVELING SALESMAN PROBLEMS

by

G. E. Liepins
M. R. Hilliard

Oak Ridge National Laboratory

J. Richardson
M. Palmer

University of Tennessee

ABSTRACT

For set covering problems, genetic algorithms with two types of crossover operators are investigated in conjunction with three penalty function and two multiobjective formulations. A Pareto multiobjective formulation and greedy crossover are suggested to work well. On the other hand, for traveling salesman problems, the results appear to be discouraging; genetic algorithm performance hardly exceeds that of a simple swapping rule. These results suggest that genetic algorithms have their place in optimization of constrained problems. However, lack of, or insufficient use of fundamental building blocks seems to keep the tested genetic algorithm variants from being competitive with specialized search algorithms on ordering problems.

INTRODUCTION

Genetic algorithms are general purpose optimization algorithms (somewhat akin to simulated annealing in that sense). They were developed by Holland (1975) to search irregular, poorly characterized spaces. Holland hoped to develop powerful, broadly applicable techniques able to handle a wide variety of problems. The genetic algorithm does not necessarily find an optimal to any one problem, but does find good solutions to problems that are resistant to most other known techniques. Holland was inspired by the example of population genetics and used crossover rather than mutation as the primary genetic operator. Thus, genetic search proceeds over a number of generations, with each generation represented by a population of current chromosomes.

"Survival of the fittest" provides the pressure for populations to develop
increasingly fit individuals. Historically, genetic algorithms applied
directly only to unconstrained problems. More recently penalty function,
multiobjective, and modified crossover formulations have been developed to
broaden the range of genetic algorithm applicability to problems such as set
covering and traveling salesman problems. These two classes of problems were
studied in this paper. For set covering problems, three penalty function formulations, two multiobjective formulations, and gene pool selection according
to ranking were investigated in conjunction with two variants of the crossover
operator. Results were encouraging and pointed to the "greedy" crossover,
tight upper bounds for cost of completion of covers (as a penalty function),
and Pareto based selection of the gene pool as promising techniques. For the
traveling salesman problems, results were arguably hardly better than with a
simple swapping rule.

BACKGROUND

Genetic algorithms were developed by John Holland (1975) who built on ideas of
Bledsoe (1961) and others. A major feature that distinguishes Holland's formulation from those of predecessors is Holland's emphasis on crossover rather
than mutation as the primary genetic search operator. Holland also formulated
the mathematical underpinnings of genetic algorithms. He showed that the
genetic algorithm searches not only among the members of the current populations, but actually searches the underlying space of schemata (hyperplanes or
templates) in an implicitly parallel manner. In fact, were the choice among
several competing schemata interpreted as a k-armed bandit problem, Holland
demonstrated that the genetic algorithm allocates its trials optimally.
Further analysis of chromosome survival lead to the schemata theorem.
Holland's students expanded on Holland's work. DeJong showed that for non-
smooth and multimodel functions, genetic algorithms were as least competitive
with PRAXIS (Brent, 1971) and the Fletcher-Powell algorithm (Fletcher and
Powell, 1963; Fletcher, 1970 Huang, 1970). Included in DeJong's studies was a
function whose evaluation was contaminated with noise. Such "noisy evaluations"
were further studied by Grefenstette and Fitzpatrick (1985) who demonstrated
that noise in the evaluations was compensated for by additional generations.
Bethke (1980) undertook a careful study of the standard genetic algorithm's
domain of applicability. He was not fully able to characterize this domain
but demonstrated that genetic algorithms work well for functions whose kth
difference approximation to the kth derivative is appropriately bounded by
certain Walsh transforms. Conversely, using Walsh transforms, he developed a
procedure to construct functions that are resistant to genetic approaches.
Goldberg (1987b) further investigated functions that would be difficult for

the genetic algorithm. Surprisingly, certain problems that might have been thought to have been misleading, exhibited combinatorics that rendered them easier than expected. In spite of their general robustness against being trapped in local optima, genetic algorithms remain prone to "premature convergence", especially for problems for which the encoding yields long chromosome strings. Wilson (1987), Bowen (1986), Booker (1987) and Schaffer and Morishima (1987) have investigated dynamic control of parameters to help alleviate this problem. An alternate solution is to dynamically modify the representation as in the Adaptive Representation Genetic Optimization Technique (ARGOT) developed by Schaffer (1985, 1987). A third possibility might be to use a real valued representation as suggested by Davis and Coombs (1987). These and other approaches are documented in Holland's book (1975) and more recently, in the work by Goldberg (1988), as well as the three volumes of Genetic Algorithm Conference Proceedings (Grefenstette (ed.) 1985 and 1987 and Schaffer (ed.) 1989).

BASICS OF GENETIC ALGORITHMS

Standard genetic algorithms are population based evolutionary search techniques whose primary search operator is crossover between two bit string representations of current "best guesses". The basic standard genetic algorithm for function optimization of a function f() is given as follows:

Algorithm SGAO

1. Select a finite, bounded domain for the search and a representation that discreticizes the search space.

2. Choose a desired population size n and initialize the starting population P.

3. Evaluate individuals according to the function f().

4. If stopping criteria met, stop; else probabilistically select individuals (according to their fitness determined by the function evaluation) to populate a gene pool of size n.

5. So long as the size of the temporary population TP is less than n, randomly choose a parent from the gene pool. With probability p_1, inject the parent into the temporary population TP. With probability $(1 - p_1)$, choose a coparent, cross parents (see below), and inject the resultant offspring into the temporary population.

6. With low probability p_2, mutate (see below) randomly chosen individuals of TP. Set P = TP and return to step 3.

(This version of SGAO is simplistic; substantial refinements are available in the literature.)

The crossover operator is straightforward. For two bit strings I_1 and I_2 of length L, choose a cut position n, $1 \leq n \leq L$. Exchange the substrings beginning at the cut position. The mutation operator simply reverses bits within an individual bit string with some low probability.

It is clear that the SGAO cannot be applied directly to constrained optimization or ordering problems. In fact, for traveling salesman problems, a common representation is an ordered list of the cities in the tour. Standard crossover does not preserve tours (see Figure 1 below). To incorporate problem specific information and handle constrained optimization problems, various representations, crossover operators, and penalty functions (Luenberger, 1965) are introduced. These increase the genetic algorithm's flexibility, but their theoretical underpinnings remain largely unexplored.

```
tour 1        1 4 5 6 2 3 | 8 7
tour 2        5 6 3 2 7 8 | 4 1
offspring 1   1 4 5 6 2 3 4 1
offspring 2   5 6 3 2 7 8 8 7
```

Figure 1. Failure of crossover to maintain tours

SET COVERING PROBLEMS

Set covering problems (SCPs) are difficult (NP complete – see Garey and Johnson, 1979) zero-one optimization problems often encountered in applications such as resource allocation (Revelle et al., 1970) and scheduling (Marsten and Shepardson, 1981). Known solution techniques for SCP include such methods as integer programming, heuristic branch and bound, and most recently, Lagrangian relaxation with subgradient variations. See Balas and Ho (1980) for a review of set covering solution techniques. Informally, one can think of a set covering problem as the problem of selecting from a collection of individuals each with a different set of skills and labor cost, a minimal cost project team that represents all the skills necessary for a certain job.

Formally, let A be an m x n binary matrix, w an n-dimensional nonnegative vector, and x an n-dimensional binary vector, with wx the inner product of the two, and $\underline{1}$ an m-dimensional column vector of ones. Then an SCP can be formulated as

$$\text{minimize (over x) } wx$$

$$\text{subject to } Ax \geq \underline{1}$$

$$x_i = 0, 1 \quad \text{for } i = 1, \ldots, n.$$

EXPERIMENTAL DESIGN

Since genetic algorithms are guided only by the representation and the evaluation function, constrained problems require the use of special representations or modifications to the evaluation function. A "natural" modification of the evaluation function is a penalty function formulation. The evaluation of a population member becomes the sum of two components; first, the function evaluation, and second, a penalty if the member does not satisfy the required constraints.

The SCP is a constrained optimization problem that can be attacked by a penalty function formulation of the genetic algorithm. A typical element of a genetic population would be a string of n binary entries such as (0, 1, 0, 1, 1, . . .) which would be interpreted as the second, fourth, and fifth columns are included in the solution, but the first and third are not. The reward is a large number M minus the sum of the cost of the columns used and a penalty function P for failure to cover:

$$R = M - \sum_{i=1}^{n} w_i x_i - P\eta$$

where x_i and η are binary with $\eta = 0$ if the solution is a cover and $\eta = 1$ otherwise. P is one of three functions of those rows that fail to be covered. The latter term

$$\sum_{i=1}^{n} w_i x_i + P\eta$$

will be referred to as "cost."

The penalty function p1 fails to differentiate among infeasible solutions. Penalty functions p2 and p3 are progressively "more graded" and use the partial information of infeasible solutions to lead the genetic algorithm to good feasible solutions. These penalty functions provide an upper bound on the cost of extending the set of columns to a feasible solution. Each requires more computation than the previous but returns a tighter upper bound. Definitions appear below:

p1: If S is a cover, cost = $\sum_{i \in S} w_i$

If S fails to be a cover, cost = $\sum_{i=1}^{n} w_i$

p2: If S is a cover, cost = $\sum_{i \in S} w_i$

If S fails to be a cover, r rows remain uncovered, and w is the maximum cost associated with any column,

then cost = $\sum_{i \in S} w_i + r\,w$

p3:   If S is a cover, $\text{cost} = \sum_{i \in S} w_i$

If S fails to be a cover, let R be the rows that remain uncovered. For each $i \in R$ let S_i be the columns that cover i. Let $w^*_i = \min \{w_j\}$ for $j \in S_i$. Then cost = $\sum_{i \in S} w_i + \sum_{i \in R} w^*_i$

This paper presents set covering results from three groups of problems. The first group of problems consist of 45 different randomly chosen 50 x 25 matrices of varying densities (five each for densities of 10% to 90%). The second group of 20 problems were intended to be more difficult than the first and were randomly generated 50 x 50 matrices with density 35% on a band within three elements of the diagonal and 5% off the diagonal. The column cost was equal to the number of ones in the column plus a random variable between 0 and 1. The third group of problems were obtained from E. Balas and correspond to problem set 1 described in Balas (1980). They were originally collected or generated by H. M. Salkin. Three types of crossover methods were investigated in combination with the three different penalty functions, and two multiattribute and one nonparametric selection mechanism. (However, not all combinations were tried on all problems.)

The multiattribute and nonparametric formulations that were investigated are the vector evaluated genetic algorithm VEGA (Schaffer, 1985), a Pareto formulation suggested by Goldberg (1987a), and the ranking selection method suggested by Baker (1985). The two attributes of the multi-attribute formulations are feasibility and cost of the (partial) cover (with no penalty). The VEGA formulation that was implemented in this study selects one-half of the gene pool on the basis of cost of partial cover, and the other half on the basis of the number of constraints violated. In each case, the number of copies an individual is expected to contribute to the gene pool is inversely proportional to the individual's cost or number of violations, respectively.

The Pareto formulation also considers the two attributes to be cost and feasibility. It differs from VEGA insofar as an individual's contribution to the gene pool is (effectively) inversely proportional to the Pareto front the individual lies on. In the experiments reported here, the Genesis code (Grefenstette, 1984) was used. Genesis performs minimizations, so the nondominated individuals were assigned to the first Pareto front and for selection purposes were given the number "1" as their evaluation. The second front was formed by individuals undominated by population members not on the first front, and the second front was assigned an evaluation of 2. Individuals on subsequent fronts were assigned their corresponding front number as their evaluation. (It can be demonstrated that VEGA is equivalent to linear fronts determined by an adaptive, linear combination of cost and feasibility.)

The ranking selection procedure was implemented as follows: Each member of the population was ranked according to fitness; the best individual was assigned a rank of 1 and so forth. The individual with rank 1 was allotted M number of expected copies as his gene pool contribution, and the poorest individual was allotted 2-M expected number of copies. The contribution of intermediate individuals was linearly interpolated. For this study, M was set at 1.1.

PERFORMANCE OF GENETIC ALGORITHMS ON CONSTRAINED PROBLEMS
IS DEPENDENT ON THE PENALTY FORMULATION

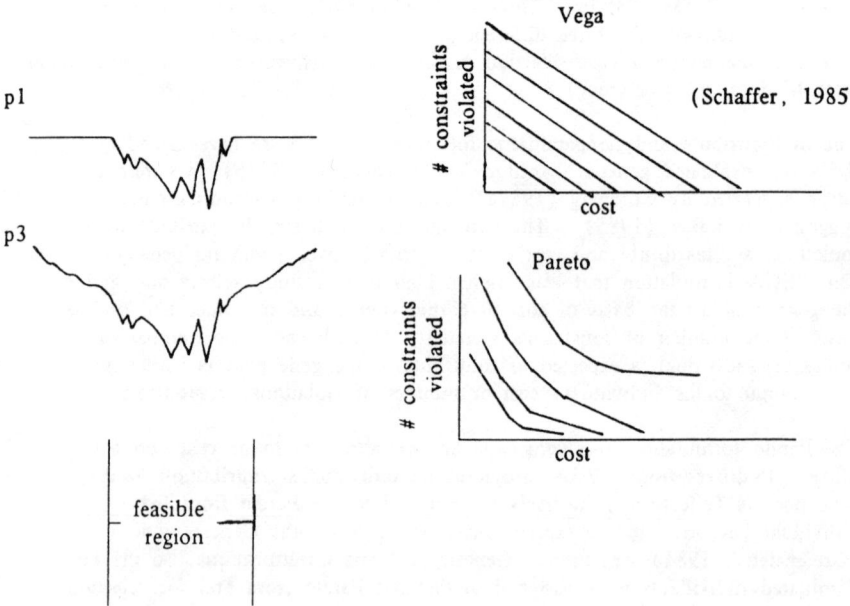

Figure 2. Constraint Formulations

Genetic Algorithms Applications

In anticipation of later comparative results, it should be noted that Baker (1985) suggested using the ranking selection only when the population was threatened to be suddenly dominated by a few individuals; the ranking selection was not intended to be routinely invoked.

Some of the concepts underlying the penalty function and multiple attribute formulations are illustrated in Figure 2, where for the Pareto and VEGA renderings, the lines represent regions of equal number of expected copies.

The three crossover operators investigated are the usual one-point crossover, two-point crossover (illustrated in Figure 3 below),

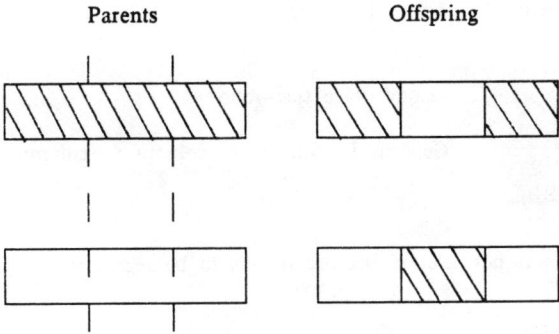

Figure 3. Two-Point Crossover

and the greedy crossover defined as follows:

For every pair of parent genes:

1. Initialize the set S to be empty and the matrix A to be the original set covering matrix with column costs (w_i).

2. For the unused columns and uncovered rows, calculate the cost-ratio (w_i/number-of-rows-covered).

3. Append to S the column (say column c) with the least cost-ratio that is included in one of the parents. (Break ties randomly.)

4. Strike column c and the rows covered by c from A and let this new matrix be A'.

5. If S is a cover or if no other columns are represented in the parents, stop. Otherwise, set A to A' and go to Step 2.

The greedy SCP crossover is illustrated below:

Parent 1: 1 0 0 0 0 $w_1=5$ $w_2=2$ $w_3=5$ $w_4=4$ $w_5=2$
Parent 2: 0 1 1 0 0   0   1   1   1   1
                           1   1   1   0   1
Initial selection for   0   0   0   1   0
child: column 2 with   1   0   0   1   0
cost-ratio 1
__ 1 __ __ __

                        second-stage cost-ratio

Second Selection: Column 1   column 3   column 4   column 5
Column 1   5   ∞   2   ∞
1 1 __ __ __

Note: Column 4 was not selected because it fails to be represented in either parent.

FIRST GROUP RESULTS

The results for a selection of the first group of set covering experiments are presented in Tables 1-3. It should be noted that for this group of problems the column costs were randomly chosen as were the positions of the "1's" in the matrices (subject only to the existence of a cover and the density constraint). Tables 1 and 2 show the clear superiority of the greedy crossover over the one- and two-point crossovers. Table 1 summarizes the performance of the techniques relative to one another; of particular note is the poor performance of the p1 penalty function formulation, the good performance of the greedy p3 formulation, and the excellent performance (relative to the other methods tested in this experiment) of the two-point crossover Pareto formulation.

	CROSSOVER METHOD					
PROBLEM	one-point		two-point		greedy	
DENSITY	wins	ties	wins	ties	wins	ties
10			1		4	
20				1	4	1
30				2	3	2
40	1			1	3	1
50		2			3	2
60				1	4	1
70		1		2	3	2
80		2		2	2	3
90		1		1	3	2
Totals	1	6	1	10	29	14

Table 1. Relative SCP Performance by Problem Density and Crossover Method for Penalty Function p2 and Group 1 Problems

		1 Pt. Best	Cross Trial	2 Pt. Best	Cross Trial	Greedy Best	Cross Trial	Pure@ Greedy	Optimal
	1	2	412	3	560	0	80	60.11	?
	2	0	593	5	593	5	56	7	?
10%	3	12	613	0	630	0	258	3	?
	4	1	702	5	663	0	90	1	?
	5	15	547	5	389	0	77	0	?
	1	3	682	0	831	0	61	13	?
	2	18	550	15	496	0	669	7	?
20%	3	22	713	40	524	0	114	10	?
	4	19	534	22	783	0	646	9	?
	5	80	614	21	641	0	75	20	?
	1	41	786	0	787	0	77	0	11.52
	2	36	918	13	733	9	93	16	14.80
30%	3	3	611	45	913	0	85	0	8.20
	4	0	763	2	431	2	106	13	?
	5	32	688	17	915	0	96	0	7.55
	1	0	937	8	788	8	81	8	10.70
	2	28	791	36	745	2	108	22	9.56
40%	3	52	949	56	1006	0	95	0	7.50
	4	24	523	0	633	0	96	6	6.42
	5	42	631	21	612	5	104	27	12.64
	1	27	703	91	623	0	78	27	2.86
	2	0	378	23	434	0	79	0	4.38
50%	3	7	855	47	684	0	81	0	7.57
	4	0	992	46	718	0	220	30	6.10
	5	3	929	3	644	1	90	3	6.19
	1	52	893	36	913	0	90	5	3.12
	2	39	687	0	581	0	1014	12	4.66
60%	3	16	723	16	962	10	77	10	5.31
	4	42	799	0	795	0	82	0	3.63
	5	84	493	7	779	0	76	36	3.77
	1	27	645	19	607	19	107	53	3.96
	2	94	663	22	795	0	94	16	1.76
70%	3	0	547	0	729	0	76	0	2.31
	4	97	727	14	610	0	97	5	2.64
	5	57	473	57	540	0	77	25	0.61
	1	0	728	127	802	0	78	0	0.92
	2	0	426	0	922	0	92	33	2.03
80%	3	73	442	52	548	0	85	30	2.25
	4	31	545	48	922	12	80	13	3.70
	5	38	819	0	686	0	80	38	2.59
	1	N/A	887	0	551	0	76	26	0.62
	2	0	979	80	787	0	86	3	0.66
90%	3	41	851	6	815	0	79	41	?
	4	4	667	4	532	0	80	42	3.65
	5	28	738	125	825	0	76	0	1.04

*Greedy Heuristic Applied Until a Cover is Generated

Table 2. Problem by Problem SCP Performance for Penalty Function p2 and Group 1 Problems

Genetic Algorithms Applications

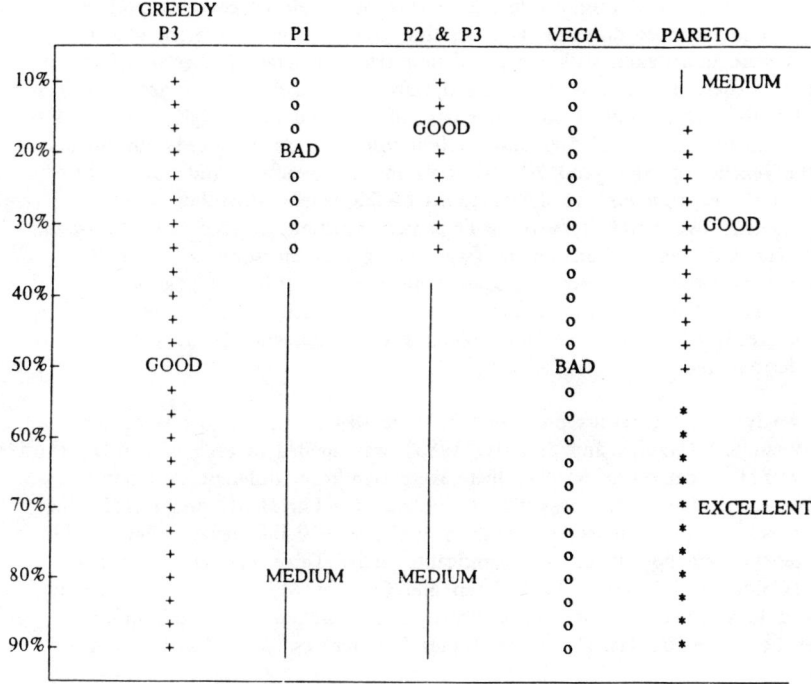

Table 3. Results for Group 1 Problem Comparison

(Ratings are relative to other methods represented in the table.)
Two point crossover was used.

SECOND GROUP RESULTS

This second group of results refers to twenty 50 X 50 set covering problems designed to be more diffucult than the first group. Three different experiments were undertaken with the population size and total number of trials varied. These experiments are denoted Hard1 with a population size of 100 and 3000 trials, Hard2 with a population size of 100 and 6000 trials, and Hard3 with a population size of 500 and 40,000 trials for the nongreedy formulations of the genetic algorithm and 20,000 trials for the greedy formulations. In each of the experiments, six formulations of the genetic algorithm were investigated, two-point crossover with penalty function p2 (scp_p2), two-point corssover with penalty function p3 (scp_p3), greedy crossover with penalty function p2 (gr_p2), greedy crossover with penalty function p3 (gr_p3), two-point crossover with ranking (rank_scp_p3), and greedy crossover with ranking (rank_greedy_p3), the latter two both in conjunction with the p3 penalty function formulation.

The analysis of the results proceeded in three steps. In the first step, the Friedman test (Golden and Stewart, 1985) was applied to each of Hard1, Hard2, and Hard3 to determine whether there were significant differences in performance among the genetic algorithm formulations. For Hard1 and Hard2, the hypothesis of no difference was rejected at the $\alpha = 0.001$ level. For Hard3, differences were not statistically significant. (See Table 4) The Wilcoxen test (Golden and Stewart, 1985; Iman and Conover, 1983) was subsequently applied to selected pairs of genetic algorithm formulations from each of the experiments, Hard1-Hard3. (See Tables 5-7 for results of the Wilcoxen tests.)

Finally, a utility of the various genetic algorithm formulations was computed as suggested in Golden and Stewart (1985). Analysis of these utilities confirmed the results of the Wilcoxen tests but shed no further light. From the Wilcoxen tests, a hierarchy of dominance can be drawn. This hierarchy is illustrated in Figures 4 for Hard1, 5 for Hard2, and 6 for Hard3, where it can be seen that the greedy crossover with the p3 penalty function dominates all other variants tested in these experiments.

Experiment	Tf	Alpha
Hard1	30.7	<.001
Hard2	11.02	<.001
Hard3	0.525	~.75

Table 4. Friedman Test Applied to Solution Cost (Statistic Tf Compared to F-table With df1=95, df2=5): H_0 = Expected Costs Equal H_A = Not All Expected Costs Equal

Wilcoxen Signed Rank Test

Hard1

Methods:	n:	Alpha:
scp_p2 vs. scp_p3	19	>.1,<.25
greedy_p2 vs. greedy_p3	14	.1
scp_p3 vs. greedy_p3	20	<.01
rank_greedy_p3 vs. greedy_p3	13	>.25,<.4
rank_scp_p3 vs. scp_-3	20	<.01

Table 5.

Hard2

Methods:	n:	Alpha:
scp_p2 vs. scp_p3	19	.1
greedy_p2 vs. greedy_p3	13	.25
scp_p3 vs. greedy_p3	20	.01
rank_greedy_p3 vs. greedy_p3	13	.5
rank_scp_p3 vs. scp_-3	19	.1

Table 6.

Hard3

Methods:	n:	Alpha:
scp_p2 vs. scp_p2	18	.25
greedy_p2 vs. greedy_p3	9	.4
scp_p3 vs. greedy_p3	17	.1
rank_greedy_p3 vs. greedy_p3	11	.1
scp_p3 vs. rank_scp_-3	15	.5

Table 7.

Tables 5–7: Wilcoxen Signed Rank Test applied to pairs of methods. With method1 vs. method2, $H_O = \{ E[\text{method1}] \leq E[\text{method2}] \}$, $H_A = \{ E[\text{method1}] > E[\text{method2}] \}$.

Statistic $T = \sum_{1}^{n} R_i / (\sum_{1}^{n} R_i^2)^{1/2}$, or in the

case of no ties, $T = \sum_{1}^{n} R_i / [n(n+1)(2n+1)/6]^{1/2}$,

where R_i, $i = 1, \ldots, n$ are the signed ranks of differences, and R is their mean (Iman and Conover, 1983).

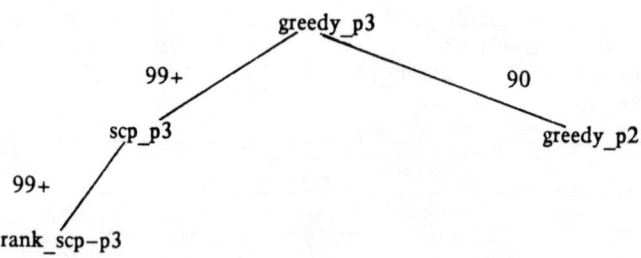

Figure 4. Domination at 90% and 99% Significance Levels for Hard1

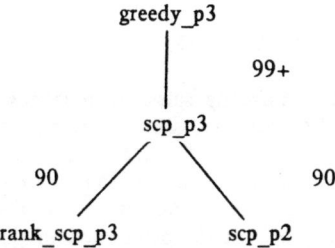

Figure 5. Domination at 90% and 99% Significance Levels for Hard2

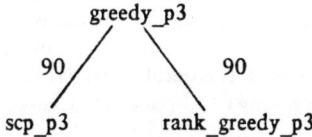

Figure 6. Domination at 90% Significance Levels for Hard3

THIRD GROUP RESULTS

The third group of set covering problems were a selection of problems shared by Balas and denoted as salk.1–salk.12. The genetic algorithm formulations used were greedy-p2 and scp-p2. Results are presented in Table 8.

Problem	Rows	Columns	Greedy_p2 Solution	scp_p2 Solution
salk.1	15	32	14	17
salk.2	30	70	14	16
salk.3	30	30	15	14
salk.4	30	50	14	15
salk.5	30	60	14	14
salk.6	30	40	15	15
salk.7	30	60	15	15
salk.8	30	80	13	16
salk.9	30	90	14	16
salk.10	30	80	15	15
salk.11	30	70	14	15
salk.12	30	90	13	17

Table 8. Set Covering Results for Selected Balas Problems

TRAVELING SALESMAN PROBLEM

Two groups of Eculidean traveling salesman problems (TSP) were studied. The first group consisted of twenty Euclidean 15-city problems with the cities randomly chosen in a 5000 by 5000 square. The second group consisted of two problems cited in Golden and Stewart (1985), a 50- and 75-city problem from Eilon et al. (1969); a close approximation to the 30-city problem investigated by Hopfield and Tank (1985) and documented in Oliver et al. (1987); and ten problems randomly generated as in the first group.

The first group of experiments provide results for the PMX and greedy crossover (defined below). The effects of anchoring the tours and seeding the initial population were investigated. (Grefenstette, 1987, has also studied the effect of seeding the initial population.) In the anchored case, all tours were constrained to a canonical form (i.e., all tours begin at one single specified 'starting' city). In the unanchored case, the starting cities were randomly initialized.

For the second group, four algorithms were investigated: a greedy crossover (Grefenstette et al., 1985), PMX (Goldberg, 1985), two-swap and a synergistic hybrid of two-swap with the greedy genetic. (For a study of a similar hybrid using two-opt, see Suh and Van Gucht, 1987.) In this group, all tours were anchored, and no seeding was used.

CROSSOVER OPERATORS

The traveling salesman problem is a pure ordering problem and the conventional crossover operators are poorly suited; they do not maintain tours. As a result, various modifications to crossover have been suggested. Two such are the greedy (Grefenstette et al., 1985) and the partially mapped crossover (PMX – Goldberg and Lingle, 1985).

The Grefenstette et al. (1985), greedy crossover was defined as follows:

1. For each pair of parents, pick a random city for the start.

2. Compare the two edges leaving the city (as represented in the two parents) and choose the shorter edge.

3. If the shorter parental edge would introduce a cycle into the partial tour, then extend the tour by a random edge.

4. Continue to extend the partial tour using Steps 2 and 3 until the circuit is completed.

The greedy crossover reported in this paper is a slight modification of Grefenstette et al. (1985). In Step 2, if the original parent's edge cannot be used, try the other parent's edge before making a random choice.

Goldberg and Lingle (1985) introduced the partially mapped crossover (PMX) for any order problem. The example Goldberg and Lingle used to illustrate PMX considers two permutations of ten objects:

$$A = 9\ 8\ 4\ 5\ 6\ 7\ 1\ 3\ 2\ 10$$
$$B = 8\ 7\ 1\ 2\ 3\ 10\ 9\ 5\ 4\ 6$$

If two random numbers, say 4 and 6, are chosen as the crossover points for these two genes, then each string (gene) is to be modified by the permutations (5 2), (6 3), and (7 10)

$$A = 9\ 8\ 4\ |\ 5\ 6\ 7\ |\ 1\ 3\ 2\ 10$$
$$B = 8\ 7\ 1\ |\ 2\ 3\ 10\ |\ 9\ 5\ 4\ 6$$

and the respective results become

$$A' = 9\ 8\ 4\ 2\ 3\ 10\ 1\ 6\ 5\ 7$$
$$B' = 8\ 10\ 1\ 5\ 6\ 7\ 9\ 2\ 4\ 3$$

The two-swap algorithm is a local improvement algorithm for tours. For any tour, the two-swap strategy simply chooses two cities at random and interchanges the cities if this reduces the tour length.

For example: $A = 9\ 8\ \underline{4}\ 5\ 6\ 7\ \underline{1}\ 2\ 10$

Replace A with

$$A' = 9\ 8\ 1\ 5\ 6\ 7\ 4\ 2\ 10.$$

The hybrid two-swap greedy algorithm simply performs the greedy crossover on a fixed percentage of the population (50 percent in these experiments) and then applies the two-swap heuristic on the remainder of the population. The number of times the two-swap heuristic was applied to a tour was equal to the tour length divided by four.

FIRST GROUP RESULTS

Both greedy crossover and PMX crossover were tested with seeded and unseeded initial populations. The seeded initial population included the 15 optimal greedy algorithm-generated tours (possibly including duplicates) beginning at each of the 15 different cities, that is, for each city, the greedy algorithm was iteratively applied until a tour was generated. This subpopulation of 15 was randomly extended to an initial population of 50. The nearest neighbor tour was the best of the 15 greedy algorithm-generated tours. Results of the performance of the greedy and PMX crossovers are given in Table 9. Discouraging is the poor performance of the PMX with the unseeded population; only for one problem did it perform better than the nearest neighbor algorithm. Among the remaining variations of the genetic algorithm, the greedy genetic seemed to perform best, and all genetic variants performed somewhat better than the nearest neighbor. Perhaps surprisingly, the unseeded greedy genetic performed nearly as well as its seeded counterpart.

prob. #	nearest neighbor	pmx unanchored and seeded	pmx anchored and seeded	pmx anchored	greedy seeded	greedy
1	19484	99%	94%	119%	99%	94%
2	16480	100%	100%	110%	100%	100%
3	17673	100%	100%	128%	100%	101%
4	15701	100%	94%	118%	91%	91%
5	16219	100%	100%	114%	100%	100%
6	13613	100%	100%	100%	100%	102%
7	17400	100%	100%	141%	100%	102%
8	15067	100%	100%	111%	100%	100%
9	19755	100%	99%	127%	99%	100%
10	21299	100%	97%	110%	98%	93%
11	18034	98%	98%	123%	96%	95%
12	18587	100%	100%	118%	100%	106%
13	17078	100%	100%	123%	100%	92%
14	19535	100%	98%	122%	95%	95%
15	16052	100%	100%	130%	100%	100%
16	21934	92%	91%	98%	88%	86%
17	16354	100%	99%	124%	99%	100%
18	21977	100%	100%	119%	99%	103%
19	17049	100%	100%	133%	100%	109%
20	17869	100%	100%	141%	100%	100%

Table 9. Shortest Tours Found for Each Problem by Technique (as percent of nearest neighbor)

Figures 7 and 8 display the results of different seeds for the randomization of initial populations and genetic mechanisms for the first problem. Discouragingly, approximately 8 percent variation in performance can be noted for the greedy genetic as either initial population or genetic mechanism is randomized differently. The counterpart variations for the PMX are 13 percent and 17 percent, respectively.

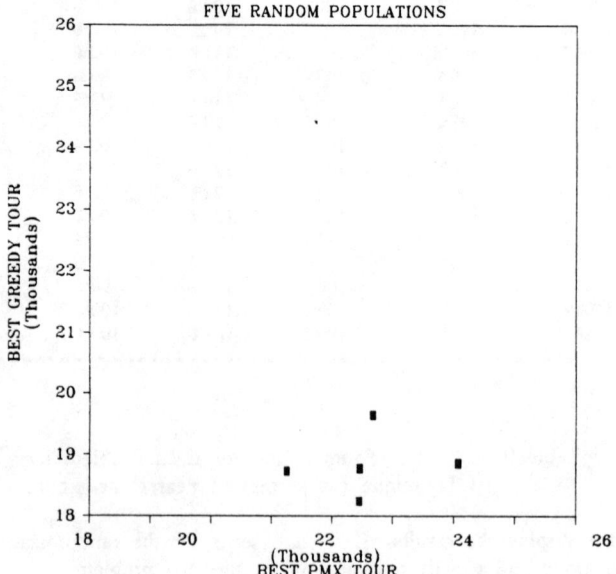

Figure 7. Random Initial Population Seeds as a Factor in Performance Differential

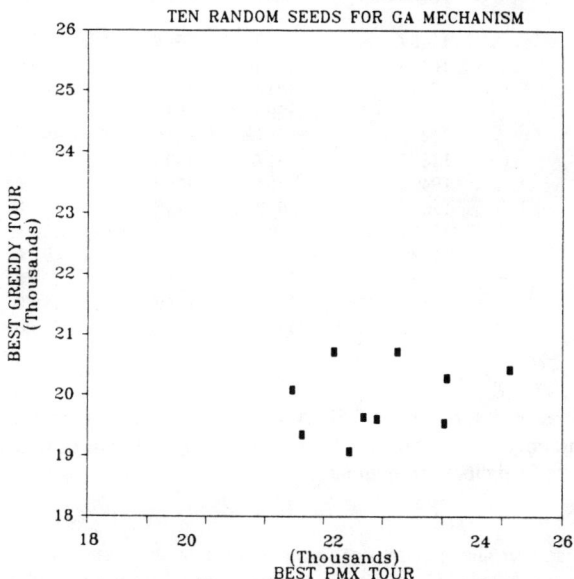

Figure 8. Random Seeds for GA Mechanism as a Factor in Performance Differential

SECOND GROUP RESULTS

In the second group of experiments, PMX, greedy crossover, two-opt alone, and greedy crossover with two-opt were run on a suite of randomly generated problems and three additional problems from TSP literature. Results of these experiments are presented in Table 10.

Problem	Nearest Neighbor or Optimal	PMX	Greedy	Two-Swap	Hybrid
1	19484	117%	94%	97%	94%
2	16480	112%	102%	100%	102%
3	17673	108%	100%	100%	100%
4	15701	130%	91%	91%	91%
5	16219	128%	100%	97%	97%
6	13613	131%	100%	106%	100%
7	17400	126%	100%	100%	100%
8	15067	105%	98%	98%	100%
9	19755	109%	99%	100%	99%
10	21299	121%	92%	92%	92%
HT 30	424	121%	105%	109%	102%
EIL 50	430	231%	115%	186%	109%
EIL 75	553	307%	107%	200%	?

Table 10. Performance for Several TSP Algorithm on a Set of Problems Both Randomly Generated and Drawn From the Literature as percent of nearest neighbor or optimal.

The results suggest that the greedy_two-swap hybrid is the best performing of the algorithms tested here. However, it remains uncertain whether the gain in performance over two-swap alone justifies the additional computation. The use of genetic algorithms in the TSP problem domain cannot be unquestionably justified yet.

CONCLUSION

This paper has provided a brief overview of current genetic algorithm results; has presented new results regarding penalty function formulations for constrained problems (especially for set covering problems); and has highlighted selected results for the traveling salesman problem. The results for set covering problems were surprisingly encouraging; those for the traveling salesman problem, less so. However, the genetic algorithm is a mulitpurpose optimization tool, and one should not necessarily expect it to compete favorably with specialized techniques that have evolved over decades of research directed at a particular class of problems.

Acknowledgements: The authors wish to acknowledge the many stimulating and fruitful discussions regarding genetic algorithms and classifier systems that they have had with Dr. David Goldberg of the the University of Alabama. Credit is extended to Dr. Goldberg for first suggesting the use of the Pareto formulation for constrained optimization.

REFERENCES

Baker, J. E. (1985). Adaptive Selection Methods for Genetic Algorithms, in Grefenstette (ed), Proceedings of an International Conference on Genetic Algorithms and Their Applications, 101-111.

Bethke, J. D. (1980). Genetic Algorithms as Function Optimizers, Ph.D. Thesis, Department of Computer and Communication Sciences, University of Michigan.

Bledsoe, W. W. (1961). The use of Biological Concepts in the Analytical Study of Systems, Presented at the November 1961 ORSA/TIMS National Meeting in San Francisco, CA.

Booker, L. (1987). Improving Search in Genetic Algorithms, in Davis (ed), Genetic Algorithms and Simulated Annealing, Pitman, London, and Morgan Kaufmann Publishers, Inc., 61-73.

Bowen, D. (1986). A Study of the Effect of Internally Determined Crossover and Mutation Rates on Genetic Algorithm Optimization, Unpublished (April 1986) Available through D. E. Goldberg, University of Alabama, Tuscallusca, Alabama.

Brent, R. P. (1971). Algorithms for Finding Zeros and Extrema of Functions Without Calculating Derivatives, Ph.D. Thesis, Computing Science Department, Stanford University.

Fletcher, R. and Powell, M. M. P. (1963). A Rapidly Convergent Descent Method of Minimization, The Computer Journal. 7:149.

Fletcher, R. (1970). A New Approach to Variable Metric Algorithms. The Computer Journal. 13:317.

Garey, M. R. and D. S. Johnson (1979). Computers and Intractability: A Guide to the Theory of NP-Completeness, Freeman, San Francisco, California.

Goldberg D. E. and R. Lingle, Jr. (1985). Alleles, Loci, and the Traveling Salesman Problem, in Grefenstette (ed), Genetic Algorithms and Their Application: Proceedings of the Second International Conference on Genetic Algorithms and Their Applications, 154-159.

Goldberg, D. E. (1987a). Personal Communications at Oak Ridge National Laboratory.

Goldberg, D. E. (1987b). Simple Genetic Algorithms and the Minimal Deceptive Problem, in Davis (ed), Genetic Algorithms and Simulated Annealing, Pitman, London, and Morgan Kaufmann Publishers, Inc., 74-88.

Goldberg, D. E. (1989). Genetic Algorithms in Search, Optimization, and Machine Learning, Addison-Wesley.

Golden, B. L. and W. R. Stewart (1985). Empirical Analysis of Heuristics, in Lawler et al. (ed), The Traveling Salesman Problem, John Whiley and Sons, New York, New York, 207-249.

Grefenstette, J. (1984). A User's Guide to Genesis, Tech Report CS-84-11, Computer Science Department, Vanderbilt University, Nashville, Tennessee. Grefenstette (1986) (1987).

Grefenstette, J. J. (ed) (1985). Proceedings of an International Conference on Genetic Algorithms and Their Applications, Texas Instruments and U.S. Navy Center for Applied Research and Artificial Intelligence.

Grefenstette, J. J. (1986). Optimization of Control Parameters for Genetic Algorithms. IEEE Transactions on Systems, Man, and Cybernetics, SMC-16(1), 122-128.

Grefenstette, J. J. (ed) (1987). Genetic Algorithms and Their Applications: Proceedings on the Second International Conference on Genetic Algorithms, Lawrence Erlbaum Associates.

Grefenstette, J. J. (1987). Incorporating Problem-Specific Knowledge Into Genetic Algorithms. In L. Davis (ed), Genetic Algorithms and Simulated Annealing (pp. 42-60). London: Pitman.

Grefenstette, J. J. and J. M. Fitzpatrick (1985). Genetic Search With Approximate Function Evaluations, in Grefenstette (ed), Proceedings of an International Conference on Genetic Algorithms and Their Applications, 112-120.

Grefenstette, J. J., R. Gopal, B. J. Rosmaita, and D. Van Gucht (1985). Genetic Algorithms for the Traveling Salesman Problem, in Grefenstette (ed), Proceedings of an International Conference on Genetic Algorithms and Their Applications, 160-168.

Holland, J. H. (1975). Adaption in Natural and Artificial Systems. Ann Arbor: The University of Michigan Press.

Holland, J. H. and J. S. Reitman (1978). Cognitive Systems Based on Adaptive Algorithms, in D. A. Waterman, and F. Hayes-Roth (ed), Pattern-Directed Inference Systems, Academic Press, New York, New York.

Hopfield, J. J. and D. W. Tank (1985). "Neutral" Computation of Decisions in Optimization Problems, in Biological Cybernetics, 52, 141-152.

Huang, H. Y. (1970). Unified Approach of Quadratically Convergent Algorithms for Function Minimization, Journal of Optimization Theory and Application, 5:405.

Iman, R. and W. J. Conover (1983). Modern Business Statistics, John Wiley and Sons, New York.

Jog, P., and D. Van Gucht (1987). Parellelization of Probabilistic Sequential Search Algorithms, in Grefenstette (ed), Genetic Algorithms and Their Applications: Proceedings of the Second International Conference on Genetic Algorithms, Lawrence Erlbaum Associates, 170-176.

Marsten, R. F. and F. Shepardson (1981). Exact Solution of Crew Scheduling Problems Using the Set Parition Model: Recent Successful Applications, Networks, Vol. 11, No. 2, 165-177.

Oliver, I. M., D. J. Smith, and J. R. C. Holland (1987). A Study of Permutation Crossover Operators on the Traveling Salesman Problem, in Grefenstette (ed), Genetic Algorithms and Their Applications: Proceedings of the Second International Conference on Genetic Algorithms, Lawrence Erlbaum Associates 224-230.

Revelle, C. D., D. Marks, and J. C. Liebman (1970). An Analysis of Private and Public Sector Facilities Location Models, Management Science 16, 12, 692-707.

Schaffer, J. David and A. Morishima (1987). An Adaptive Crossover Distribution Mechanism for Genetic Algorithms, in Grefenstette (ed). Genetic Algorithms and Their Applications: Proceedings of the Second International Conference on Genetic Algorithms, Lawrence Erlbaum Associates, 36-40.

Shaefer, C. G. (1985). Directed Trees Method for Fitting a Potential Function. Proceedings of an International Conference on Genetic Algorithms and Their Applications, 207-225.

Shaefer, C. G. (1987). The ARGOT Strategy: Adaptive Representation Genetic Optimizer Technique, in Grefenstette (ed), Genetic Algorithms and Their Applications: Proceedings of the Second International Conference on Genetic Algorithms, Lawrence Erlbaum Associates, 50-58.

Suh, J. Y. and D. Van Gucht (1987). Incorporating Heuristic Information Into Genetic Search, in Grefenstette (ed), Genetic Algorithms and Their Applications: Proceedings of the Second International Conference on Genetic Algorithms, Lawrence Erlbaum Associates, 100-107.

Wilson, S. W. (1987). Classifier Systems and the Animat Problem, Machine Learning, 199-228.

DISCOVERING AND REFINING ALGORITHMS THROUGH MACHINE LEARNING

Michael R. Hilliard, Gunar E. Liepins and

Mark Palmer

ABSTRACT

The development of solution methods for classes of operations research problems involves formulating the problems mathematically, developing algorithms to solve the abstracted formulations and evaluating the solutions. One possible contribution of artificial intelligence to this process is the application of machine learning to algorithm discovery and refinement. This paper presents several genetic algorithm and classifier system based experiments to discover and refine algorithms for simple scheduling problems. The discovered algorithms can be considered to be rule bases that are modified and adapted through training with examples. The quality of the resultant algorithm is investigated as a function of the training.

INTRODUCTION

Operations research concentrates on developing solution methods for problems arising in business, industry, and government. The solution methods are formulated as combinations of modeling, algorithm and evaluation techniques [Hillier and Lieberman, 1986]. The development of a solution method proceeds in three steps: the problem is abstracted as a mathematical model, an algorithm is developed and refined to solve the mathematical model, and the solutions produced are evaluated in terms of the original problems. The development of a solution method for a specific problem is typically an iterative process. Although the literature (in good scientific style) does not usually report on all of the failures and false starts in the process, typically the mathematical model and algorithm are modified numerous times before a methodology is finalized. One of the possible contributions of artificial intelligence to this process is the application of machine learning to algorithm discovery and refinement.

Several genetic algorithm and classifier system based experiments provide

a start at understanding how machine learning might support heuristic scheduling algorithm development.

THE METHODOLOGY DEVELOPMENT PROCESS

Modeling the Problem

The first step in solution methodology development for a particular class of problems is to mathematically abstract the problems. The abstractions must fit the notions of sets, constraints, functions, networks, or whatever formalism seems appropriate. This abstracted structure is used in the second step, the discovery and refinement of an algorithm.

Developing the Algorithm

Formally, algorithm development and refinement can be viewed as a search problem in a space of possible algorithms. While not usually formulated as a search problem, algorithm development does consist of trying different parameters, different control structures, different bounds, or, in the case of expert systems techniques, different rules. Automating the search process through the use of machine learning techniques requires formally defining and bounding the space of possible algorithms and determining an intelligent method of searching that space. Algorithms which can be implemented in a data driven form are particularly amenable to this type of approach. By data driven, we mean algorithms which can be separated into a general procedure and adaptable data which controls the algorithm's behavior on a specific type of problem. These algorithms range from heuristic procedures defined by a small set of parameters, to rule based systems in which the inference engine constitutes the procedural portion and the entire rule base is the adaptable data.

Evaluating the Solution Technique

Learning requires feedback. Unless a system (human or machine) nows how well it is performing there is little chance of improving. An

DISCOVERING AND REFINING ALGORITHMS

important aspect of a machine learning system is the evaluation determined by some form of critic. The critic may be omniscient and provide accurate evaluations of the proposed problem solutions based on full knowledge of the optimal solution, or the critic may be able to evaluate the problem solution only in terms of some theoretical bound or on the basis of the best solution seen so far. The evaluation is linked to a mechanism which modifies the algorithms adaptable data. This adaptation of the algorithm to a specific problem type brings the benefit of machine learning to algorithm development.

Overview of Learning Experiments

This paper describes two experiments on algorithm development for scheduling problems through the use of genetic algorithms and classifier systems. The scheduling algorithms are rule based systems with simple inference engines. The form and relative importance of the rules is modified through a learning process. The reminder of this paper describes the classifier system and the genetic algorithm, formulates three solution techniques for simple job shop scheduling problems, and highlights the results of a series of experiments.

GENETIC ALGORITHMS AND THE CLASSIFIER SYSTEM

Classifier systems were developed by Holland and Reitman (1978) to automatically discover rules to perform desired actions. In contrast to traditional expert systems where rules are handcrafted by knowledge engineers, classifier systems use the genetic algorithm as a discovery operator to generate rules. These rules are tested in an environment and are automatically assigned relative strengths based on the system's performance and a technique for allocating credit. An understanding of the genetic algorithm provides a useful starting point for discussing the details of the classifier system.

The Genetic Algorithm

The genetic algorithm is a mechanism with a probabilistic component that provides a means to search poorly understood, irregular spaces. The genetic algorithm has been successfully applied to various problems (for example, tuning of expert systems) that could not have been readily solved with more conventional computational techniques. John Holland developed the genetic algorithm and provided its theoretical foundation in his book *Adaptation in Natural and Artificial Systems* (1975). Holland's formulation was motivated by the observation that sexual reproduction in conjunction with the pressure of natural selection has enabled nature to develop species remarkably well adapted to their environment. The breakthrough contribution that set his formulation apart from those of predecessors was the central role played by the crossover operator as the underlying discovery mechanism; mutation is an infrequent operator, used primarily to preserve population diversity. Although Holland was not able to prove convergence theorems for genetic algorithms, he did prove one theorem about the efficiency of the technique--if the choice of search hyperplanes is viewed as a k-armed bandit problem, then the genetic algorithm optimally allocates trials to the hyperplanes and, in this sense, it is optimal.

The Genetic Algorithm Cycle

Although there are many possible variants of the basic genetic algorithm, the fundamental underlying mechanism is relatively standard and consists of three basic operations: (1) evaluation of individuals in the population, (2) formation of a gene pool, and (3) recombination and mutation. The individuals resulting from these three operations form the next generation's population. The process is iterated until the system ceases to improve. (Usually at this time, the population has converged to a few well performing individuals.) Generally, each individual in the population is represented by a fixed length binary string which encodes a solution to the problem. The population size remains fixed from generation to generation and is typically between 50 and 200 individuals. Individuals contribute to the gene pool in proportion to their relative merit (function evaluation divided by average evaluation) on the function being optimized, that is, well performing individuals contribute multiple copies, and poorly performing individuals contribute few (if any) copies. (Typically, each individual contributes its allotted integer number of copies. Whether or not an individual contributes an additional copy corresponding to the fractional part of its relative merit is determined probabilistically.) The recombination operation is the crossover operator. The simplest variant of the crossover operator selects two parents at random from the gene pool as well as a crossover position within the binary encoding. The parents exchange tails, the portion of the string to the right of the crossover point, to generate two offspring. Mutation is a probabilistic operator which changes each position on a string with a small probability, usually in the .01 range. The strings generated by crossover and mutation form the new population. Sometimes it is useful to generate only a fraction of the subsequent population by crossover. In these cases, the offspring probabilistically replace the poorest performing individuals in the prior population. The fundamental cycle and basic operations are illustrated in Figures 1-4. A thorough introduction to genetic algorithms is provided in [Goldberg 1989], current research is surveyed in the proceedings of two conferences [Grefenstette 1985, 1987] and public domain code is available [Grefenstette 1984].

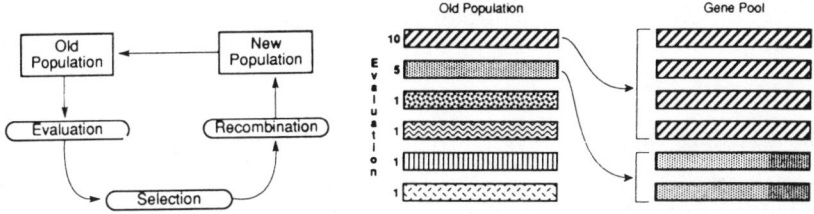

Figure 1. The Genetic Algorithm   Figure 2. Evaluation and Selection

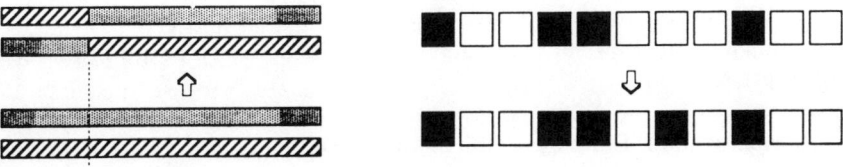

Figure 3. Crossover   Figure 4. Mutation

An Example of the Genetic Algorithm

As a simplistic example of a genetic algorithm application, consider maximizing the function $f(x) = 1-x^2$ where the variable x is restricted to belong to the interval [0 , 1]. Assume that the population size is fixed at four and that a four-bit binary representation is chosen. Then one possible cycle is illustrated in Figure 5.

X-VALUE	OLD POPULATION	FUNCTION EVALUATION	RELATIVE MERIT	GENE POOL	NEW POPULATION
1/16	0 0 0 1	.996	1.27	0 0 | 0 1	0 0 0 0
1/4	0 1 0 0	.938	1.20	0 1 | 0 0	0 1 0 1
3/16	0 0 1 1	.965	1.23	0 | 0 1 1	0 0 1 1
7/8	1 1 1 0	.234	0.30	0 | 0 1 1	0 0 1 1
		Average = .7833			

Function Evaluation at X = 1/16.   Relative Merit at X = 1/16
$1-(1/16)^2 = .996$ .996/.7833 = 1.27

Figure 5. One Genetic Algorithm Cycle for $f(x) = 1 - x^2$

The Classifier System

The classifier system is a data-driven discovery mechanism developed by Holland and Reitman (1978). It differs from a pure genetic algorithm insofar as it is a production rule system. While the full implementation of the classifier system allows for much more flexibility, we have utilized a simpler implementation referred to as a stimulus-response or immediate evaluation classifier system. The classifier system cycle is similar to most production rule inference mechanisms. Rules match information about the current state of the system and suggest actions, conflicts are resolved, and the system moves to a new state. The classifier system adds an additional step in which the new state is evaluated and the active rules are reinforced appropriately.

Matching and Bidding. Holland and Reitman's formulation of the classifier system uses a message list to store the current environmental state and any internal messages. For our purposes these messages are represented as fixed length binary strings which encode numerical or logical

DISCOVERING AND REFINING ALGORITHMS

values. The position of a bit on the string determines its meaning. Each classifier is an "if-then" rule, with a condition part and an action part. The condition part of a classifier is encoded as a string of the same length as the strings on the message list but created from a ternary alphabet (0,1,#) where the # symbol is considered a "don't care" symbol and matches either a 1 or a 0. The action part of the classifier specifies the suggested action and is also encoded as binary string but not necessarily of the same length as the messages. Each classifier has an attribute called its strength which takes on a numerical value and represents the merit or expected performance of the rule. If the conditional part of a classifier matches a message(s) on the message list, the classifier competes to pay a portion of its strength (its bid) for the privilege of acting. The bid is calculated as strength times a bid constant.

Conflict Resolution, Action and Evaluation. Should competing classifiers specify incompatible actions, conflict resolution is based on the size of their bids. Once an action is chosen by the conflict resolution mechanism, the acting classifiers pay out their bids, the system moves to a new state, and the evaluation of that new state provides an appropriate reward or punishment for the acting classifiers of the current cycle. This "bid-payment" cycle is called distribution of strength is repeated for a predetermined number of cycles. System development is based on the assumption that the strengths of the rules after a large number of cycles reflect the merit or average reward that the rule can expect. In fact, a rule which receives a consistent reward reaches a steady state strength when the reward is exactly offset by the bid payment and any taxes. As the system begins to approach steady state, the only possibility for increased performance is the discovery of new rules.

Discovery of New Rules. The genetic algorithm provides the discovery mechanism for the classifier system by recombining the individual classifiers to form a new population. (A classifier's strength determines its contribution to the gene pool.) The encoding of classifiers as binary strings is particularly suitable for this process. Thus, a stimulus-response-reward classifier system closely resembles the genetic algorithm; the major differences are that the underlying primitives are classifiers with a conditional part and consequent part, the evaluation of a classifier requires a separate process, and the system is searching for a set of co-adapted rules rather than a single optima. Figure 6 illustrates the full classifier system cycle, including the genetic algorithm for discovery. A thorough example of a classifier formulation can be found in Holland (1986), and in Goldberg (1989). A review of the state of the art is available [Liepins et. al., 1988].

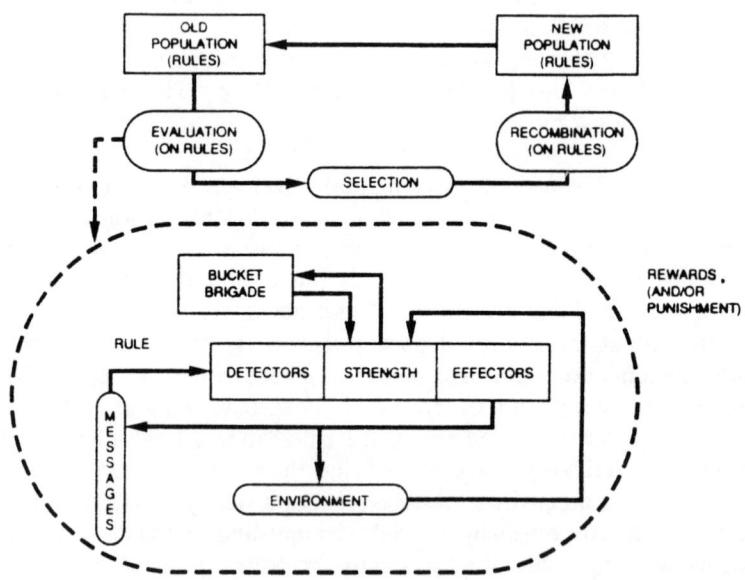

Figure 6. The Classifier System

EXPERIMENTS

The Scheduling Problem

Three implementations of the classifier system were designed to discover solution algorithms for a simple class of scheduling problems. The problems were to schedule sets of eight jobs on a single machine without precedence constraints and to minimize the cumulative lateness of the jobs.

DISCOVERING AND REFINING ALGORITHMS

Each job has a deterministic run time and due date so a problem can be modeled as:

(1) $\quad \text{minimize} \sum_{i=1}^{n} (\text{run}_i + \text{run}_{i-1} + ... + \text{run}_1 - \text{due}_i)$

Obviously this is a simple problem, and the optimal algorithm is to sort the jobs by increasing run time [Conway, 1967]. Knowing the optimal algorithm, however, allowed for the omniscient evaluation of the algorithms discovered by machine learning. Three implementations provided different spaces of algorithms for the learning technique to search. The first two implementations forced the system to work in a primitive representation, while the third allowed the system to deal with a higher level of abstraction.

The Primitive Representation

The classifier system was presented with a set of eight messages with each message encoding the necessary data about a single job. Two different encodings of the data provided a measure of the system's ability to deal with the data. The first representation encoded each job's run time in binary. Run times were randomly chosen integers between 0 and 127, that is seven bit messages. The second representation encoded the ranking of each job's run time and then concatenated the rank of a job with four extraneous bits. Thus the third shortest job might be encoded as 00 011 10. The first two bits were always 00, and the last two bits were randomly generated. This provided a seven bit message compatible with the first representation and a test of the system's ability to ignore extraneous information. In all cases, the due dates were generated randomly between 0 and 255 but were not incorporated into the messages.

In both representations the actions available to a classifier were to place the job(s) it matched into a particular queue position (1st, 2nd, 3rd, ..., 8th). The second representation, the ranking representation, could be solved by a simple set of deterministic rules, which the system generally discovered. The minimal rule set:

000 ##; 000
001 ##; 001
010 ##; 010
.
.
.
111 ##; 111

consisted of eight rules embodying the "sort by run time" heuristic.

The first representation presented a much more complex task. The system had to develop a set of rules which probabilistically performed well. Thus the rule set could not be expected to produce optimal orderings for every possible queue of eight jobs that it encountered; it could only be expected to determine generally good orderings for queues with jobs whose parameters were chosen stochastically from the same distributions as the training set. In short, the system was required to accomplish the difficult task of inferring approximate correlational relationships on the basis of a limited (thirty-two different job queues out of a possible 128^8) training set. The intent of this run time representation was to experiment with a solution space which was not already constrained by knowledge of the optimal algorithm.

Conflict Resolution

The conflict resolution scheme for the classifier system implementation employed a noisy auction mechanism which added an independent identically distributed random variable (mean=0) to each bid to create a noisy bid. This allowed rules with initially low strengths to still be explored. The noise was scaled so that as the performance of the system improved, the noise would attenuate. The intention was to allow the system to exploit its better rules as the search proceeded, in a manner similar to the cooling function used in simulated annealing. This was implemented by scaling the standard deviation of the noise to be proportional to a dynamically controlled fraction of the bid range, that is

DISCOVERING AND REFINING ALGORITHMS

(2) noise = $\dfrac{\text{Best - Average}}{\text{Best - Worst}}$ * bid_range

where
Best = highest reward received so far
Worst = lowest reward received so far
Average = average of the last 20 rewards

The conflict resolution scheme is best described by means of an example. Consider the partial queue and partial rule set in Figure 7.

CURRENT QUEUE

POSITION	JOB	RUN MESSAGE	DUE TIME	DATE	LATENESS
1	A	0100	4 days	day 9	-5 days
2	B	1010	10 days	day 15	-1 day
3	C	0111	7 days	day 8	13 days

Total Lateness 7 days

RULE SET

RULE NUMBER	CODING	NOISY BID	INTERPRETATION
1	0111;10	62	Job C -> Pos. 2
2	010#;01	86	Job A -> Pos. 1
3	1###;11	125	Job B -> Pos. 3
4	###0;01	79	Job A -> Pos. 1
			Job B -> Pos. 1

Figure 7. Conflicting Rules

Each of the rules matches a job(s) in the current queue and suggests a new queue position for that job(s). Rule 4, for instance, recommends placing either job A or job B into position 1.

The bid matrix for an n-job queue is an nxn matrix with the rows indexed by the jobs and the columns indexed by the queue positions. For each job and for each queue position, the system enters into the corresponding matrix cell the highest noisy bid made by any rule suggesting that the job be placed in that queue position. Figure 8 illustrates this for the rule set of Figure 7.

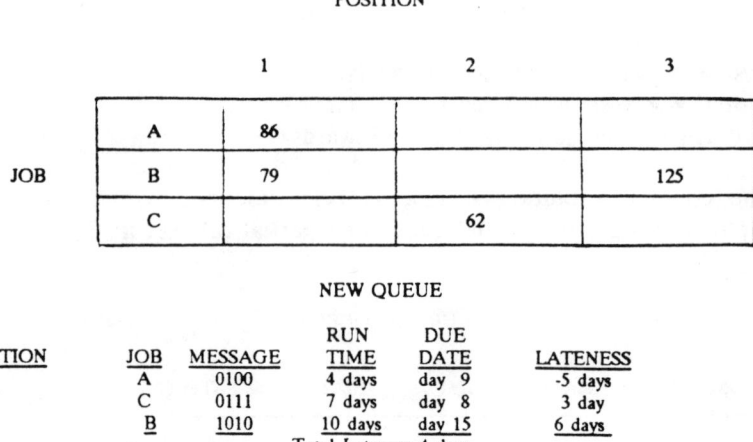

Figure 8. Example Bid Matrix and New Queue

To generate the new queue, choose the largest entry in the matrix (break ties randomly) and place that job in the suggested position. Strike the row and column corresponding to the selected cell and choose the next largest entry. Continue in this way until all queue positions are filled. In our example, job B is placed in position 3, job A is then placed in position 1, and finally job C is placed in position 2.

Reward

Because the system was required to learn general stochastic rules for the problem and not just solve individual problems, the system was trained on a set of 32 randomly generated training problems. After the system had produced schedules for each problem in the suite (the problems were presented in random orders), the rule strengths were updated. Each queue was evaluated for lateness and the resulting number was scaled to fall between 0 and 500, with 500 representing perfect ordering. Each rule's strength was augmented by the average of the evaluations for queues which it participated in forming and decremented by its average bid during those queue formations. To discourage rules which never act from forming, all bidding rules pay a small tax to bid whether they win the conflict resolution or not.

DISCOVERING AND REFINING ALGORITHMS

Generation of New Rules

In addition to the new rules generated by the genetic algorithm, the classifier system created rules to match jobs with no current matching rules. This encouraged exploration of the full search space. These rules were generated by concatenating the unmatched job's message with a random action and then generalizing the rule by randomly (with 50% probability) substituting #s for specific entries in the message.

Results for Primitive Representation

The experiments on these two formulations consisted of three steps.

(1) Establish benchmark performance for a randomly initialized population of rules on a suite of ten job queues. (Call the population of rules resulting from this stage P_B.) (2) Train the system (with the rules resulting from the benchmark stage) with a suite of 32 training queues. (Call the population of rules resulting from this stage P_T.) (3) Test the relative performance of populations P_B and P_T by running P_T against the ten-queue benchmark suite.

In these experiments, the system was run for 1,000 cycles for the benchmarking stage, 40,000 cycles for learning, and another 1,000 for testing. Two strategies were used for benchmarking. In the first, the only population modification was strength modification due to bids and rewards -- called discovery off. In the second, the genetic algorithm was invoked between cycles -- called discovery on. (It should be noted that with discovery on, the benchmark set actually undergoes learning on the very queues on which it is tested. Hence, a comparison of the benchmark with the trained set effectively evaluates the usefulness of additional training on different queues.)

All strategies and experiments were run with five different initializations of the random number generator to minimize stochastic effects. The results are presented in Table 1 in terms of average performance and best performance. The average performance is the average over all 10 test problems of all queue evaluations during an epoch

(100 cycles) averaged over the 5 runs. This represents what performance might be expected from the solution technique on a single problem at that point in time. The best performance is the average of the best performance on each of the 10 test problems during the epoch averaged over the five runs. This represents an estimate of the best performance if the solution process had 100 trials on a problem. It is important to note that the test set is very small compared to the possible problem sets (10 problems out of 128^8).

Table 1 summarizes the resulting improvements in performance. Note that the average performance improves between 15% and 85% while the best performance improves between 2% and 9%.

	IMPROVEMENT	
	AVERAGE	BEST
RUN TIME - DISCOVERY ON	15%	2%
RUN TIME - DISCOVERY OFF	52%	8%
RANKING - DISCOVERY ON	37%	4%
RANKING - DISCOVERY OFF	85%	9%

Table 1. Results of Non predicate Learning Experiments

The Predicate Representation

Rather than only provide the classifier system with the raw data, it was decided to allow the system to choose which of a number of predicates and functions it should use to process the data. The system's actions would then be a function of the preprocessed data. The three logical predicates to be evaluated on a pair of jobs were: (1) run time (jobi) < run time (jobj), (2) due date (jobi) < due date (jobj), and (3) present queue pos (jobi) < present queue pos (jobj). Rules matched on the outcome of these predicates and suggested one of two actions: (1) leave jobi and jobj in their present positions in the queue or (2) invert the positions of jobi and jobj. By use of a bubble sort, rules with this syntax could properly arrange any job queue. It is important to note that inversion and maintenance rules need to cooperate to generate and maintain optimal queue orderings.

DISCOVERING AND REFINING ALGORITHMS

As described in the above paragraph, a queue of jobs was iteratively presented to the system as pairs of jobs. The pairs were presented in the order that they would be examined in a simple bubble sort:

```
DO    I = 1to I=N
  DO  J = I+1 to J = N
    Compare JOB (I) with JOB (J)
    IF action is invert
      Swap JOB(I) with JOB(J)
    ELSE maintain order
  END DO
END DO
```

As in the non predicate approach, all rewards were scaled to lie in the interval [0, 500], and rewards (and corresponding adjustment of strength) were delayed until all the comparisons necessary for a bubble sort were completed. At the end of the sort, each classifier's strength was increased by its average net reward.

The conflict resolution strategy was modified to favor more specific classifiers by increasing their bid values. No taxes were imposed, and no rule generation mechanism was used. The convergence factor in the determination of noise was similar to that defined in equation (2), but the noise was modified only at the end of each epoch.

Results from the Predicate Formulation

The results of these experiments were encouraging. The system did discover the optimal rules and generate reasonable on-line queue arrangements. (A typical list of rules and their associated strengths at cycle 20001 is listed in Table 2.)

Current Classifiers Cycle 20001							
ID	Condition	Act	Str.	ID	Condition	Act	Str.
10	011	1	4263	143	0#1	0	942
19	110	1	4138	140	10#	1	10
79	001	1	4129	142	010	1	10
1	100	1	4089	141	1#1	1	10
11	101	0	4023				
15	010	0	3883				
122	1#1	0	3854				
3	000	0	3852				
53	100	0	1510				
8	0#1	1	1417				
12	111	1	1193				
42	110	0	1164				
121	00#	1	1069				
103	000	1	1002				
37	1##	1	995				
137	101	1	974				

Table 2. A Typical Population of Classifiers for the JSS Predicate Implementation at Cycle 20001

With the noise set to 0 (thus producing a deterministic system), the system also identified the optimal rule set. Additional experiments indicate that such performance can be realized in less than 5000 cycles (approximately 180 training examples).

Parameter Tuning

Many algorithms can be formulated to depend on a small set of parameters. These parameters may determine rankings of alternatives, step sizes, sample size, search depth, or numerous other controls for the algorithm. One logical application at the machine learning task is to discover settings of the parameters which produce good results over a problem class. The basic algorithm remains unchanged; however, the choice of parameters makes the algorithm perform better, on average, on the particular problem class. In a separate effort, we have shown how a genetic algorithm might assist in discovering appropriate parameters for a complex multi-objective scheduling algorithm. A parallel implementation of a multi-objective genetic algorithm was developed and initial experiments indicate that a collection of good parameter sets can be found [Hilliard et. al. 1989a]. This system takes advantage of natural parallelism by evaluating 64 parameter sets simultaneously. The system also employs a non-omniscient critic which evaluates solutions based on the current best solution rather than on an absolute optimum.

CONCLUSION

Future Work

These initial job shop scheduling experiments are obviously limited by the simplicity of the problem being solved and by the sparse information being processed. The predicate formulation seems to be the most fruitful candidate for expansion into more complex realms, and initial experiments are being designed to test the formulation on harder (NP complete) problem classes. The predicate representation allows for great flexibility in expressing the relationship between jobs; however, it requires the designer to make a decision about which logical comparisons will be allowed. If the designer has a strong background in the problem area, or good intuition, this can prove successful, but the design limits the capability of the system to create novel solution techniques not envisioned by the designer. Introducing a large class of predicates, however, would allow the system greater flexibility in discovering algorithms.

Summary

The use of machine learning to develop and refine algorithms could provide a unique contribution to operations research. Typically research in algorithms has either been directed at general techniques which perform adequately on a variety of problems or development of specialized algorithms which surpass the generalized techniques in specific domains. The use of machine learning to adapt general algorithms to special problems may provide a compromise position. Practical implementation of this methodology will require the development of robust solution techniques designed to be adapted. Some parameter based techniques are easily adapted but may not be broadly applicable. Rule based approaches, similar to those described above, may provide greater flexibility. Investigations into combining machine learning with rule based representations of algorithms might provide the necessary insight to develop practical implementations of machine learning for algorithm discovery and refinement.

REFERENCES

Conway, R.W., W.L. Maxwell, and L.W. Miller (1967), Theory of Scheduling, Addison-Wesley, Reading, MA.

Goldberg, D.D. (1989), Genetic Algorithms in Search, Optimization, and Machine Learning, Addison-Wesley, Reading, Mass.

Grefenstette, J. (1984), "A User's Guide to Genesis," Tech Report CS-84-11, Computer Science Department, Vanderbilt University, Nashville, Tennessee.

Grefenstette, J.J. (1985), (ed.), Proceedings of an International Conference on Genetic Algorithms and Their Applications (Carnegie-Mellon Univ., Pittsburgh, Penn., July 24-26).

Grefenstette, J.J. (1987), (ed.), Genetic Algorithms and Their Applications: Proceedings of the Second International Conference (MIT, Cambridge, Mass., July 28-31) Lawrence Erlbaum Associates.

Hilliard, M.R., G.E. Liepins, M. Palmer, and G. Rangarajan (1989),"The Computer as a Partner in Algorithmic Design: Automated Discovery of Parameters for a Multi-Objective Scheduling Heuristic" in R. Sharda, B. Golden, E. Wasil, O. Balci, and W. Stewart (eds.) Impacts of Recent Computer Advances on Operations Research, North-Holland, New York,pp. 321-331.

Hillier, F.S. and G.J. Lieberman (1986), Introduction to Operations Research, Holden-Day, Oakland, CA, pp 16-25.

Holland, J.H. (1975), Adaptation in Natural and Artificial Systems, University of Michigan Press, Ann Arbor, Mich.

Holland, John H. and Judith S. Reitman (1978), "Cognitive Systems based on Adaptive Algorithms," in D. A. Waterman and F. Hayes-Roth, (eds.), Pattern Directed Inference Systems, pp. 313-329.

Holland, John H., K. J. Holyoak, R. E. Nisbett, and P. R. Thagard (1986), Induction, MIT Press.

Holland, John H. (1986), "Escaping Brittleness: The Possibilities of General-Purpose Learning Algorithms Applied to Parallel Rule-Based Systems, in Michalski, Carbonell, and Mitchell," (eds.), Machine Learning II -- An Artificial Intelligence Approach, pp. 593-623.

Liepins, G.E., M.R. Hilliard, M.Palmer, and G. Rangarajan (1988) "Credit Allocation and Discovery in Classifier Systems," IJCAI-89.

Pearl, J. (1984), Heuristics, Addison Wesley, Reading, Mass.

Smith, Stephen F., Mark S. Fox and Peng Si Ow (1986), "Constructing and Maintaining Detailed Production Plans: Investigations into the Development of Knowledge-Based Factory Scheduling Systems", AI Magazine, Vol VII, Number 4, (Fall 1986), pp. 45-61.

II. UNCERTAINTY MANAGEMENT

Three papers comprise this section. The first paper ("Using Probabilities as Control Knowledge to Search for Relevant Problem Models in Automated Reasoning" by Bhatnagar and Kanal) examines the problem of reasoning with uncertain knowledge from a perspective not adopted by many of the existing approaches. This paper examines the alternative of hypothesizing causal models that result in inferences of interest rather than determining inferences in predetermined causal models. The capability to hypothesize interesting causal structure is necessary for any reasoning system, and further research in this area would help build more powerful mechanical reasoners.

The paper by Schum ("On the Marshalling of Evidence and the Structuring of Argument") is concerned with issues that arise when arguments are constructed from a mass of evidence and when the evidence is weighed or assessed in the process of drawing conclusions from it. Of major concern are problems associated with the productive integration of structural and weight-related analyses. In addition, the author addresses some common misconceptions about theories of evidential reasoning and attempts to set the record straight by mentioning the legacy of research on the evidential foundations of probabilistic inference that has been available for many years. All of these matters should be of concern in attempts to provide a productive integration of research in OR and AI in situations in which people must draw conclusions from a mass of evidence that is incomplete, inconclusive, and that comes from sources with every gradation of credibility.

The final paper of the section by Pate-Cornell and Lee ("Hybrid Systems for Failure Diagnosis") concerns the application of a probabilistic representation of uncertainty to failure diagnosis. Failure diagnosis is an integral part of speedy and effective field service support for large and complex industrial systems or machines. The complexity and the interactions of the many components of these systems have made failure diagnosis a difficult task to perform. The contribution of this paper is to develop a hybrid system that combines the advantages of probabilistic risk analysis models, heuristic methods, and optimization techniques as decision support for failure diagnosis and repair. The paper discusses the effectiveness of these methods as a function of the nature of the failures and the decision circumstances. While there have been published results in this area using a single method or approach, this paper is a first attempt to develop such a hybrid system to take advantage of the strengths of several techniques. As a result, the system can be more responsive and flexible to the needs for managing industrial systems or machines. Based on this framework, the authors are interested in the development of prototypes leading to actual implementation in real industrial settings.

Using Probabilities as Control Knowledge to Search for Relevant Problem Models in Automated Reasoning

Raj K. Bhatnagar and Laveen N. Kanal

Machine Intelligence and Pattern Analysis Laboratory
Department of Computer Science
University of Maryland, College Park, Md. 20740.

Abstract

A decision for taking an action is based on problem related evidence and the domain knowledge of the reasoner. The aspect of the problem situation on which the decision is to be based may not be part of the observable evidence. The domain knowledge provides a connection between the evidence and the aspect of interest. The model of the problem world that provides the link between what is observed and what is of interest is of crucial importance. In many approaches to computing uncertainty a model of the problem situation is predetermined by a human and made available to the mechanical reasoner. Construction of such models of interest by mechanical reasoners themselves will certainly make them more effective reasoners. We present a methodology whereby a mechanical reasoner can hypothesize interesting models of a problem situation by using his knowledge of various domain related variables and their inter-relationships. The knowledge about alternate scenarios and their predictions about the items of interest can provide guidance for seeking further evidence in order to either corroborate or refute the various scenarios of interest. The main problem encountered in considering various scenarios is that of efficiently constructing only those that may be of interest. Our methodology involves a graph based formalism that searches for such appropriate scenarios of the problem world.

This research has been supported by NSF grants ECS-83-00799 and DCR-8504011 to MIPAL, the Machine Intelligence and Pattern Analysis Laboratory, University of Maryland, College Park.

1. Introduction

Reasoning with uncertain knowledge has been studied and formalized in many different ways. The desire to provide this capability to mechanical reasoners has led to an examination of existing formalisms and new investigations along some hitherto unexplored lines. Most of the current work has focused on computing the uncertainties associated with propositional inferences. The model of the problem world that provides the context for such computations is provided by humans. For example, the determination of probabilities by using Bayesian networks [7] for some problem instance assumes that the network structure containing relevant aspects of a situation is already available. Various instances of determining probabilities are performed using the same network structure. For the simple case where the network is a tree, the root node, representing an aspect of interest is connected to the leaf nodes for which observations can be made. As noted by Lauritzen and Spiegelhalter in [4, p216] and also presented in [12], many graphical representations of probabilistic knowledge presented in the literature actually represent the influence relationships and not causal relationships in the literal sense. We propose to view the graphical links as purely causal links and want to interpret the graphs as causal structures. In many situations it is easy to elicit the cause effect relationships from the domain and then construct trees with the aspect of interest at the root and the observable aspects at the leaves. However, it may be possible to construct a number of such causal trees for the fixed root and the leaf nodes. This is possible because a number of different causal chains may connect one aspect to another aspect of the problem situation. Which one of these trees shall be used as representing the real situation? This needs to be decided. The method sketched in [5] also uses a causal physiological model for representing the domain knowledge. The algorithm mentioned (but not described) by the authors of [5] is based on signal flow analysis for predicting effects of therapy and identifying causal chains responsible for these effects. This algorithm seems to be quite different from our approach. We present a method for a relatively general problem of selecting preferred hypotheses about the causal structure of a situation that also explain the observations. The selection of preferred hypotheses is achieved through a search algorithm and is guided by one of several possible preference criteria.

A similar situation arises in the case of default logics [11]. Each default rule represents a relationship among variables but it may not be consistent to include all the relationships in any one extension. The question is again to select one of the many possible consistent subsets of all the known default rules. In the case of causal relationships, it may not be necessary to consider all of them for modeling any situation because only a few of them may be present in any particular situation. In this paper we examine a methodology suitable for mechanical reasoners that lets them determine the interesting

Searching for Relevant Problem Models 85

subsets of causal relationships to construct the trees modeling the problem situation. The probabilistic knowledge associated with individual causal relationships and the criterion of interestingness are used to control the search for the interesting trees.

2. Knowledge Needed for Reasoning

A reasoner's domain knowledge may contain alternative sequences of cause-effect relationships between any two aspects of a problem situation. The available evidence may also be uncertain and only about a few aspects of interest. In such a case one of the problems faced by a reasoner is to select a unique cause-effect model of the problem world from among many that may be possible. Probabilities for the aspect of interest are then computed in the context of this selected model of the problem situation. Ideally one would like to identify the actual subset of the possible cause-effect relationships active in the problem situation, but this is generally not possible due to limited observability of the situation. For example, a physician may observe some symptoms of a patient, but the actual chain of physiological cause-effect relationships is a matter of hypothesizing. These hypotheses need to be refuted or corroborated, but before that the reasoner needs to construct some interesting and consistent causal structures as hypotheses. One way of dealing with this situation is to consider all possible models of the problem situation and decide upon that action which, from among all possible actions, has the maximum expected utility. An alternative approach that we present here is to determine those models of the problem situation that are of interest, given some criterion of interestingness. We consider such an approach to be more useful because it

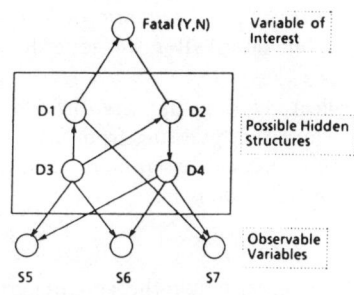

figure-1

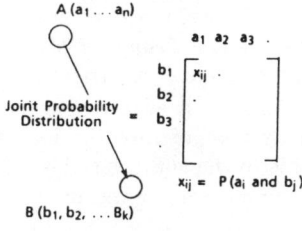

Conditional probabilities derived from x_{ij}'s and known probabilities for one variable can be used to compute the probabilities for the other variable.

figure-2

provides the knowledge about the causal structures of the interesting models of the problem situation. Knowing the model of interest one can design queries for either refuting or corroborating the model and thereby continue one's search for actual causal structure active in the problem situation. The causal structure also provides a good basis for selecting an action because the causes against which such an action must be directed are explicitly known.

One simple formalism for describing the state of a problem world is in terms of a number of variable-value pairs. Each variable may have a number of possible values, but can take only one of them in any particular situation. If all these variables are made the nodes of a graph, then the reasoner's knowledge about the causal relationships can be represented by edges of this graph. A reasoner's complete knowledge may consist of a large number of multi-valued variables and a large number of edges even though only a subset of these causal relationships may be active in any particular problem situation. Also, only a very small subset of the graph's nodes may correspond to the variables observable in any problem situation. Assignments for some unobserved variables of interest may have to be inferred based upon the assignments of the observable variables. In attempts to link the unobserved and observed variables, a number of other variables and edges may have to be involved. In the example of Figure-1, some of the variables $D1$, $D2$, $D3$, and $D4$, and only some of the edges may be included in a cause-effect model that attempts to relate variable *Fatal* to the observed symptom variables. The assignment inferred for *Fatal* may depend on the causal structure that is chosen as the model for the situation. A graph that includes all the nodes relevant to the problem situation and includes edges for all the cause-effect relationships known to the reasoner represents the reasoner's complete knowledge. Any subgraph of this graph that includes the aspect of interest and all those observed nodes that can be linked to it (following some constraints of consistency as discussed later) represents one scenario of interest. A more precise definition of a scenario is given later, after we have defined the concept of causal correctness.

Since each edge represents a cause effect relationship, the direction of causality is represented by making the edge directed; pointing from the cause variable to the effect variable. This edge represents the assignment caused at the head node whenever an assignment at the tail node is fixed.

3. Knowledge of Uncertainty

We now consider the situation in which upon fixing the assignment at the tail variable there is some uncertainty about the assignment *caused* at the head variable of an edge. We represent the knowledge about this uncertainty by associating with the edge the joint probability distribution for the two variables connected by the edge. This probability distribution is based only on those events in which the assignment for the tail variable is known to

Searching for Relevant Problem Models 87

have caused the assignment for the head variable. For example let us consider two variables *Play(board-game, court-game, field-game)* and *Injury(None, minor, severe)*. Let us say the set of known events from which we are going to determine the joint distribution contains a number of pairs $<Play_i, Injury_j>$. Since we know that the kind of game a person plays causally influences the kind of injury he may get and also an injury a person may already have causally influences the kind of games he may be able to play, we associate with each known event (pair of $<Play_i, Injury_j>$) a flag and set it to "1" if $Play_i$ is known to have caused the $Injury_j$. For determining the joint distribution to be associated with the edge from *Play* to *Injury* we count only those events for which the flag is "1". For an edge directed from *Injury* to *Play* we set the flag to "1" only if the $Injury_j$ is known to have caused $Play_i$. The joint distributions associated with the two edges between these two variables may be completely different from each other. The probabilities associated with an edge can be viewed as conditioned on the fact that only the causal influence represented by the edge is active and all other causal influences are absent.

Let us say, we construct some tree out of the graph representing the reasoner's knowledge. This tree is such that the aspect of interest is at the root and the observed nodes are at the leaves. With each edge we can assign a conditional probability matrix containing the values P($Child_i \mid Parent_j$). These values can be determined from the joint probability distributions associated with the edges. The assumption here is that it is these conditional probabilities that remain constant across various problem instances for which the observations at the leaf nodes may vary. How we define a tree to be consistent, to be of interest, and the method for determining it from the graph are the focus of the remaining discussion of this paper. An example is illustrated in Figure-2.

Let us briefly examine the nature of a typical instance of a decision making problem. A decision may involve selection of an action that optimizes the utility. The utility or the expected payoff of an action can be defined only in the context of a conjecture of the model of the problem world. In particular, the utility value may be determined by the assignments made to or the probability distribution assigned to some variable contained in the causal structure of the conjectured scenario. In the example of Figure-1, it may be desired to determine only those causal structures according to which the probability of the patient dying may be highest. The utility of an action may depend solely on its capability to counter the possible death-causing scenarios. Before the action can be selected, it is necessary to determine those scenarios that a physician may consider death-causing. In the case of default reasoning, the value taken by *Fatal* in an extension depends on assertions known to be true (observations) and the selected default-rules. Here a parallel can be drawn and we can say that the probability distribution for

Fatal depends on the observations and the choice of cause effect relations assumed to be active. The edges of the graph represent causal relations and one now needs to select one of the many possible trees to link the observations at the leaves to the aspect of interest at the root.

An elementary path from a node corresponding to an observable variable to the node corresponding to the variable of interest can be used to compute the probability distribution for the latter. When there are many observable variables and each can be used to infer a probability distribution for the same variable, we should find a way to determine their combined influence. This is one of the main problems that faces most methodologies that use numeric measures for representing uncertainty. In the framework of probability theory it is not possible to ascertain P(A,B,C) from the knowledge of P(A,B) and P(A,C). In the absence of the knowledge of P(A,B,C), a reasoner must make some assumptions to be able to say something about A from the knowledge of P(A,B) and P(A,C). An assumption that has been made in many systems is that of conditional independence, that is, P(A.B | C) = P(A|C)*P(B|C). This, or close variations of this assumption have been used in PROSPECTOR and MYCIN etc. In most formalisms that use graphical structure for representing probabilistic knowledge, absence of an edge between two nodes denotes the conditional independence of these nodes, given the set of nodes S, such that if S is removed the two nodes become disconnected [4,5]. In our formalisms an edge represents a causal influence and the absence of an edge between two nodes does not mean any thing more than the absence of a causal relationship between the two. An assumption of the conditional independence of these nodes is only one possible assumption. However, if causal influences are assumed to be independent of each other, then this assumption may hold in many situations. The consequences of the assumption of conditional independence have been addressed in [4, pp 206,217] (in the discussion by Dubois and Prade). Many issues relating to the aggregation of probabilistic information have also been addressed in [2].

If we assume such conditional independence for the tree of causal influences described above, then the method described in [7] can be used to compute the probabilities at all the non-leaf nodes of this tree. However, if we can not assume conditional independence and consider some other rule of obtaining P(A|B,C) from P(A|B) and P(A|C) as more suitable, we should use it for aggregating the effects of B and C on A . Such a rule may be different for different nodes of the tree. When such assumptions have been decided upon, we can determine the probabilities at each of the non-leaf nodes of the tree.

4. Knowledge of Causality

Another important problem that has received some attention relates to the appropriate representation and usage of knowledge of causality [8]. For example, let us consider the graph shown in Figure-3. It shows that disease D1 causes symptoms S1 and S2 and Disease D2 causes symptoms S2 and S3. If we have information about S2, we can hypothesize about D2. From this hypothesis about D2, we can infer about other symptoms of D2. A different situation results when the disease D1 is the observed variable. We can then infer the probability of symptom S2 knowing the probability of D1. But should we now use this probability of S2 to infer the probability of D2? We should not. Knowledge of some effects of a cause makes other effects likely and we can hypothesize about their existence, that is, if we hypothesize the common causal event. But direct observation of a cause of an effect does not make other causes of the same effect likely. This difference results because of the difference between the explicit nature of the knowledge relating to observed variables and implicit nature of the hypothesized causes or effects. This constraint should be obeyed by any description of a scenario that is constructed based upon some observed variables and may be stated as follows :

From explicitly known variables, we can hypothesize either about their effects or about their causes. From a hypothesized cause variable, we can hypothesize about further causes or effects. However, from hypothesized effects, we can not further hypothesize about their other causes.

At best, we can say that knowledge about one cause of an effect makes other causes less likely. The decrease in the confidence associated with competing causes is difficult to discern. The principle stated above will be followed in the methodology described below for constructing model-subgraphs.

The joint probability distribution associated with an edge should also be viewed as representing only the causal influence represented by the edge. It is possible that two variables may have causal influence upon each other, only one being active in any particular scenario. In that case we would have two edges between the nodes corresponding to these variables, each node being pointed to by one of the edges. The joint probability distributions associated with the two edges may be entirely different from each other. If a directed edge points from variable A towards variable B, then the probability distribution associated with the edge represents that relationship in which assignments for A are allowed to cause B to take corresponding assignments and not vice versa.

When we want to discern the belief in one variable given the observed knowledge about some other variable, we need to find a path in our knowledge graph that connects the two variables. The nature of this path may be one of the following: 1. One variable is the causal parent of the other, that is, there is a directed path from one node to the other. 2. Both the variables have a common ancestor node, which has causal influence on the two variables, that is, there are directed paths from the ancestor node to both the nodes. In the second case, the path from an observed node to the ancestor node is the diagnostic part and is traversed against the direction of the arrows. From the ancestor node to the node of interest it is the predictive part and is traversed along the direction of the arrows. Any particular edge of the graph is capable of being used in either of these two directions and the use depends on the problem instance being tackled. The constraints for the correct use of causal knowledge can be summarized in terms of conditions for elementary paths between observed variables and the variables of interest. The basic condition that should be met is that no path between the observed variable and the variable of interest should have a head-to-head connection of edges. Figure-4 summarizes these conditions on the permissible paths.

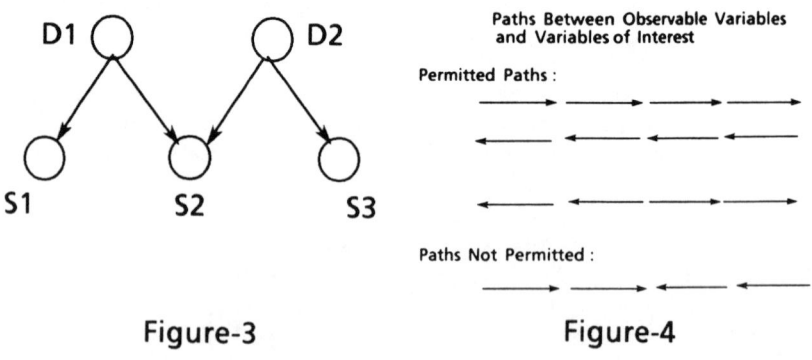

Figure-3   Figure-4

When either the observed variables or the variables of interest are two or more in number then a more general set of conditions that ensures consistency with causality knowledge must be applied. The basic idea underlying the causality constraints is that we must not use the information about a hypothesized effect to make an inference about other (possibly competing) causes of the same effect.

5. Models of Problem Situation

A description of the problem can now be restated as follows. A graph whose nodes represent multi-valued variables and whose edges represent the joint probability distributions for the connected variables, represents a reasoner's knowledge. Probability distributions for some variables of this graph may be directly computed from the observations. A decision making

situation may require the decision to be based on some variable other than the observed ones. Each subgraph (of the graph) that satisfies the following represents a consistent model of the problem domain.
- It satisfies the constraints shown in Figure-4, that is, it is consistent when viewed from the point of view of the causality information.
- There should exist only one path between any two nodes of the subgraph, that is, based on an observation about one variable, only one conclusion is drawn about another variable.
- It includes the node corresponding to the variable of interest. It also includes each such observed node to which a path from the node of interest exists and this path does not violate the previous two conditions.

The above definition forces any scenario to be only a tree structure. There may be a large number of trees that include all the observed variables and the variable of interest and are correct according to the above definition. It may not be possible to select only one of these model-subgraphs for deciding on the action to be taken. The underlying principle of classical decision theory is to find that action which optimizes the utility in the long run if all the model-subgraphs were to occur according to their respective probabilities. In many situations, the maximization in the long run is not as important as making the best decision for the particular problem instance at hand. Scope for acquiring further knowledge may be there but one must know what to ask from the external environment. Our strategy is to determine those model-subgraphs that are of interest (given the criterion of interestingness). The variables corresponding to the nodes present in the structure of a model-tree provide the needed guidance for obtaining new evidence for either corroborating or refuting any particular model-subgraph. Each model-subgraph is like a different line of reasoning for inferring the assignments for the variable of interest. Figure-5 shows an example of a graph and various possible lines of reasoning embedded in it. A graph may have a number of lines of

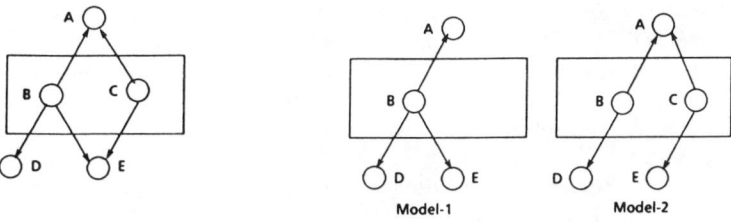

Figure-5

reasoning, each yielding a different probability distribution for the variable of interest. According to the knowledge embedded in the graph, any of the scenarios is equally good and we need some other information for preferring one model-subgraph over the other.

Let us consider the example of Figure-1 again. A large number of scenarios may be possible, each yielding a different probability distribution for the variable *Fatal*. A physician, though aware of the existence of other scenarios, may still like to focus on only those lines of reasoning in which the probability of *Fatal*, i.e., the chances of the patient dying, is very high. His action, which may be either to order further tests to confirm or refute the scenario, may depend on the knowledge of the underlying causal structure of the scenario of interest. Another example is the case of a legal battle between two attorneys.

Given the same set of facts, one attorney may be interested in determining that model of the world or the possible line of reasoning in which the probability for the client being guilty may be minimum and the other attorney may be interested in determining a different line of reasoning or scenario such that the same probability may be maximum. The guidance for obtaining further evidence is provided by the inferred nodes included in the two scenarios. Thus, a criterion of optimality can be specified in terms of the probability distributions for the variables of interest. So, now the problem is to identify that line of reasoning from the graph that connects the observable variables to the variable of interest and optimizes some function of the probability distribution of the variable of interest. The idea of viewing reasoning as selecting appropriate theories that can be built from the known facts has been examined in the framework of formal logic. Some aspects of such an approach, in the context of circumscription have been discussed in [10].

6. Identifying Relevant Models

If A represents the adjacency matrix for the entire knowledge graph, and it has n nodes in it then the matrix A^{n-1} can be computed in such a way that the element $A^{n-1}(i,j)$ of the matrix contains all the directed paths that leave node i and reach node j. We can obtain such a matrix if the matrix A contains the names of the nodes connected by an edge rather than the usual 1's as the entries. Multiplication of A^{k-1} by A to obtain A^k can then be described as follows.

```
For i = 1 to n do
  For j = 1 to n do
    For l = 1 to n do
      A^k(i,j) ← A^k(i,j) ∪ Product(A^(k-1)(i,l), A(l,j))
    End
    A^k(i,j) ← A^(k-1)(i,j) ∪ A^k(i,j)
```

Searching for Relevant Problem Models

End
End.

Where $Product(A^{k-1}(i,l), A(l,j))$ is defined as :

$A^{k-1}(i,l)$ is a set (possibly null) of paths from
   node i to node l, say p_1 to p_m
$A(l,j)$ is either path $l \to j$ or $null$
concat (p_x, p_y) is defined as :
   $= null$ if either p_x or p_y is $null$
   $=$ concatenation of the two paths, that is,
      concat (i-a-b-c-l, l-j) $=$ i-a-b-c-l-j if j is not included in i-. . .-l;
      $= null$ otherwise.

$$Product(A^{k-1}(i,l), A(l,j)) = \bigcup_{i=1}^{m} concat(p_i, A(l,j))$$

F	1	2	3	4	5	6	7
F							
1	1F						17
2	2F			24	245	246	247
3	31F 32F	31	32	324	35 3245	36 3246	317 3247
4					45	46	47
5							
6							
7		3245 = (path) 3 → 2 → 4 → 5					

F can be reached from nodes 1,2, and 3
Find all the paths that connect nodes 1,2 and 3 to observable nodes

MATRIX OF ALL DIRECTED PATHS
An-1 : where A is the Adjacency Matrix of the Graph

Figure-6

The above algorithm can compute the path matrix for the graphs with cycles. All the cycles are avoided by performing a check during *concat* to see if the node being added already exists in the path. The matrix for the example of Figure-1 is shown in Figure-6. Computation time and storage space for the path matrix may be large because theoretically an exponential (in number of vertices) number of paths may exist in the graph. However, the knowledge graphs are expected to be sparse and the path matrix needs to be constructed only once. The computation required for constructing this matrix is a one time requirement and should be considered as part of knowledge acquisition effort. The path matrix contains all the knowledge about the structure of the graph. All problem instances to be solved refer to this same path matrix without having to do any more path related computations. Since the graphs are expected to be very sparse, the storage requirements for this matrix would also be reasonable for any problem situation.

Let us consider the example shown in Figure-1. It has one variable of interest *Fatal*, and three observable variables *S5,S6, and S7*. Four hidden variables *D1 - D4* and their relationships are represented in the graph. We are interested in selecting a subset of the hidden structure such that it connects the symptoms to the variable *Fatal*.

The selected subset of the hidden structure constitutes the assumed model for the problem instance. We may then infer the probability distribution for *Fatal* in the context of the selected model.

There are three different ways in which *Fatal* may be related to an observed symptom variable. 1. *Fatal* and the observed variable are effects of a common unobserved cause. 2. The observed variable is an effect, caused by the variable of interest *Fatal*. 3. The observed variable has causal influence on *Fatal*. Our search for the optimal scenario should explore the graph for all these types of paths connecting the observed variables to the variable of interest.

When an instance of a problem is presented, it includes the observed variables and the variables of interest. Our first step is to identify that subgraph which contains only those paths between variables of interest and the observable variables that are in accordance with the constraints imposed by the knowledge of causality. For the example of Figure-1, this can be performed as follows.

Step 1.   In the path matrix shown in Figure-6 look under the column *Fatal* and find all those variables that have at least one path leading to *Fatal*. *D1, D2, D3* is such a subset in this case. These variables are the candidates for being the common causal ancestors of the observed variables and the variable of interest.

Step 2.   This step is to determine the set of paths that start from the above identified candidates for ancestors of *Fatal* and lead up to any of

Searching for Relevant Problem Models

the observed variables. This is achieved by looking in the rows for *Fatal* and ancestors of *Fatal* and columns corresponding to the observed variables. Steps 1 through 4 mentioned in section 6 are used to determine this set of paths.

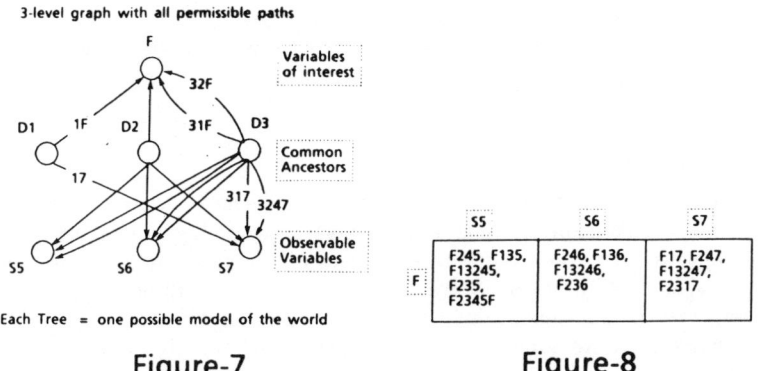

Figure-7   Figure-8

We can now draw a three level graph, as shown in Figure-7. The first level has the variable of interest *Fatal*, the second level has the nodes that are common ancestors of the observed variables and the variables of interest. Let us refer to this set as CA (Common Ancestors). The third level has the nodes for the observed variables. Each edge in this graph corresponds to an elementary path in the original graph. The edges in this graph are directed towards the first and the third levels from the middle level. We should keep in mind that many nodes of the original graph may appear in a number of paths in this reduced graph. Any subset of the set CA that has paths leading to the variable of interest and all the observed variables is sufficient to construct a model of the problem world. There may be a large number of such subsets. Moreover, from any such sufficient subset, there may be a number of ways in which paths can be found to relate them to the observable and other variables. The choice of the common ancestors and the paths defines the model of the problem world. The choice of the paths also adds some new nodes to the subgraph. All the nodes thus included provide the context in which the utility of some proposed action may be measured. Below we see how we can select the model of interest from among many that may be feasible.

Having obtained the graph of Figure-7, the task now is to select a unique path between the variable of interest and each of the observed variables. In each model tree that results, the probability for the event $Fatal = Y$ can be computed by using the assumed method of combination. The objective is to determine the model-tree which maximizes this probability. We need to efficiently search the space of possible model-trees to determine the desired tree. What we need is a heuristic to guide us as to which subset of paths should be chosen. If there are k observed variables then we need to select one from each of them leading up to the root node. In Figure-8 we show a table that lists all the elementary paths that connect the root node to the observed nodes.

7. Searching for Optimal Models

The above discussion has addressed the issue of constructing models from the graph representing a reasoner's knowledge. The uncertainty knowledge that is associated with the edges of the graph is used for controlling the search of the most desirable scenario. The desirability of a scenario is determined by the uncertainty measures associated with the nodes included in the corresponding subgraph. We present below the algorithm for accomplishing such a search process.

Associated with each path there is a matrix of conditional probabilities that links the probabilities for the variables at the observed end to the probability of the variable at the other end of the path. This matrix is fixed for each path and can be computed along with the path matrix. Using the matrix for each path we can compute the probability distribution resulting at the root node if that path were to be the only path in the tree. In each column of the table of Figure-8 the paths can now be ordered according to their merit. In this case the merit is measured simply by the probability for the event $Fatal = Y$. Now our objective is to select one path from each column of the table in Figure-8. However we are constrained by the fact that any subset of paths selected arbitrarily from the columns may form cycles and thus not result in a tree. So, one alternative we have is to search through the space of all possible subsets and determine the one that forms a tree and maximizes the probability for the desired event. The algorithm for this search, based on the A^* algorithm described in [6] is as follows.

A state for the search process includes a partially constructed tree and the set of all those observed variables and their paths that have not yet been included in the tree (similar to the table of Figure-8).

Initial state = null-tree T and the table of Figure-8.

In each of the columns of Figure-8, order the paths according to their merit and place the initial state on the Search Open-List.

Perform the following Steps :

1. Select and remove the top state description S_t from the Open-List.

2. If all the includable observed nodes have been included in T then this is the desired model-tree. Stop the Search.

3. For each unincluded observed node, select the path that has maximum probability for the event $Fatal = Y$. From among the paths selected in step 3, select that path P which has the maximum probability for the event $Fatal = Y$.

4. Generate a new state S_1 as follows :
   T ← T (of S_t) ∪ P;
   Delete all the paths from the column of P.
   If T is a Tree then
   [update the merit values of those paths in table that have a node in common with P ;
   place S_1 on the Open-List according to its merit.]

5. Generate a new state S_2 as follows :
   T ← T (of S_t);
   Delete P from the table of unincluded paths.
   If the table of unincluded paths is not empty then
   Place S_2 on the Open-List according to its merit.

6. Return to Step 1.

The heuristic information depends on the reasoner's objective, that is, on the nature of the model-tree sought. The heuristic information is used at two places during the above procedure. First for selecting a state description from the open list and second for selecting a path in the selected state description. The latter choice corresponds to selecting one of the many possible expansions of the selected node. The state descriptions are ordered on the Open-List according to decreasing heuristic merit value.

For the example under consideration, the preference criterion is to determine that scenario in which the path selected to connect an observed node and the needed node $Fatal$ is the one which results in maximum probability being assigned to the event $FATAL = Yes$. For this case, we define a simple merit function to evaluate the suitability of a scenario from the point of view of the above mentioned preference criterion. We say that the merit of a scenario is measured by the average of the probability values assigned to the event $Fatal = Yes$ by various paths included in the scenario. This merit value is not intended to replace the computation of the probability for the event $Fatal = Yes$, but to facilitate the search of a scenario that satisfies the above preference criterion. An estimation of the upper bound on the merit value for a state can be computed as following :

P_t = probability for the event $Fatal = Y$ in the partially constructed tree. It is 0 when the tree is a null tree.
$P_k = \max_j P[Fatal = Yes \mid a_j]$ where a_j s are the events corresponding to the variable at the other end of the edge containing variable $Fatal$ for a path in the kth column of Figure-8, and P is the joint probability distribution associated with this edge.

$$H = (m * P_t + \sum_{unincluded-nodes} P_k) * 1/n$$

where m is the number of observed nodes already in the tree and n is the total number of all the observed nodes.

The value computed by function H can be shown to be an upper bound on the merit values of the trees that can be generated from the search-state description. That is, H is greater than the merit value for a tree that can be obtained by adding to the partial tree any paths leading up to all the unincluded nodes.

For computing the probability value for the event $Fatal = Yes$ we can assume conditional independence of child nodes when conditioned on the parent node. The computation of probability value for a variable which is the child of more than one edges may require more information or assumptions, but we concern ourselves here only with determining the tree structure that satisfies the above mentioned preference criterion. The merit value defined above represents one type of preference criterion. For each different type of merit value function we would need to find a different function that is an upper bound on the merit value of trees that can be generated from a search state. The merit of a tree in this case is given by the probability of the event $Fatal = Yes$.

Proposition : If the heuristic function that measures the merit of a state description is an upper bound on the actual merit of the state description,

Searching for Relevant Problem Models

then the above algorithm will terminate with the optimal model-tree for the particular problem instance.

Proof : Let us say H represents the heuristic estimate of the merit of a state description. Let H^* represent the maximum of the merit values of all the trees that can be generated from a state whose merit is H. The search terminates in step-2 when the state description selected in step-1 has all the nodes included in the tree. If H_t is the merit of this terminal state then it must be same as H_t^* because the selected state description consists of only a tree containing all includable observed nodes. We are given that the nature of function H is such that $H_i \geq H_i^*$. Since the state with maximum merit is selected by the search process, H_t must not be less than the merit of any other state H_i on the open list. So $H_t^* \geq H_i$, and therefore also $H_t^* \geq H_i^*$. Hence, the search terminates with the optimal tree.

Once we have obtained the optimal model tree the search is terminated. However, if we are interested in obtaining alternate model-trees that are in decreasing order of merit then we may continue the search process and obtain as many top ranking models as we desire. The decision about an action many a times is made based on the top few model-trees.

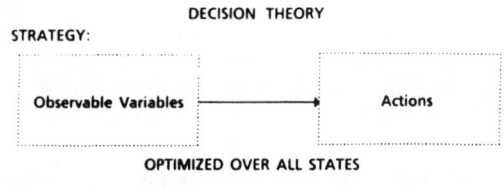

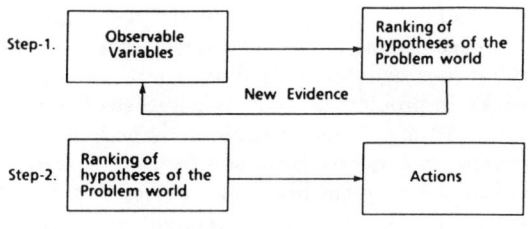

OUR METHODOLOGY - A TWO STEP PROCESS

Figure-9

This is an important consideration for a decision making process. We start with some evidence and obtain a number of possible scenarios as pointed to by this evidence. The search for further evidence is guided by the need to corroborate or refute any of these scenarios. This methodology provides such guidance by making available to a reasoner the alternate scenarios. It seems to us that this situation is certainly much better than a formulation based on classical decision theory where a strategy provides a mapping from the domain of observed values to the domain of actions. That strategy optimizes over all possible scenarios of the world and does not provide the reasoner an opportunity to ascertain the state of the world more precisely by asking new questions. A comparison of our methodology with classical decision theory is shown in Figure-9.

8. Extensions of the Method

The optimality criterion of the reasoner may vary with the problem instance and a large number of criteria are possible. For example, instead of specifying a particular event, we may be interested in determining a scenario which has maximum probability for any one of the elementary events corresponding to the variable of interest. We may also be interested in either specified events composed of a fixed number of elementary events or in events of larger cardinality (number of elementary events required to compose the event) having the probability value above some threshold value. The heuristics that are upper bounds for the merits of state descriptions and for selecting a path in a state description depend on the criterion of optimality for the model-trees. These heuristics would also depend on the assumed method of combining the evidence. There is a large number of ways in which a reasoner may specify the criterion of optimality for determining the model of the problem world. If we classify various ways of specifying the optimality criteria and determine heuristic functions that are upper bounds for the merits, then this constitutes a specification of strategies in a domain independent manner.

The example considered above is a very simple one and all computations are easy to perform. However, this may not be the case when some other methods for combining evidence are used. Here the causal influences are represented by edges connecting two nodes at a time. However, causal influences of simultaneous or interactive influence of a number of aspects on some other aspect of a problem domain requires edges that connect more than two variables at a time. This gives rise to hypergraphs for representing knowledge. The search in this case is for subgraphs of this hypergraph. This generalized version of the problem has been addressed in [1]. In this paper, since we are primarily interested in explaining the basic ideas of our approach, we have not discussed alternate more sophisticated search strategies which would allow more efficient search. Examples of such alternate strategies which can be used here include SSS^* and other algorithms which are discussed in the book [3].

Let us consider the case where we are interested in determining the scenario where the resulting probability for any one of the elementary events is maximum. An example of the situation is the case where the possible values of the variable of interest consist of n person-names. The observable variables may be the known facts of a murder and each person-name may be linked to the facts (symptoms) in various differing ways (model-trees). The merit of a model-tree may be defined as the maximum of probabilities assigned to each of the the n person-names. The objective is to ascertain model-trees in order of decreasing merit. The search algorithm described above can be used with the following modifications for step-3.

3. For each unincluded node, select the path that has maximum probability for that person-name which has maximum probability in the partially constructed tree. In case of a *null* partial tree, select the path with maximum probability for any of the n person-names.
From among these paths, select that path P which has the maximum probability for that person-name which has maximum probability in the partially constructed tree. In case of a *null* partial tree, select the path with maximum probability for any of the n person-names.

9. Conclusion

One of the main problems in handling uncertain and incomplete knowledge is determining an underlying causal model for relating the variables of interest to the observable variables. In current methodologies, these models are thought of by humans and made available to mechanical reasoners. Mechanical reasoners then compute measures of uncertainties associated with the variables of interest in the given model. For example, in methods based on probability theory or any other uncertainty calculus, the structure of the problem domain must be specified by a human interface. This structure includes information about which variable should be inferred from which other variable. Hypothesizing a number of possible models that explain the observations can be achieved by performing abductive reasoning. A methodology for abduction based on the set covering model has been presented in [9]. The problem that arises due to abduction is that a large number of hypotheses may become possible and we need some criterion to select one or some of these. In [9], criteria based on parsimonious covering have been discussed. In our methodology, a traversal of an edge against its direction (representing causality) is an abductive step of reasoning. The preference criterion that we use for preferring one hypothesis over the other is the inferred probability for some pre-selected event in each hypothesis. In addition to abduction, the uncertainty also contributes to a number of causal structures being possible. In our methodology, the knowledge of relationships of causality among various variables provides the basis for constructing various causal structures for a problem situation. Our objective is to provide a mechanical reasoner the capability to construct problem models of interest using a reasoner's domain knowledge consisting of known variables and their interrelationships and the available evidence. The criterion for interestingness of a scenario may vary with each problem instance and our methodology takes this into account. The probability knowledge is used to select that scenario which may be of maximum interest. In many situations it is the structure of the scenario and not the probability value associated with a variable, which provides the basis for selecting the action to be taken. Given a set of observed variables and a variable of interest, we have presented a methodology for use by mechanical reasoners for identifying various possible candidates for models, ranked by their level of interest.

References

1. Raj Bhatnagar, Construction of Preferred Causal Hypotheses for Reasoning with Uncertain Knowledge, Ph.D. Dissertation, Computer Science Department, University of Maryland, 1989.

2. Benjamin N. Grosof, Evidential Confirmation as Transformed Probability: On the Duality of Priors and Updates, *Uncertainty in Artificial Intelligence*, ed. L. N. Kanal and J. F. Lemmer, North Holland, 1986, pp. 153-166.

3. L. N. Kanal and V. Kumar (eds.), *Search in Artificial Intelligence*, Springer Verlag 1988.

4. S. L. Lauritzen and D. J. Spiegelhalter, Local Computations with Probabilities on Graphical Structures and their Application to Expert Systems, *The Journal of the Royal Statistical Society*, Series B(Methodological), volume 50, No. 2, 1988, pp. 157-224.

5. W. J. Long, S. Naimi, M. G. Criscitiello, S. Kurzrok, Reasoning About Therapy from a Physiological Model, *MEDINFO 86*, eds. R. Salamon, B. Blum, M. Jorgensen. Elsevier Science Publishers B. V. (North Holland), pp. 756-760.

6. Nils Nilsson, *Principles of Artificial Intelligence*, Tioga Press, 1980.

7. Judea Pearl, Fusion, propagation, and structuring in Bayesian networks.*Artificial Intelligence* Vol. 29, 1986, pp 241-288.

8. Judea Pearl, Embracing Causality in Formal Reasoning. *Proceedings of AAAI Conference, 1987, pp. 369-373*

9. Yun Peng, A Formalization of Parsimonious Covering and Probabilistic Reasoning in Abductive diagnostic Inference. Doctoral Dissertation, TR #1615, Dept. of Computer Science, University of Maryland, 1986.

10. Donald Perlis, Autocircumscription, *Artificial Intelligence*, Vol. 36 (1988), pp. 223-236.

11. R. Reiter, A Logic for default Reasoning, *Artificial Intelligence*, vol. 13, pp. 81-132.

12. Ross D. Shachter, DAVID: Influence Diagram Processing System for the Macintosh, *Uncertainty in Artificial Intelligence 2*, ed. J. F. Lemmer and L. N. Kanal, Prentice Hall, 1986, pp. 191-196

ON THE MARSHALLING OF EVIDENCE AND THE STRUCTURING OF ARGUMENT*

David A. Schum

Department Of Operations Research And Applied Statistics
George Mason University
Fairfax, Virginia

Abstract

This paper concerns issues that arise when arguments are constructed from a mass of evidence and when the evidence is weighed or assessed in the process of drawing conclusions from it. Of major concern in this paper are problems associated with the productive integration of structural and weight-related analyses. In addition, the author addresses some common misconceptions about theories of evidential reasoning and attempts to set the record straight by mentioning the legacy of research on the evidential foundations of probabilistic inference that has been available for many years. All of these matters should be of concern in attempts to provide a productive integration of research in Operations Research and Artificial intelligence in situations in which people must draw conclusions from a mass of evidence that is incomplete, inconclusive, and that comes from sources with every gradation of credibility.

* This research was supported by the National Science Foundation under grant # SES-8704377 to George Mason University.

1.0 Some Issues In The Analysis Of Probabilistic Reasoning

My task in this paper is to examine two related issues that arise in the analysis of inferential reasoning based upon a mass of evidence. The first issue is structural in nature and involves the process of forming arguments from evidence to hypotheses or possible conclusions we entertain. The second issue is probabilistic and it concerns the means by which we assess and combine the inferential "weight", "value" , or "force" of the evidence in drawing a conclusion from it. The history of research on probabilistic reasoning reveals a somewhat curious separation of these issues. Some studies focus on the details of a probabilistic system and ignore the often-complex argument structures to which this system will be applied. Other studies are concerned about the structure of inference and ignore probabilistic matters in assessing the force of evidence. Concern about both issues is necessary in any attempt to make sense out of masses of evidence that, on the surface at least, may appear not to make any sense. We must be able to construct defensible arguments [that will often be very complex]; but we must also be able to express what we believe to be the force of the argument we seek to defend when the evidence is considered in the aggregate.

The trouble is that, although structural and probabilistic analyses can be symbiotic activities, the productive integration of both of these forms of analysis presents us with difficulties that, I believe, add considerable frustration in our efforts to develop computer-based systems for assisting the performance of complex probabilistic reasoning. Human inference is a remarkably rich intellectual activity having many attributes and dimensions; it seems far too much to expect that any particular combination of structural and probabilistic analyses will capture all of this richness. Selecting a combination of analyses that best captures the necessary elements of a problem at hand requires us to integrate the understanding that alternative views of inference structuring and probabilistic analysis provide. So, another part of my present task is to discuss what I believe to be the major difficulties encountered in the useful integration of structural and probabilistic analyses.

In inferential or diagnostic tasks we often find it necessary to consider large collections of evidence in order to draw conclusions. Lacking one or more items of evidence that would be nearly conclusive on matters of interest to us, we must somehow evaluate and combine often massive amounts of weaker evidence in the hope that, in the aggregate, the evidence will point with reasonable strength to some specific possibility or hypothesis we are entertaining. Even if we did have one or more items of evidence we thought compelling we would ordinarily seek to buttress our conclusion on this evidence by attempting to show that the weaker items of evidence we also have are consistent in favoring this same conclusion. Thus, for example, a trial attorney would ordinarily hope to buttress the testimony of a single eye-witness [however credible this witness seems to be] with a consistent pattern of less direct or circumstantial evidence that may also be available. One trouble is that we are far better at the tasks of collecting, transmitting, storing, and retrieving information than we are at the task of drawing defensible conclusions from it. Not all information we collect is relevant or significant in the problem at hand; information judged relevant in some inferential problem is commonly called "evidence" within the context of this problem. However massive is our collection of evidence in some inferential task, it will usually be incomplete in its coverage of matters relevant to possible conclusions we entertain. Upon examination of evidence items either separately or in the aggregate we usually find the evidence to be inconclusive in the sense that it does not make necessary some particular possibility or hypothesis. Finally, we may note that the evidence we have comes to us from sources having any gradation of credibility.

How we can close the obvious gap existing between our well-advanced methods for collecting, transmitting, storing, and retrieving information and our not so well-advanced methods for drawing conclusions from information is a matter of considerable interest to persons in many disciplines. Quite naturally, we expect that computers may assist us in closing this methodological gap. Much has been promised by our colleagues in the area of artificial intelligence who have an interest in the development of various systems for assisting us in the performance of inferential reasoning and decision tasks; as a result, much is now expected by the public. The trouble is that, upon close examination, even "simple" inference tasks reveal many interesting and important evidential and inferential difficulties; these difficulties get compounded when there is a mass of incomplete, inconclusive, and unreliable evidence to be evaluated and combined. It is also true that matters we do not consider can hurt us in inferential reasoning: hypotheses or possibilities we did not consider become embarrassing in post-mortem analyses of an inferential failure; also embarrassing is evidence we did not consider or subtleties we overlooked among the evidence items we did consider. I believe that success in closing this methodological gap rests to no small extent upon recognition of particular features of human inference in natural settings that I will discuss. These same features act to prevent ready integration of structural and probabilistic analyses.

2.0 Structuring An Inference Based On A Mass Of Evidence

It would be very helpful indeed if every human inference problem emerged in well-posed form in which all of the necessary reasoning linkages in one's arguments from evidence to hypotheses were apparent and uncontroversial. It would be similarly helpful if there was just one form of structure that suited every inferential purpose. Good reasons exist for believing that there is no such thing as a uniquely correct argument from any evidence to any hypothesis; thus, it is very difficult to justify saying that one has formed the "correct" structure or cognitive model for some inferential problem. Further, there are often quite different but related tasks that are associated with the drawing of a conclusion from evidence; each task requires somewhat different structural considerations. Following is a brief account of some important and interesting structural issues; much can be gained from the study of structural relations between evidence and hypotheses and between one item of evidence and another.

2.1 A Bit Of History

Scholars of inferential reasoning soon learn to take all the help they can get. In seeking such assistance it is often discovered that others have tred the very same path that now seems so novel and exciting. There is much discussion these days about inference "networks" and other graphical schemes for depicting a structure or, if you like, a cognitive model of some complex reasoning task. This seems natural since one form of inferential structuring seems to resemble a complex network of interrelated elements. This is precisely what the American jurist John H. Wigmore discovered 75 years ago when he discussed a method for charting complex masses of evidence in preparation for a trial at law [Wigmore, 1913, 1937]. Figure 1 shows a small portion of one of Wigmore's evidence charts for a particular case ; this one comes from the 1937 edition of his work The Science Of Judicial Proof [Wigmore, 1937, p. 876] [1]. Wigmore noted that an overall argument from evidence to major hypotheses or possible conclusions is usually composed of one or more subarguments. Further, he noted that arguments themselves usually involve what he termed "catenated" reasoning, i.e., chains of reasoning in which there are often many links or reasoning steps from evidence to possible conclusions. In modern terms we describe such inference as being "cascaded" or "hierarchical" in nature. He distinguished between directly relevant evidence, that can open up a chain of reasoning, and ancillary or indirectly relevant evidence that can bear upon the strength or weakness of any link in this chain. Along with Venn [1907, p 506], Wigmore noted that there is arbitrariness in the construction of any reasoning chain; different persons might justifiably perceive different reasoning routes from the same evidence to the same conclusions. They might also, of course, reach different conclusions from the same evidence.

In order to decypher a Wigmore evidence chart one must make reference to the elaborate collection of symbols Wigmore used to distinguish among different forms of evidence he identified and to show the strength of probative or inferential relationships among elements of a reasoning chain. As noted elsewhere [Tillers & Schum, 1988], Wigmore's evidence charts can be described in modern terms as directed acyclic graphs whose vertices are propositions and whose edges are "fuzzy" inferential or probative force qualifiers. For various reasons, some quite obvious, Wigmore's schemes for charting evidence and structuring argument were never popular among the advocates for whom these schemes were intended; William Twining [1985] discusses several reasons why Wigmore's schemes went over "like a lead balloon". One obvious reason is that diagrammatic schemes such as Wigmore's, by themselves, can only rarely suggest a conclusion. Although some of Wigmore's symbols concern probative or inferential force he offered no mechanism for combining, propagating, or aggregating these probabilistic ingredients in order to draw a conclusion from the evidence items and the arguments they suggest. In short, Wigmore's graphic analyses are structurally strong but probabilistically weak. We would do well not to dismiss Wigmore's efforts simply on the basis of the difficulty of applying his structural analyses; what he had to say about evidential and inferential issues in evaluating masses of evidence is at least as profound as what we hear from inferential theorists in our own day.

2.2 Forms Of Structural Analyses

In deliberating upon available evidence it becomes apparent that it can be marshalled or organized in different ways. One reason is that there are often different inferential ends to be served by the evidence; this was also obvious to Wigmore as he distinguished between his chart method for organizing evidence and another method he called the "narrative" method [Wigmore, 1937, p 821-857]. In many situations evidence is arranged in such a way that it allows someone to tell a story or to offer an explanatory account of what appears to have happened or what might happen; such accounts typically rest upon a chronological ordering of the events for which there is some evidence. So, one obvious structural form involves the believed order in which the events reported in evidence occurred. Such <u>temporal</u> structuring can lead to the formation of "narratives" or "scenarios" [von Winterfeldt & Edwards, 1986, 163-176]. Of course it is true that the order in which events have occurred can, itself, be a source of inferential value. But the trouble is that many different scenarios or stories might be constructed on the basis of the same event chronology; in a trial at law the prosecution will tell one story and the defense another. Different scenarios arise when we consider the inevitable gaps in any sequence of reported events; even if we agree about the actual occurrence of some collection of events and about their ordering, we might attempt to fill in the gaps in entirely different ways each of which may provide an importantly different account of what happened.

Given several different plausible stories based on the same evidence ordering, how do we decide which story to believe ? Wigmore argued that preference for one story over another is based upon specific arguments we make relative to the major elements or points in a story; he argued that his chart method or network analysis was simply a device for marshalling evidence in terms of the major points at issue upon which the evidence bears. Thus, for example, we might sort out the evidence in a civil case in terms of its apparent bearing on the several elements or points necessary to prove that plaintiff's scenario is one in which defendant is indeed negligent. Figure 1 above shows Wigmore's analysis in a criminal case of the prosecution and defense arguments about whether or not defendant knowingly gave poison to the victim in this case; this is just one element of a proof that defendant committed murder with malice aforethought. Network analyses tend to be <u>relational</u> in nature since they involve the juxtaposition of evidence and elements of arguments the evidence suggests. There are, of course, other forms of inferential analysis apart from the temporal and relational analyses just mentioned. I have argued that structural and probabilistic or "weight-related" analyses are symbiotic; a bit later on I shall mention how alternative views of probabilistic reasoning suggest other forms of structural analyses.

Not all network or relational analyses exactly resemble the ones Wigmore discussed years ago. Of great current interest are the causal inferential networks captured in the analyses of Judea Pearl [e.g., 1986, 1987] and of Spiegelhalter [1987; Lauritzen & Spiegelhalter, 1988]. In some cases we can conceive of networks that represent the interplay of both inference and choice processes; the various current applications of "influence diagrams" provide ready examples [Howard & Matheson 1980; Shachter, 1986, 1987]. Other network analyses are similar to Wigmore's but have been cast in terms that facilitate probabilistic analyses [e.g Schum, 1980,; Schum & Martin,1982]. My point here is that concern about the structuring of complex inferential processes is not a recent phenomenon; scholars of evidence in jurisprudence have, for a very long time, accepted both structural and weight-related elements of complex inference as their stock-in-trade. They have left us a legacy of recorded experience and scholarship on human inference that ought to be both acknowledged and made use of in current research.

2.3 Structural Analyses And The Study Of Evidence

In many important contexts people frequently wish to know what questions they might ask of existing evidence in the process of drawing a conclusion from it; they may also wish to know what questions to ask in the discovery of other relevant evidence and in the generation of other possible conclusions they might entertain. Structural considerations, by themselves, can supply questions to ask of evidence and can assist the process of discovery. Attending only to structural matters, and considering nothing at all about probability and the "force" of evidence, we can learn a substantial amount about evidence and its role in inference. Thus, there are many important elements of a comprehensive theory of evidence that do not require commitment to any particular view of probability, as is sometimes supposed. For example, it is meaningful to ask how many logically-distinguishable kinds of evidence there are and to inquire about the various ways in which we can form combinations of evidence. We do not need Bayesian, Baconian, Shafer-Dempster, or other systems of probabilistic reasoning in order to answer these inquiries, as centuries of experience in jurisprudence testify. So, it should not be supposed that a theory of "evidential reasoning" begins within the confines of any particular formal system of probabilistic reasoning.

A person confronted with a complex reasoning problem often ponders the following question: "what questions should I be asking of my evidence" ? One possible response to this question is another question: "what kinds and combinations of evidence do you have" ? On the surface, this latter question may seem meaningless since evidence varies substantively in a near infinite fashion. However, the work of evidence scholars in jurisprudence, as well as others, gives us some hope that there is a finite and manageable number of different logically-distinguishable species of evidence, regardless of the substance of the evidence. Figure 2 shows one possible categorization of evidence and it is based only upon structural considerations. It also rests upon two very necessary distinctions that are not always made [to the detriment of many probabilistic analyses of the weight or value of evidence]. First, there is a distinction to be made between evidence about an event and the event itself; the necessity for this distinction was obvious years ago to Wigmore [1937, p. 45] and to Keynes [1957, p 181]. The reason is that evidence about some event does not ordinarily entail that the event actually occurred. So, for example, neither my testimony that I saw you running from the house where the crime was committed nor a photograph of you running from this house entails that you were indeed running from this house; I might be untruthful or have been mistaken, or the photograph doctored, misinterpreted, or taken at a different time than is presently at issue as you now stand trial. Second, as Wigmore and Venn realized, most if not all human reasoning involves chains of reasoning and is therefore "catenated", "cascaded", or "hierarchical" in nature.

There appear to be two basic dimensions for categorizing evidence; the first involves the nature of the relevance of the evidence. As I noted above, evidence can be either directly or indirectly relevant on matters at issue in an inference. Directly relevant evidence sets up a plausible chain of reasoning to major possible conclusions; indirectly relevant evidence either strengthens or weakens the links in this chain. There appear to be two forms of directly relevant evidence; direct evidence and circumstantial [or indirect] evidence. Direct evidence settles the matter if the source of the evidence is perfectly credible; my testimony that you were running from the scene of the crime is direct evidence that you were running from the scene of the crime. Circumstantial evidence provides just inconclusive grounds for believing some event even if the source of this evidence is perfectly credible. Thus, even if you were running from the scene of the crime, this event would be just circumstantial evidence that you did commit it; you could have been running away from the person who actually committed the crime. One complication here is that a given item of evidence can be direct evidence at one level of an inference and just circumstantial evidence at another. My testimony is direct evidence about your running from the scene, but just circumstantial evidence that you committed the crime.

The other dimension of the categorization shown in Figure 2 basically concerns the _form_ of the evidence and how its source stands in relation to the user of the evidence. So-called "real" or demonstrative evidence is that which can be examined by the person using it to draw a conclusion; this person uses his/her own senses in order to determine the occurrence or nonoccurrence of an interesting event or events. Testimonial evidence, on the other hand, is that which has been recorded by the senses of a person other than the one using the evidence to draw a conclusion. As Figure 2 shows, both "real" and "testimonial" evidence can be "positive" or "negative" evidence. Positive evidence records the occurrence of some event; "negative" evidence records the nonoccurrence of an event. Suppose a source of evidence is queried and no answer is given; or suppose that no source can be found that has recorded the occurrence or nonoccurrence of some event of interest. It has long been recognized that missing evidence can, itself, be evidence; there are, of course, different possible explanations for missing evidence. Perhaps the source(s) have nothing to tell us and we have simply queried the wrong ones. On the other hand, perhaps a source does have information about the event that interests us and prefers not to provide it.

The behavior of a source of evidence can often be as inferentially interesting as what the source tells us. One well-known species of interesting source behavior is commonly called "stone-walling" or equivocation. Asked whether or not some event occurred, the source says: "I can't remember", "I don't know", "I couldn't tell" , and so on. Any of these responses may be expressions of honest self-impeachment; the trouble is that they may also reflect refusal to divulge information that the source does have. Quite often, we have to evaluate evidence that comes to us from a chain of some number of sources; such evidence is commonly called "second-hand" or "hearsay" evidence. In a very wide variety of circumstances we must use evidence that does not come from "the horse's mouth", i.e., from the person or other source immediately privy to the occurrence or nonoccurrence of the event that interests us. In some cases, what we take to be evidence is simply an inference or a conclusion reached by someone else; this is commonly called "opinion" evidence. A certain source tells us that event E occurred; asked if he observed E, the source says: "no, but I observed the occurrence of events C and D from which I inferred the occurrence of event E". Finally, there is a category of evidence that represents "accepted facts" in the sense that the events reported are accepted without further proof. For example, suppose an element of your argument rests upon the time of high tide in Miami, Florida on 4 March, 1987. You report the time as recorded on an appropriate page of the tide tables, not expecting anyone to require further proof of the time these tables record. You might, of course, expect scrutiny about whether or not you consulted the correct page of the tide tables.

It is true, of course, that the categorization of some item of evidence is always relative to a particular inferential problem and, further, to a stage in the problem development. A given item of evidence can have its status changed during the life cycle of a problem. For example, evidence deemed of only ancillary importance at one stage of an inference problem can emerge as directly relevant evidence in another. Finally, the source-related dimension in Figure 2 has some items that are not mutually exclusive. For example, second-hand or hearsay evidence can have a base in either "real" or testimonial evidence and can be either "positive" or "negative" evidence.

On purely structural grounds we can identify various ways in which two or more evidence items can be taken in combination; Figure 3 illustrates these combinations and helps us to sort out the distinctions among combinations of evidence that are sometimes confused. In this figure we distinguish between evidence about some event and the event itself; for example, F* is evidence that event F occurred. Parts A and B of Figure 3 illustrate the distinctions between "corroborative" and "confirming" [or "convergent"] evidence and between "contradictory" and "conflicting" [or "divergent"] evidence. One

Marshalling Evidence 115

species of evidential corroboration occurs when two or more sources report the same event; evidence is contradictory when one source reports the occurrence of some event and another source reports its nonoccurrence. As Part A shows, for multiple sources we may have any pattern of corroboration and contradiction. Evidence is confirming or convergent when sources report different events that favor the same hypothesis or conclusion. Evidence is conflicting or divergent when sources report different events and the events favor different hypotheses. Thus, the events reported in contradictory evidence cannot both be true but the events reported in the conflicting case may both be true.

There appear to be two species of redundant evidence as Part C shows. In one case two or more sources report the same event [i.e. they give corroborative reports]; in the other case the sources report different events that have the following characteristic. If we knew, say, that event D had occurred, then knowing event E would have less value [i.e., D acts to inhibit the value of E]. The trouble, of course, is that D* and E* are just evidence about events D and E; how redundant E* is depends, in part, upon the credibility of the source reporting D*. The same is, of course, true with corroborative redundancy; how much value the second report of F has depends, in part, upon the credibility of the source giving the first report of F. There also appear to be two species of nonredundant evidence. In one case we may suppose, with justifiable reasons, that the evidence given is independent both in the sense that the sources of evidence have not interacted in any way and that the events they report seem to be independent. The remaining case in Part C illustrates an instance in which knowing event D would act to enhance the value of knowing event E. Thus, some combinations of evidence can exhibit a synergistic effect; in some instances two or more items of evidence taken together seem to have more inferential value than they do when considered separately.

Categorizations such as those just discussed are indeed helpful in determining questions that might be asked of evidence as I recently explained in some detail [Schum, 1987]. But, it should not be thought that the only virtue of structural considerations is simply to help us identify ways of categorizing evidence. Care in identifying stages of reasoning that link evidence and possible conclusions assists in the process of tracing dependencies among evidence and among elements of arguments we make; this is one very important step in recognizing and then exploiting the many evidential subtleties that can exist. Another virtue of careful argument structuring concerns the all-important tasks of generating evidence and new or revised possibilities. To lay out carefully what you believe are the stages in an argument from your observed evidence to conclusions you entertain is also to identify potential evidence that may be better than what you already have. In addition, careful structural analyses can assist in the process of generating new or revised possibilities or hypotheses. Careful structural thinking about evidence that does not seem to fit into your existing cognitive model of an inferential problem is part of the process of generating other possible conclusions you might entertain.

So, to suppose that a theory of evidential reasoning must rest upon a particular view of probability is to display considerable innocence of the work of the very many scholars of evidence in several disciplines for whom probabilistic considerations are just one element in a much larger enterprise.

3.0 Probability And "Evidential Reasoning"

As is commonly known, there are alternatives to the standard calculus of probability and Bayes' rule for thinking about probabilistic reasoning. Some of the better known alternatives are Shafer's belief measures [1976], L. J. Cohen's Baconian or inductive probabilities [1977], and Zadeh's fuzzy probabilities and possibilities [1978]. These formal systems have a common characteristic: they all concern what current jargon calls "evidential reasoning". In the contemporary literature in some areas this term is frequently used just with reference to Shafer's theory of belief based upon evidence [e.g. Lowrance & Garvey, 1983; Lowrance, Garvey, & Strat, 1986]. But no part of Professor Shafer's profound and influential work [1976] suggests that the inferential use of evidence is peculiar to his formal probabilistic system. In addition, I have already noted that extremely valuable study of the inferential use of evidence has been going on for a very long time by learned persons who have had no awareness of the alternative formal views of inference we are so fortunate to have today.

In one way or another, each of these formal systems concerns the inferential "weight", "value", or "strength" of evidence and each suggests a method for combining assessments of the value of evidence in the act of drawing a conclusion from it.

Persons having an interest in various formal systems of probabilistic reasoning are sometimes asked: "which of these systems do you prefer" ? This is rather like asking if you prefer your hammer to your saw. When you want to cut wood you prefer a saw over a hammer; when you want to pound a nail you prefer a hammer over a saw. Further, you may sometimes want to pound a nail and to cut wood and so you find both hammer and saw useful; but, although these tools complement one another, you find it very awkward to use them both at the same time. Each one of these formal systems of probabilistic reasoning, as well as others that could be mentioned, capture some but not all of the richness evident in human inferential reasoning. Thus, they offer us some complementary tools for use in our attempts to make sense out of masses of evidence.

Quite apparently, Professors Shafer, Cohen, and Zadeh, as well as others having an interest in formal studies of probabilistic reasoning, have examined such processes and have observed different features of them. One factor involved in our ability to see objects in three dimensions is the fact that an object casts slightly disparate images on the retina of each of our eyes. In addition, of course, one observer of a scene will note important features of it that escape the perceptions of another equally attentive observer. Thus, we can think about the nature of probabilistic inference in different ways and, upon examination of the evidence we have, we note that it has many properties to which we should attend. There are algorithms or processes in each of the formal systems mentioned above for combining our beliefs based on the evidence we have; Bayes' rule, Dempster's rule, and the combinatorial processes supplied by the connectives in the Baconian and fuzzy systems are examples. Thus, each of these systems supplies a mechanism for valuing evidence whose relation to possible conclusions is specified by structural analyses in a particular case. Algorithmic processes are interesting but not helpful in particular inference problems unless they are implemented within the context of some problem structure. One interesting additional fact is that these alternative formal systems suggest different ways of marshalling evidence and structuring argument.

The alternative formal systems being discussed today differ quite significantly about what is meant by the inferential "weight" of evidence. In a Bayesian formulation the inferential value of evidence is commonly graded in terms of what we believe to be the relative likeliness of this evidence under various hypotheses being considered. In the Shafer-Dempster system of belief combination the "weight" of evidence is related to the support we believe the evidence provides various subsets of hypotheses we are considering. In L. J. Cohen's system of Baconian probabilities the weight of evidence is related to the number of evidential tests performed in an effort to eliminate hypotheses; the more evidential tests some hypothesis passes without being invalidated by the evidence, the higher is its grade of Baconian probability. Zadeh tells us that, in considering the weight of evidence, both the directly relevant evidence that sets up a chain of reasoning and the ancillary evidence that supports or diminishes the strength of each link in this chain frequently involve fuzzy or imprecise propositions. Thus, the "weight" of evidence is itself a fuzzy concept.

The integration of the structural and probabilistic elements of inferential "networks" is not a product of current interest in the development of artificially intelligent systems. For the past twenty years or so I have had an interest in applying Bayesian formalizations to all of the species and combinations of evidence shown in Figures 2 and 3 and to various inferential networks involving mixtures of these species and combinations of evidence; there are several current summaries of this work [Schum & Martin, 1982; von Winterfeldt & Edwards, 1986; Schum, 1987]. Such analyses supply information of considerable heuristic value in generating questions that might be asked of a mass of evidence as well as some suggestions about the implications of answers that are obtained when these questions are asked. The concept of conditional independence/nonindependence in Bayesian analyses can capture a wide assortment of evidential subtleties. In some instances two or more of the formal probabilistic systems can be applied side by side and their complementary nature noted in the analysis of an argument structure [e.g. Schum, 1988].

Throughout Shafer's work appears an emphasis on the process of constructing a probabilistic argument [e.g. 1986]; indeed, his work has been particularly stimulating to persons interested in structural issues. He has helped many of us to see that in the "growth" of a cognitive model of some inferential problem the necessary elements of it will undergo important changes. For example, the character of the hypotheses or possibilities we entertain depends critically upon the evidence we have. If the evidence is vague or imprecise in its focus, we are

Marshalling Evidence 119

only able to entertain imprecise or undifferentiated hypotheses. Thus, an investigator who arrives at the scene of some incident may initially obtain evidence that only allows him/her to discriminate among the vague possibilities: "accident", "suicide", or "foul play". Asking appropriate questions leads to the discovery of other evidence that may be more precise in its focus and that allows the investigator to entertain more specific hypotheses. It seems that one element in the productive integration of structural and probabilistic analyses should involve enhancement of the critical process of asking appropriate questions; I shall later return to this matter.

Cohen's work on Baconian probability makes us aware of the generalizations we must make in the formation of reasoning chains and of the nature of the evidential basis for them; generalizations form the inferential "glue" that holds an argument together and, therefore, are crucial elements in the defense of arguments. If we are asked to infer B from A in some particular instance we might inquire about why such an inference is licensed or warranted; we might be told in response that "if a thing is an A it is [always, usually, often] a B". Now, we might be able to identify many ways in which this generalization could be undermined in any particular situation such as the one we are now facing. The more of these ways we rule out by evidential test, the stronger is the warrant for inferring B from A in this particular case. In short, we must subject our generalizations to a _variety_ of evidential tests; this is precisely the role of the ancillary evidence shown in Figure 2. The more factors we can rule out on the basis of ancillary evidence that is relevant in testing this generalization involving A and B, the more confident we can be in inferring B from A in this particular instance. Cohen's system of Baconian probability is very informative about this eliminative aspect of our probabilistic reasoning.

As indicated in the above example and as noted elsewhere [Tillers & Schum, 1988], the generalizations we make are frequently stated in fuzzy terms and the ancillary evidence is itself frequently fuzzy in nature. For example, the generalization "if something is an A it _usually_ is a B" involves a fuzzy qualifier. Here is an example of fuzzy evidence. Suppose we wish to infer that person W actually believes what he just told us; such an inference would be involved in assessing the veracity of W. Now, suppose someone asserts the following generalization: "if a person reports an event, then this is _probably_ the event this person believes to have happened". We begin to think about all of the ways in which this generalization could be undermined as far as W and his present testimony are concerned. One test of this generalization involves W's reputation for truthfulness. In a test of this veracity-relevant factor we query several persons who have known W for quite some time. One person says: "W <u>usually</u> tells the truth"; another says: " W would <u>rarely</u> lie"; and another says: " W is a <u>very honest</u> person". So here we have an instance of a fuzzy generalization put to the test using fuzzy ancillary evidence. The work of Zadeh and his numerous colleagues is of obvious value throughout structural and probabilistic analyses of evidential reasoning.

I have given just a very brief account of some of the instances in which there is a natural integration of structural and probabilistic analyses. But, if it is so natural to think of integrating probabilistic and structural analyses, why is this integration so difficult in work on actual inferential tasks as they occur in many important contexts such as in law, medicine, science, business, and intelligence analysis ? I believe there are some answers to this question.

4.0 Probabilistic Inference In Contrived And In Natural Settings

To illustrate the workings of a theory of probabilistic reasoning one usually contrives a situation in which interesting elements of the theory will be exposed. Similarly, to illustrate an approach to the structuring of complex inference one either contrives some situation involving a collection of evidence or uses some already-existing collection of evidence. Wigmore, for example, used existing collections of evidence from actual cases to illustrate his methods of structural analyses. However comprehensive contrived inferential problems may appear to be they typically do not sample all of the elements of inferential behavior that are required in many natural settings. A careful examination of certain characteristics of human inference in natural settings is useful in evaluating the adequacy of existing structural and probabilistic methods of analysis; as I asserted earlier, such examination will help us to see why the methodological gap I mentioned at the outset continues to be quite wide.

4.1 Mixtures Of Inferential Reasoning

The word "assumption" appears regularly in accounts of any formal system; thus, for example, we might say: "assuming a set of hypotheses and a collection of evidence [having certain properties] we will illustrate the differences between Bayes' rule and Dempster's rule in the combination of probabilistic beliefs". Similarly, we might say: "assuming the existence of the documented trial evidence in the Sacco-Vanzetti case, we will illustrate the apparent structure of the prosecution's argument in this case". Unfortunately, such assumptions are no part of inferential tasks as they commonly occur in any natural setting. Neither hypotheses, evidence, nor plausible arguments linking them are typically provided for physicians, intelligence analysts, attorneys, or any other persons who routinely draw conclusions from emerging masses of evidence; they have to be discovered or generated by some means. One result of this rather obvious characteristic of inference in natural settings is that there appear to be no "pure" inferential reasoning tasks; the inferential reasoning required in contexts such as those mentioned appears to be a mixture of three recognized forms of reasoning: deduction, induction, and what has been termed "abduction".

Following Hanson [1981], we may say that deduction shows that something is necessarily true, induction shows that something is probably true, and abduction shows that something is possibly true. Though there is some argument about the distinction between abduction and induction [both involve reasoning from evidence to hypotheses], abductive processes lead to the generation of novel hypotheses and to the further generation of appropriate evidential tests of a novel hypothesis as well as other hypotheses being entertained [e.g., see Rescher, 1978; Eco & Sebeok , 1983; Tursman, 1987]. Since hypotheses and evidential tests of them are rarely "given" and certainly do not materialize out of thin air, we have to suppose that probabilistic inference in natural settings involves a distinctly creative element. Indeed, considerable imagination is required in the formation of a structure or cognitive model of some inferential problem in terms of hypotheses at various levels, evidence, and generalizations linking hypotheses and evidence. By imaginative or abductive reasoning we generate new possibilities, by deductive reasoning we generate possible evidential tests of them, and by inductive reasoning we draw probabilistic conclusions from the evidence we actually gather. The point here is simply that inference in natural settings should not be construed as, say, a purely inductive process; "pure" inductive reasoning tasks are encountered only in contrived situations and in classroom exercises.

4.2 The Acts, Scenes, And Actors In Inferential Tasks

When we examine the life cycle of an inference problem we will not be too far wrong in imagining this life cycle as a "play" having discernible "acts" and "scenes". Further, we observe that there are frequently several or many "actors", each of whom has particular inferential roles to play. My first illustration of an inferential problem's "life cycle" involves events resulting in a trial at law; I have provided a more extensive analysis of this situation elsewhere [Schum, 1986]. In legal matters that result in litigation three general acts can be identified: discovery, proof, and deliberation and choice. Discovery seems to involve the generation of possibilities, hypotheses, or theories about what happened before, during, and after some "moment [or interval] of substantive importance" [labeled MSI in Figure 4 below]. It is at or during the MSI that a potential inferential problem is recognized [Binder, & Bergman, 1984, p 191]. Discovery also involves the generation of evidence on possibilities being considered; and, it involves the elimination of possibilities that do not appear justified by the evidence as it emerges. Finally, legal discovery seems also to involve the structuring of argument on those possibilities that are retained and taken seriously.

The act of "proof" involves the presentation of argument by both sides of the matter in contention; this presentation takes place before some tribunal or factfinding body sanctioned by law to hear the evidence in the case according to procedures that are also legally sanctioned. Following the presentation of argument by both sides on the matters at issue, the factfinders deliberate upon the evidence and the arguments, render a verdict, and decide upon consequences to the parties in contention. As Figure 4 illustrates, in a very general way, there are different actors that appear in these acts and scenes and they perform inference tasks involving various mixtures of the three reasoning methods discussed above. Litigation is an example of situations in which inferential reasoning tasks are embedded in further decision tasks.

The situation depicted in Figure 4 has a most interesting and important feature; it is the only situation known to me in which the acts of discovery, proof, and deliberation/choice occur in a definite sequence. This feature of legal inference allows us to place within the context of specific inference-related activities the various insights provided by alternative theories of probabilistic reasoning. Though not itself a theory of discovery, Shafer's theory of beliefs is attentive to the inevitable changes in the whole structure of our beliefs as we generate and refine new possibilities and discern new evidential tests of them. Cohen's Baconian system of probability is especially attentive to the means by which we rule out possibilities in the process of determining which one we will seek to defend; how completely we have covered matters of relevance to such elimination is all-important in this process. Bayesian formulations are especially valuable in testing the belief-related consequences of complex argument structures and offer specific ways in which we can capture a very wide assortment of evidential subtleties. Finally, Zadeh reminds us that, throughout all of these acts and scenes, we may have to rely upon fuzzy possibilities, fuzzy evidence, and fuzzy generalizations in drawing conclusions that can only be fuzzy in nature.

But the general course of litigation depicted in Figure 4 differs significantly from the course of events in other inferential contexts which I shall illustrate using Figure 5. In science, medicine, intelligence analysis, business, and in other natural settings, more than one of the three acts depicted in Figure 4 can be on stage at the same time; they can also recur in various cycles. In science and in intelligence analysis, for example, discovery is an ongoing process that [presumably] never ceases; in medicine a physician is faced with cycles of inference and choice. The mixing together of the inference-related discovery, proof, and deliberation/choice elements occurring during the life cycle of inference problems in many natural settings is what makes implementation of alternative theories of inference so difficult; each theory attends to some but not all of the necessary elements of inference-related acts and scenes that can be mixed together in various ways. Unfortunately, it will not do to suppose that we can easily integrate these alternative theories of inference; they have quite incommensurable properties. This we expect since these alternative theories examine different elements of a very complex process.

4.3 Time And The Structuring Of Argument

One of the most frequently-overlooked elements of human inference is **time**. Inference in most natural settings takes place in a world that changes with the passage of time and the march of events. The very carefully-reasoned argument you constructed and that seemed so persuasive yesterday now suddenly seems obsolete in light of an event that happened just an hour ago. Typically, we acquire evidence over time; this process depends to a great extent upon possibilities we entertain; new or revised hypotheses suggest new kinds of evidence we should examine. In turn, new evidence suggests new possibilities. We may also experience changes over time in generalizations we make that license stages in our reasoning; the inferential linkage between events E and F, that appeared so strong in the past, may now appear weaker in light of new information we have or in light of new possibilities we entertain. In short, one's structure or cognitive model of an inferential problem grows and changes in light of new insight as well as new evidence. Some theories of probabilistic reasoning seem better tuned than do others to these dynamic characteristics of human inference. Shafer's theory of belief assumes that the process of discerning possibilities is dynamic and not static; Cohen discusses mechanisms for the legitimate purchase of support for some hypothesis by making revisions in it.

Time passes, evidence is gathered and organized in various ways, an argument is constructed, and a conclusion is reached and successfully defended. After all of this labor we later discover that our conclusion was incorrect; where did we go wrong ? A critic, now armed with outcome knowledge we did not have during our deliberations, may appear and be eager to tell us where we went wrong. As Fischhoff relates [1982], this hindsight critic may seek to add insult to our injury by telling us that what actually happened was "relatively inevitable". In other words, this person can usually, after the fact, find reasons why we should have reached a different conclusion than we did. Now, our thinking about this inferential problem developed over time, during the passage of which we may have changed our minds about many of the problem ingredients we thought necessary to consider. The evidence emerged over time and in response to questions we asked; and we entertained different possibilities at different points in time. If we were able to reconstruct our thinking at various stages, as well as to reconstruct what the evidential situation was at each stage, we might be able to take much of the sting out of the arguments made by our hindsight critic. Perhaps this person would have reached the same conclusion we did. The further trouble, Fischhoff says, is that our own reconstruction of our thinking as it developed over time may be similarly biased; we can always find some way to make ourselves look better now that we ourselves have hindsight. Peter Tillers and I are now at work on the development of ways of forming "intellectual audit trails", by means of which we can preserve records of our structural thinking as it develops over time. Such records can add more objectivity to post-mortem analyses and may also have important pedagogical uses [Tillers & Schum, 1988].

4.4 Causal And Non-causal Linkages In The Structure Of An Inference

In offering an explanation for some pattern of events we naturally seek to identify at least proximal or immediate causes; on occasion we may also seek to trace more remote or distal causal relationships in some chain of events that, we believe, has a particular explanation. Reference to various standard works on the topic of causality [e.g. Mackie, 1974; Skyrms, 1980; Hart & Honore, 1985] reveals that it is no easy matter to answer the question: "What does it mean to say that A 'caused' B" ? As I mentioned, here is a substantial level of effort now being applied to the analysis of causal inferential networks. I have noted above that human inference in natural settings appears to involve mixtures of reasoning activities; I shall now argue that such inference also involves various mixtures of causal and non-causal elements. But, by a "non-causal" relationship between events, I shall not mean that this relationship is necessarily "accidental".

A witness at trial asserts that he saw defendant's car at the scene of the crime just a short time before the crime was committed. The relevance of this testimony is defended on the following grounds. If defendant's car was at the scene, it is certainly possible that defendant was at the scene; and if defendant was at the scene it is certainly possible that he was the one who committed the crime. Working in either direction in this chain of inferences we experience some difficulty in determining strict causal relationships. We might be no more inclined to say that defendant's car being at the scene caused defendant to be there than his being at the scene caused his car to be there. Similarly, we would be no more inclined to say that defendant's being at the scene caused him to commit the crime any more than his commiting the crime caused him to be there.

Suppose the credibility of this witness has not been seriously impeached; if so, it would not seem improper for a juror to reason that his testimony about defendant's car being at the scene has at least some value in an inference about the guilt or innocence of the defendant. But if the events in the reasoning chain set up by his testimony had only an "accidental" association, it is difficult to see how a court would allow such testimony, which it routinely would. What the court supposes is that people can apply common-sense generalizations in reasoning from one proposition to another. Thus, a person's car being at a particular location can sometimes mean that the person is also there and the presence of a person at the scene of some crime can sometimes mean that this person committed the crime. The function of ancillary evidence introduced by either side of a matter in contention is to show why some generalization holds or fails to hold in the particular case being examined. Often, these generalizations concern associations that, if they are not causal in nature, are also not accidental. The result here is that formal mechanisms based only on the charting of suspected causal influences cannot capture all of the richness of inferential reasoning that occurs when structural considerations reveal both causal and non-causal elements.

4.5 Structural Forms And Their Relationships

Quite possibly, there are as many different ways of marshalling and organizing evidence as there are persons who perform such tasks. I have mentioned two that the jurist Wigmore recognized: those I referred to as temporal and relational structuring. There are certainly others, some of which are suggested by alternative theories of probabilistic reasoning. For example, there are some schemes suggested by Shafer's discussion of the process of discerning possibilities and others suggested by Cohen's discussion of eliminative inference based upon variative evidential tests. One trouble is that it is not always clear how one structural form relates to another and how different structuring activities can be made mutually facilitative. The research Tillers and I now have research in progress is designed to provide some answers to these questions . As I mentioned earlier, it seems that attention to structural methods is useful not only in the construction of defensible argument but also in the stimulation of questions that have not yet been asked; some of these questions concern new possibilities, others concern new categories of potentially valuable evidential tests of all possibilities being entertained. Having better structural means for showing what one has in the way of hypotheses, evidence, and their apparent relationships is, perhaps, one important element in the discernment or discovery of hypotheses and evidence that one does not have but ought to have.

4.6 The Amount Of Evidence And The Burden Of Assessment

No theory of probable reasoning that I am aware of promises to simplify the process of inference; if one did, such a promise could not be taken seriously in most natural settings such as the ones I have mentioned. As I noted, having access to large amounts of evidence is not our problem; what does present a problem is the appropriate use of it in drawing defensible conclusions. With more evidence about more matters available from more sources, the possibilities for inferentially valuable evidential subtleties increase. But the tracking of evidential subtleties during the emergence of a mass of evidence and the construction of an argument takes time and considerable judgmental effort; and there is no assurance about what the payoff for such effort will be. All we know is that ignoring an evidential subtlety can be harmless in some situations and catastrophic in others.

Many existing artificially intelligent systems do not appear to be very "intelligent" after all; one reason is that there is often a focus on just the obvious ways in which elements of an inferential argument can be related. The metaphor of an argument as a "fabric" is not such a bad one; both warp [vertical] and "weft" [horizontal] threads of an argument can be discerned [e.g. see Schum, 1987, Vol. II, p . 43-45]. Quite often, attention is given only to the warp and not the weft of an argument; but it is in the weft that many important evidential subtleties can be discerned. The weft of an argument concerns the various ways in which the elements of different lines of argument can be related or elements of different parts of the same argument can be related. Here is just one example. Two sources give contradictory reports; the credibility of each source is an obvious consideration in our efforts to resolve this evidential inconsistency. What is not so obvious is that we also have a potential evidentiary redundancy problem on our hands as well. If we believe the first source to be lying to us, we may legitimately conclude the opposite of what this first source says; but, this is exactly what the second source has reported and so there is some potential redundancy in their testimony.

To construct arguments at the level of detail suggested years ago by Wigmore is to examine inference under a microscope; in the process we discern evidential subtleties that are not obvious under less careful scrutiny. At the same time, however, a fine-grain analysis may result in inferential paralysis when we begin to consider the number of separate probabilistic assessments or judgments such analysis suggests. There is another problem that both Cohen [1977] and Shafer [1986] have noted; we may easily construct an argument whose details far outrun the evidence we have. We seek to test a generalization that, we believe, licenses an inference from E to F at some stage in an argument we make. We may discover that there is no evidential backing for this generalization in our particular instance and so, if we rely only upon the reasonableness of this generalization, we will be quite far out on a limb when we consider the variety of ways this generalization could be undermined in the particular situation we now face. Shafer is correct in saying that we must often adopt simpler arguments, ones whose details do not outrun the evidence we have. Part of the research Tillers and I now have in progress concerns development of strategies or heuristics, if you like, for at least deciding upon the order in which we might cope with the evidential subtleties revealed in structural analyses of complex reasoning.

5.0 In Conclusion

A colleague of mine has a sign on his office door that reads: "Operations Research Is The Science Of Informed Choice". I know of other disciplines in which a similar identity might be announced. We have choices to make in many different kinds of situations; and the ways in which we seek to inform ourselves in making these choices are many and varied. In an array of important inferential contexts people must consider masses of incomplete, inconclusive, and unreliable evidence; arguments they might construct on the basis of such evidence can be extremely complex. The major focus in the present collection of essays is on the integration of problem-solving strategies found useful in Operations Research and Artificial Intelligence. If there can be a productive integration of ideas from these two disciplines, as well as from others, I can think of no collection of persons in which such integration would be more welcome than among those persons who must inform their choices on the basis of complex inferential activity.

Each one of us has a somewhat different agenda in our research on these matters; my own agenda involves attempts to identify what seem to be the requisites of complex inference and to investigate how these identified requisites can assist us in deciding what evidential and inferential questions we might well be asking in the process of drawing a conclusion. For insight into these matters I have had to draw upon research from many disciplines. This study has forced me to draw several conclusions regarding the integration of structural and probabilistic analyses of complex inference; I take such integration to be one of the important issues facing those who attempt to integrate OR and AI problem-solving strategies in the analysis of complex inference tasks.

(1) It might be argued that structural and probabilistic issues are always bound together in any analysis of probabilistic reasoning and so my separation of them in this essay is misleading. For example, even in the most elementary applications of Bayes' Rule we have to make a structural assumption; we assume a single-stage reasoning linkage between conditioning event E and hypothesis H in determining the posterior probability $P(H|E)$. What I have argued is that there are valuable ideas obtainable from the works of persons interested in the structure of argument who have had no commitment to any particular system of probability. Much can be learned about possible evidential forms and relationships without any numerical consideration of the weight or the value of evidence. Some forms of structural analyses merely assist us in the process of laying out an argument in order that we might recognize sources of uncertainty or doubt in our argument that someone else might discover and exploit.

(2). There are different structural forms that suit particular inferential purposes; theories of probability assist in the analysis of some structures but not others. For example, the Baconian, Bayesian, and Shafer-Dempster systems can each be applied in a relational or network analysis involving chains of reasoning; but none of these systems tells us how best to fill in the evidential gaps in a temporal analysis in which we seek to present a persuasive scenario or provide a compelling explanation.

(3). The alternative views of probabilistic reasoning that exist today are each uniquely informative about some elements of the behaviorally rich task of probabilistic inference. The trouble is that the implementation of a particular theory in an applied or natural setting is often accompanied by disappointment when it is discovered that there are elements of inference in this setting to which the probabilistic theory does not attend. As I noted in Section 4.0, there are several features of inference in natural settings that add considerable difficulty to the task of integrating structural and probabilistic analyses. These same difficulties do not vanish when the analysis of complex reasoning is construed in other non-probabilistic terms.

(4). There are some very good reasons why current computer-based methods for assisting human inferential reasoning deliver less than is frequently expected. Any method based on just a single view of probability cannot capture all of the behavioral richness evident in the inferential reasoning tasks performed in so many natural settings. Although of obvious importance, the expression and combination of our uncertainties in inferential reasoning is just one element in a much more complex enterprise. Various mixtures of reasoning are required in the generation and defense of arguments set forth in the structure or cognitive model of an inferential problem at hand; the imaginative processes required in such generation are not well-understood. One's structure of a problem will change over time and will reveal mixtures of causal and non-causal elements; different forms of evidence marshalling and argument structuring are required to suit different inferential purposes. The analysis of even the "simplest" inference problems reveals a very wide array of potentially important evidential subtleties; the exploitation of these subtleties places enormous judgmental burdens on persons whose task is to try to make sense out of the _masses_ of evidence our current technologies can provide. We may make various attempts to simplify our inferential procedures and calculations by ignoring these subtleties; but we do so at our own peril. Taken together, these characteristics of human inference in natural settings show how much we have yet to learn about the task of drawing conclusions from a mass of evidence.

Footnotes

1). Figure 1 is just a small portion of Wigmore's analysis of the case: *Hatchett vs Commonwealth* [1882, Court of Appeals of Virginia, 76 Va. 1026]. I know of Wigmore evidence charts that measure 37 feet in length [see Twining, 1984, p 31]. The numbers on the symbols in Figure 1 correspond to items on what Wigmore called a "key list". Wigmore's process of argument structuring consisted of both analysis and synthesis. The analytic part consisted of compiling a key list of all evidence items, events or propositions at intermediate stages in an argument from evidence, and generalizations linking one proposition and another; the synthetic part consisted of the evidence chart itself.

References

1). Binder, D., Bergman, P., Fact Investigation: From Hypothesis To Proof, St. Paul, West Publishing Co., 1984
2). Cohen, L. J., The Probable And The Provable, Oxford, The Clarendon Press, 1977.
3). Eco, U., Sebeok, T., The Sign Of Three: Dupin, Holmes, Peirce, Bloomington, The Indiana University Press, 1983
4). Fischhoff, B., For Those Condemned To Study The Past: Heuristics And Biases In Hindsight, In: Judgment Under Uncertainty: Heuristics And Biases, Cambridge University Press, 1982
5). Hanson, N., Patterns Of Discovery, Cambridge University Press, 1981
6). Hart, H., Honore, T., Causation In The Law, Oxford, The Clarendon Press, 1985
7). Howard, R., Matheson, J., Influence Diagrams, SRI International Report, January, 1980.
8). Kaynes, J., A Treatise On Probability, London, Macmillan & Co., 1957
9). Lauritzen, S., Spiegelhalter, D., Local Computations On Graphical Structures And Their Applications To Expert Systems, Journal Of The Royal Statistical Society, (B) Vol. 50, No. 2, 1988,
10). Lowrance, J., Garvey, T., Evidential Reasoning: An Implementation For Multisensor Integration, SRI Technical Note, 30 December, 1983
11). Lowrance, J., Garvey, T., Strat, T., A Framework For Evidential Reasoning Systems, Proceedings Of The Fifth AAAI Conference, 1986
12). Mackie, J. L., The Cement Of The Universe: A Study Of Causation, Oxford, The Clarendon Press, 1974
13). Pearl, J., Fusion, Propagation, And Structuring In Belief Networks, Artificial Intelligence, Vol. 29, No. 3, September, 1987
14). Pearl, J., Evidential Reasoning Using Stochastic Simulation Of Causal Models, Artificial Intelligence, Vol. 32., No. 2, May, 1987
15). Rescher, N., Peirce's Philosophy Of Science, Notre Dame University Press, 1978
16). Schum, D., Current Developments In Research On Cascaded Inference Processes, In: Cognitive Processes In Choice And Decision Behavior, [ed. T. Wallsten], Hillsdale, N. J., L. Erlbaum Press, 1980

17). Schum, D. Probability And The Processes Of Discovery, Proof, And Choice, Boston University Law Review, Vol. 66, Nos. 3 & 4, May/July 1986
18). Schum, D., Evidence And Inference For The Intelligence Analyst, [Two Volumes], Lanham, Md., University Press Of America, 1987
19). Schum, D., Jonathan Cohen And Thomas Bayes On Chains Of Inferential Reasoning, In: Rationality And Reasoning: Essays In Honor Of L. Jonathan Cohen, [eds. Eells, E., Maruszewski, T], Amsterdam, Dodopi Press [forthcoming]; also available as Report # 32, Center For Computational Statistics And Applied Probability, George Mason University, 1988
20). Schum, D., Martin, A., Formal And Empirical Research On Cascaded Inference In Jurisprudence, Law & Society Review, Vol. 17, No. 1, 1982
21). Shacter, R., Evaluating Influence Diagrams, Operations Research, Vol. 34., No.6., 1986
22). Shacter, R., Thinking Backward For Knowledge Acquisition, AI Magazine, Fall, 1987
23). Shafer, G., A Mathematical Theory Of Evidence, Princeton University Press, 1976
24). Shafer, G. The Construction Of Probability Arguments, Boston University Law Review, Vol. 66, Nos. 3 & 4 , May/July 1986
25). Skyrms, B., Causal Necessity, Yale University Press, 1980
26). Spiegelhalter, D., Coherent Evidence Propagation In Expert Systems, The Statistician, Vol. 36., 1987
27). Tillers, P., Schum, D., Charting New Territory In Judicial Proof: Beyond Wigmore, Cardozo Law Review, Vol. 9, No. 3, Feb. 1988
28). Tursman, R., Peirce's Theory Of Scientific Discovery: A System Of Logic Conceived As Semiotic, Indiana University Press, 1987
29). Twining, W., Taking Facts Seriously, Journal Of Legal Education, Vol. 34, 1984
30). Twining, W., Theories Of Evidence: Bentham & Wigmore, Stanford University Press, 1985
31). Venn, J., The Principles Of Inductive Logic, New York, Chelsea Press, 1973 reprint of 1970 edition
32). von Winterfeldt, D., Edwards, W., D ecision Analysis And Behavioral Research, Cambridge University Press, 1986

33). Wigmore, J. The Problem Of Proof, Illinois Law Review, Vol. 8., 1913.
34). Wigmore, J., The Science Of Proof, Boston, Little, Brown, & Co., [3rd ed], 1937
35). Zadeh, L., Fuzzy Sets As A Basis For A Theory Of Possibility, Fuzzy Sets And Systems, Vol. 1, 1978.

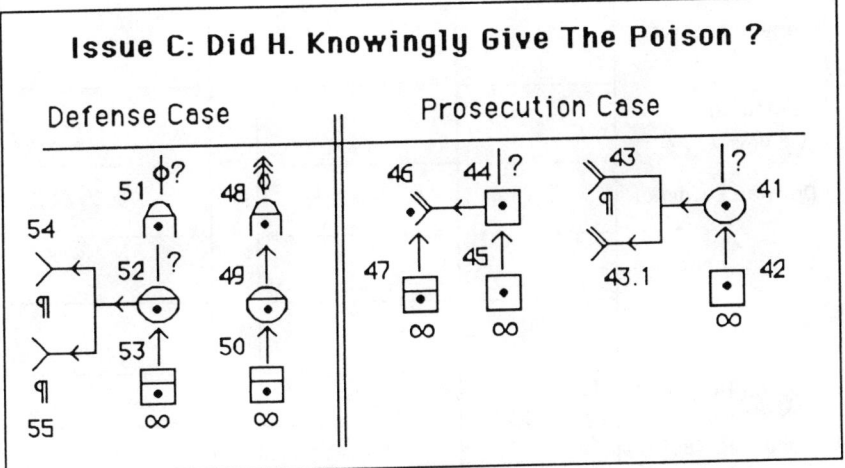

Figure 1. A portion of a Wigmore evidence chart.

	Direct Relevance		Indirect Relevance
	Direct	Circumstantial	Ancillary
"Real" or Demonstrative (+ or –)			
Testimonial (+ or –)			
Opinion Evidence			
"Second-hand"			
Missing			
Source Behavior			
Accepted Facts			

Figure 2. A categorization of evidence forms.

A. Corroborative vs Contradictory Evidence:

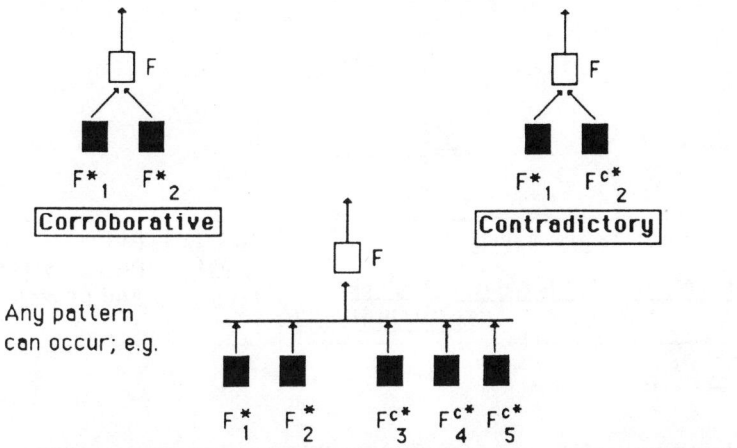

B. Confirming [Convergent] vs Conflicting [Divergent] Evidence:

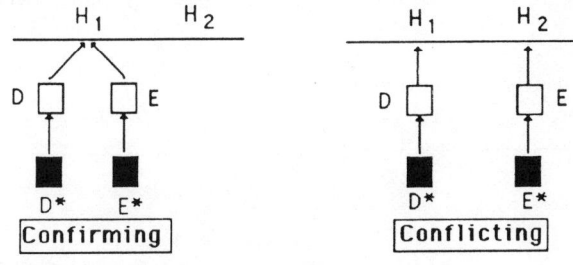

C. Redundant vs Nonredundant Evidence:

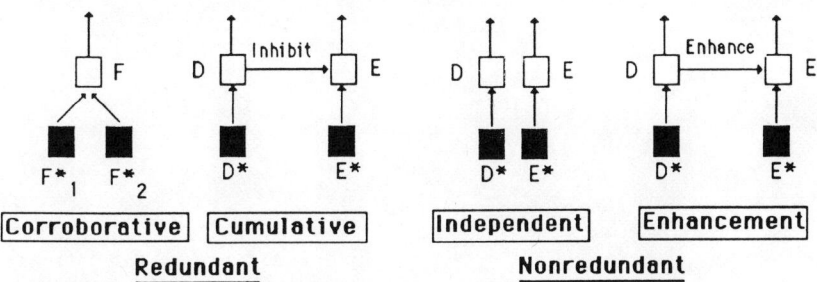

Figure 3. A categorization of evidential relations.

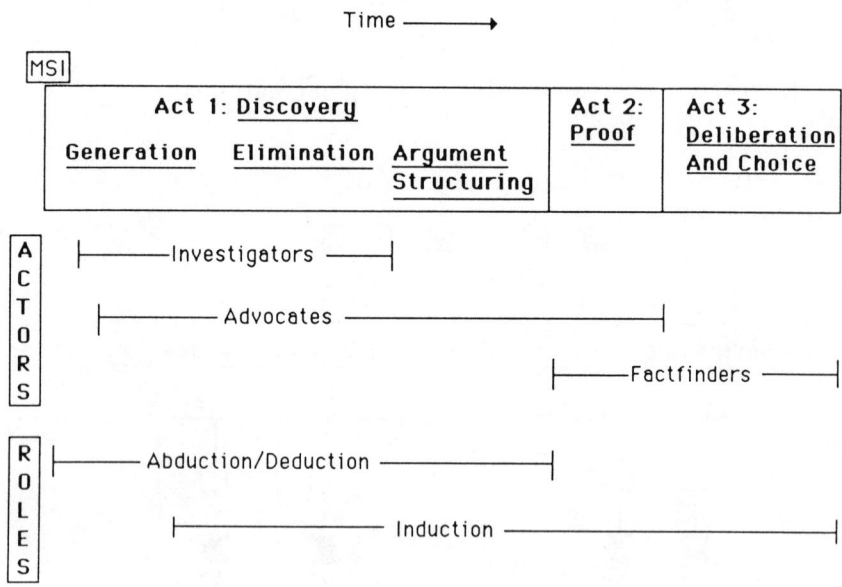

Figure 4. The "acts" and "scenes" of inference in jurisprudence.

```
                          Time ⟶
_____
──────────────── Ongoing Discovery ────────────────
─────Proof ──────── Choice ────── Proof ────── Choice ──────
```

Figure 5. The processes of discovery, proof, and choice in other inferential contexts.

HYBRID SYSTEMS FOR FAILURE DIAGNOSIS

by Elisabeth Paté-Cornell and Hau Lee
Department of Industrial Engineering and Engineering Management
Stanford University

ABSTRACT

Optimization as well as heuristics can be used as a basis for decision support systems for failure diagnosis and repair. The object of this paper is to discuss the desirability and the characteristics of hybrid decision support systems, designed to operate in three different modes either on the basis of probabilistic risk analysis (PRA) models, or on the basis of heuristics, or in a simple information mode, according to the nature of the failures and the decision circumstances. This paper presents first an overview of the capabilities of current approaches to failure diagnosis (in particular, their treatments of uncertainty) and a discussion of the basic issues in the diagnosis and repair of complex systems. The argument then focuses on a PRA-based method for the optimization of inspection and repair procedures when the objective is "minimum repair" (restoring the system to an operational mode under time or resource constraints) and on the inclusion of this analytical model in an extended expert system, partly model-based and partly rule-based. This method is illustrated by the case of a hypothetical discrete engineering system.

INTRODUCTION

Optimization as well as heuristics can be used as the basis for decision support systems for failure diagnosis and repair. The object of this paper is to discuss the desirability and the characteristics of hybrid decision support systems, designed to operate either on the basis of probabilistic risk analysis (PRA) models, or on the basis of heuristics, or in a simple informational mode, according to the nature of the failures and the decision circumstances.

PRA was originally designed to compute the failure probability of physical systems made of components in series or in parallel (for example, engineering systems such as nuclear power plants). The PRA method is based on a logical analysis of the functions performed by the physical system and on a probabilistic analysis of the different failure modes accounting for interdependencies and external events (Paté-Cornell, 1984). The PRA model can be imbedded in an artificial intelligence framework and turned into a diagnosis tool to identify and fix the sources of a system's malfunction. Whenever possible, this search is done following an optimal procedure under resource constraints. Artificial intelligence (AI), in addition to its capacity to treat automatically repetitive problems, offers a convenient medium of interface with the user. It also allows inclusion in the knowledge base of different types of information, some of which are PRA-related, and some of which are more loosely related to components' failures. The AI framework thus allows support of a large range of diagnosis and repair decisions that may or may not fit an optimization model.

Hybrid failure diagnosis systems whose principles are presented here are particular examples of intelligent decision systems as described, for instance, by Holtzman (1989). Instead of attempting to clone systematically the human thought, such a system is designed to operate on different modes according to the circumstances. In cases where the problem is well defined, the failure diagnosis system is designed to operate on the basis of PRA, thus following a normative, analytical approach to sequential decision making (first mode). When the problem does not fit the PRA model or when optimization is unnecessary, it is proposed here to design the diagnosis system so that it can shift either to a second mode that provides heuristics for action without attempt of optimization, or to a third mode in which it provides relevant information even if there is no attempt to handle formally the complete decision problem.

In the first mode, AI is used to formulate the search and repair problem and explain or interpret the recommendations. Probabilistic optimization techniques are used to process information because they generally perform better than the human mind for the treatment of well defined, complex problems involving uncertainty and rare events. Furthermore, the Bayesian techniques generally perform better than current AI methods of treatment of uncertainty (Henrion, 1987). In the second mode, the system is designed to function like a classic rule-based expert system. In the third mode, the system is designed to provide information rather than recommendations. The focus of this paper is primarily on the first mode, and therefore on the structure of PRA-based diagnosis and repair systems.

This paper presents first an overview of the capabilities of current approaches to failure diagnosis (in particular, their treatments of uncertainty) and a discussion of the basic issues in the diagnosis and repair of complex systems. The argument then focuses on a PRA-based method for the optimization of inspection and repair procedures when the objective is "minimum repair" (restoring the system to an operational mode under time or resource constraints) and on the inclusion of this analytical model in an extended expert system, partly model-based and partly rule-based. This method is illustrated by the case of a hypothetical discrete engineering system.

UNCERTAINTIES AND THE CURRENT METHODS OF COMPUTER-AIDED FAILURE DIAGNOSIS

Diagnosis of failure has been one of the first uses of artificial intelligence and expert systems. Most of the existing computerized systems for diagnosis of failures in complex industrial settings are based (1) on the automation of the thought process of a particular expert, or (2) on existing procedures, or (3) on logical, functional analysis coupled with experts' recommendations. For example, in the domain of hydroelectric power generation, a rule-based expert system was designed to replicate as faithfully as possible the thought process of an expert in dam safety (Rose, 1988). The usual problem encountered by knowledge engineers, in such cases, relates to the structuring of the information provided by the expert. In the nuclear power industry, a knowledge-based expert system (named ESCARTA) has been developed for the diagnosis of boiler tube failures on the basis of the information and procedures contained in the manuals (EPRI, 1987). It is assumed in such a system, that the information is correct and appropriate and has been tested against PRA. In the aerospace industry, NASA developed the Knowledge-based Automatic Test Equipment (KATE) system designed to perform "system monitoring, signal validations, fault location and diagnosis, and automatic control and reconfiguration" (EPRI, 1987). KATE is based on the functional logic of Failure Modes and Effect Analysis (FMEA) and on additional rules provided by experts to reflect their perception of priorities, but it does not involve concepts of probability. The electric power industry is currently working on the adaptation of the KATE system to power plant applications (ibid.).

Because of the way these expert systems, whether knowledge-based or rule-based, are often designed, they have limitations that may seriously affect their ability to perform an accurate diagnosis in the first place, and to do so efficiently if at all. One major limitation of purely heuristic methods, as currently implemented, is that they perform poorly in the face of uncertainty (AAAI, 1985). An efficient diagnostic method must rely on the probability of failure of each element of the system given the available information (i.e., observed failures and previous knowledge of the system). During the diagnosis process, this information must be updated when additional information becomes available. Heuristic methods often treat this updating mechanism on the basis of experts reasonings, that is, implicitly rather than explicitly. This means that the experts intuitively base their recommendations on perceived failure likelyhoods based on their experience which does not ensure internal consistency in the probabilistic reasoning. Different approaches to the treatment of uncertainties used in popular expert systems include certainty factors and other belief levels (Shortliffe and Buchanan, 1975) based, for example, on the "Theory of Evidence" (Shafer, 1976). These approaches have the advantage that they appear to fit the thought processes of most experts, but they also raise serious questions concerning their fundamental logic, particularly in their way of incorporating new information.

Another limitation of rule-based expert systems is that the recommendations that they provide are guided, in some cases, not only by experts' knowledge but also by their preferences. This is true, for instance, of decisions involving trade-offs among risks, costs, and delays. In this respect, the value judgments of the experts are not necessarily those of the decision maker (the final user of the system). One solution is to include in expert systems explicit information about the preferences that are implicit in the rules. This allows the user to check that these preferences correspond to his own, and if not, to make sure that his own criteria prevail. In the medical field, where this question of values is often unescapable, such a system was developed, using methods of decision analysis, to advise infertile couples about available medical treatments (Holtzman, 1985). In problems of diagnosis of technical failures, the confusion of facts and values in experts recommendations can lead to severe inefficiencies in the control of diagnostic depth as well as the choice of a repair policy.

Artificial intelligence, however, has some major advantages over strict optimization methods because it allows processing of information that does not bear any strong relationship to failure of the elements, but may be critical to making decisions under unexpected circumstances. The question is to structure this "loose" information in a retrievable manner so that it can be efficiently used under time constraints. PRA can then provide a sound base for causal reasoning while artificial intelligence can provide the interface and the framework necessary to expand the knowledge base. For well defined decision circumstances that can be anticipated at the time of the system conception, Bayesian techniques coupled with utility theory allow the design of a decision support system that can simultaneously guide a user in the search for basic failures and make appropriate recommendations along the diagnosis path for the repair or the replacement of parts and subsystems as failures are detected. The characteristics of the failed technical system as well as the decision circumstances are relevant both to the choice of the inspection path and to the repair decisions. Of particular interest is the case where the immediate goal is to restore the system to operating condition through "minimal repair" rather than (or prior to) systematically fixing all basic failures.

The key to the use of PRA in a diagnosis system is the solution of a dynamic programming model involving the updating of failure information for the yet-uninspected parts as the inspection procedure progresses. Some practical difficulties may arise in this process due to the possible complexity of the analytical problem which grows with the complexity of the physical system itself. New methods based on the manipulation of influence diagrams can help solve this updating problem under reasonable time constraints. A shift to simpler heuristics may be required when the power of computer systems does not allow updating to be fast enough or when the optimization is unnecessary. There are also some purely theoretical difficulties in the optimization approach that require a shift to heuristics of different kinds. These difficulties can be due, for example, to systems' features that

PRA does not handle well such as particular types of partial failures, or simply to unexpected events that may affect all or parts of the system but cannot be anticipated at the time where the diagnosis system is constructed, for example, because they have never yet been observed.

DIAGNOSIS PROBLEMS IN COMPLEX ENGINEERING SYSTEMS

Hybrid decision support systems for diagnosis and repair

A system may fail in such a way that there is no uncertainty as to which parts of it caused the failure. In other cases, it may be unclear from the symptoms how the failure occurred and what repairs are needed. The latter is true, in particular, for large complex systems in which failure can be attributed to many different subsystems and individual parts. The question is then to identify the failure cause and to decide which part(s) should be fixed or replaced. Most of the times, the problem of optimal inspection and repair can be described as one of sequential decision under uncertainty. At each inspection step, a decision has to be made (what to inspect next, and whether to repair a failed part) before all information has been gathered about the state of all subsystems.

Analytical, OR-based methods to address these questions have existed for a long time (see for example, Gluss, 1959). They generally involve probabilistic methods coupled with the economic analysis of failure detection, for example, models based on dynamic programming (Bellman and Dreyfus, 1962) under cost and time constraints. These models combine logic and probability to provide the most effective search algorithms when the information needed is available and sufficient, and when the goals and criteria can be clearly formulated (e.g., minimization of costs or minimization of time). In the published literature, however, the choice of an inspection sequence and the repair decision are treated separately. The purpose here is to couple these two functions.

Practical problems that can arise in the analytical approach are due to the complexity of the physical system. A key question is how to handle a large number of potential failures and how to treat probabilistic dependencies among failures of the different parts. An implication of these dependencies is that along the diagnosis path, the probability that a yet-unchecked part has failed depends on the information obtained so far, including (1) the symptoms observed and (2) the results of the inspection (and possibly, repair) for the parts that have already been checked. The process of updating the information at each step may be complex and time consuming. It is proposed here to use influence diagrams that allow compact representation and effective treatment of the updating problem (Shachter, in Paté, Lee, Tse, and Shachter, 1986).

Sometimes the optimization is infeasible or unnecessary. The problem formulation itself can be difficult in cases where there is little information on which to base functional and probabilistic dependencies. Future circumstances may still be so vague that the decision itself is unclear. In this case, the hybrid diagnosis system must be able to switch to a second mode of operation. The design problem is thus to choose, gather, structure, and store in a retrievable manner general information about the physical system's characteristics (such as the physical properties of some of the materials) that may be relevant to the decision but do not enter the classical framework of reliability and optimization. Also, the system's complexity can make analytical methods impractical. Simple (and "greedy") heuristics may then be attractive because they can be quicker (even though they may not be optimal) and time may be key to avoiding potentially catastrophic delays. Finally, the effort involved in the analytical formulation and resolution may simply be unnecessary given the simplicity of some decision problems and for these cases, one can adopt the classical artificial intelligence approach, for example, under the form of rule-based expert systems. The knowledge base can then rely on existing procedures assuming that these procedures are correct and, for example, that they have been tested against a PRA-type of system analysis.

In the design of a hybrid diagnosis system, one issue is thus to decide when to shift from one mode to the other. The adequacy of logical versus heuristic methods for fault diagnosis and repair relies on two sets of criteria: (1) the technical characteristics of failures and (2) the decision situation as anticipated at the time when the decision support system is constructed.

Complexity

Engineering systems' complexity may be related to several types of characteristics that result in difficulties of inspection and maintenance: difficulty to conduct quickly an exhaustive search for failures, to characterize failure itself, or simply to use one's intuition to perceive causal and probabilistic links between the failures of the different elements and the symptoms observed.

The characteristics of complex systems that are of particular interest are therefore those that may constitute discriminants in the choice of optimization versus heuristics when faced with a problem of diagnosis and repair. For this purpose, complex systems are defined as follows:
  ° They have a large number of components.
  ° The failures of the basic components involve significant probabilistic dependencies.
  ° The physical properties and the physical characteristics of the components and their failure mechanisms are complex, or poorly known. The information contained in the classical inputs of reliability models (functional diagrams, failure probabilities etc.) may be insufficient to guide the diagnostic process and the repair decisions. Other kinds of information may be necessary to understand and repair the failure.

° The system may be subjected to external events, such as massive loss of electric power, fires, floods, etc. These phenomena may constitute common causes of failure that modify the failure probabilities, the probabilistic dependencies of element failures, and possibly the failure mechanisms themselves.
° The components of the system may be in a state of partial failure.
° The basic components of the system may not be clearly identifiable as such; for example, the system may be continuous (i.e., the elements of the system are not discrete) or unknown beyond a given level of detail.

An example of a complex system is a computer network with a large number of terminals that can be subjected (in total or in part) to a common cause of failure, for example, a sudden increase in the electrical voltage, or an earthquake in seismic regions such as California. In addition, some poorly understood phenomena may occur in the connections between the elements of the network, and some of these elements may be difficult to access and to test.

Optimization vs heuristics for fault diagnosis in complex systems

Analytical models of optimization are attractive because they are systematic and rely on clear optimization criteria. Yet, they can only accommodate a limited set of systems. First, their resolution can be time consuming which might make them unusable in real time or emergency situations. Second, they involve almost exclusively the limited information provided by functional diagrams, failure probabilities, and diagnostic costs and times. They rarely include the physical characteristics of evolution and deterioration mechanisms. Third, these methods do not easily involve partial failure states and they are difficult to use in the case of continuous systems whose basic elements cannot be clearly identified and discretized. In addition, probabilistic dependencies among element failures and the effects of external events present practical--if not theoretical-- difficulties.

In spite of these limitations the optimization of search paths based on probabilities can be powerful when coupled with methods of decision analysis (Raiffa, 1968; Howard, 1968). For instance, the decision maker's preferences in the repair decision can help determine the appropriate level of diagnostic depth: if replacing a whole subsystem instead of part of it involves only marginal additional cost, further inspection beyond the subsystem level may unnecessary. The minimization of expected costs is only one possible criterion. More complex multiattribute utility functions (Keeney and Raiffa, 1976) can be used when considering not only the search costs, but also the detection and repair costs and times, and, possibly, the risks of subsequent failures for different levels of repair and upgrading.

Heuristic methods (see, for example, Nilsson, 1980) have been more recently developed to address these problems from the perspective of artificial intelligence and expert systems. These *ad hoc* methods are based on rules that can accommodate a less structured body of information than the knowledge and data processed by classical reliability methods (Genesereth, 1978). The heuristic aspect of the method may alleviate considerably the search procedure and allow the decision maker to find a satisfactory solution more quickly. This solution may not be the optimal one, but the optimum may not be much better, and, in some cases, may be unreachable within reasonable cost constraints. These techniques can thus be quicker and more flexible than analytical method and allow the use of a knowledge base that extends beyond reliability data and that includes fundamental notions about the system's physical characteristics.

The completeness issue

One of the classical problems of fault tree and event tree analyses is the issue of completeness. This term refers to the impossibility of ascertaining that all possible failure modes and failure mechanisms have been considered in the analysis. The problem is not so much one of ensuring that all possible combinations of failure events have been considered. Rather, it is one of facing the fact that deterioration mechanisms and physical phenomena may occur that were either unknown at the time of the analysis or were overlooked in the design of diagnosis and repair procedures. Obviously, this issue is more likely to arise with new emerging technologies than with well-known ones, and is often raised in the nuclear industry where probabilistic risk analysis is used in safety decision making (Lewis, 1984).

The completeness issue arises in expert systems as well but in a different form. Rule-based and PRA-based expert systems are obviously subjected to shortcomings similar to those of PRA due to the limitations of the experts' knowledge and experience. The AI framework, however, can provide a partial solution to the completeness issue because the knowledge base can be extended beyond obvious procedures for diagnosis and repair. It can be built to include additional information that may be necessary to understand the system's operation, and, therefore, to address a new failure problem. This information might be, for example, the description of the physical effects of ground shaking on the behavior of some electronic components. By definition, however, one cannot be sure that the additional information provided by extending the knowledge base is the information that will be needed in an unpredictable situation. Furthermore, heuristics alone may well overlook phenomena (e.g., rare events) that exhaustive sytematic analysis would have considered. In other words, the completeness problem exists whether one uses analytical or heuristic techniques, but under different forms. In this respect, hybrid method may be helpful to reduce if not resolve the completeness problem.

Decision situations and criteria

Expert systems that rely on experts' opinions often result in recommendation for action. Therefore they involve not only the experts' knowledge but also their preferences (e.g., minimize repair time as opposed to repair costs). Analytical methods, such as reliability analysis coupled with decision analysis give a description of the systems' state and leave the decision criterion to the decision maker. In the design of a hybrid system, the question of value judgment versus pure information is thus important: in the end, if there is any potential conflict, the values of the user must be allowed to prevail. If he relies on the recommendation of an expert system, he must be able to check that the decision criterion involved corresponds to his own.

Although the link between failure diagnosis and repair decisions can be considered as a relation between information gathering and decision making (Holtzman, 1985), both rely, in fact, on value judgments. The depth of an investigation in the diagnosis and repair of a technical system may be determined by the costs of replacing a whole subsystem without further investigation as opposed to looking for a more specific failure cause and fixing a smaller component. The choice of a desired level of depth and precision in the diagnosis is based on the value of additional information. This value depends (1) on the availability and desirability of alternative repair possibilities and (2) on the diagnostic "costs" associated with a particular decision criterion (e.g., overall minimization of the diagnosis and repair time). Therefore, since diagnosis is never free, this essentially descriptive task includes in fact a normative element. Indeed, in case of emergency, the question of time versus cost is critical. For more routine repairs, cost minimization may be the principal objective. Diagnostic methods and repair procedures are thus closely interrelated and must be developed concurrently. We refer to a particular instance of diagnosis and repair as an intervention. Clearly, once the intervention is completed, the state of the system will depend on the quality of both operations. In the proposed hybrid system, the choice between first and second mode requires the explicit use of a decision criterion. Furthermore, in either case, an elicitation of the user's preferences is needed, either for an explicit model of decision analysis (mode 1) or to check that the experts' recommendations for action correspond to the user's preferences at the time when the procedure is activated.

Facts, values, and disagreements

As mentioned above, state-of-the-art rule-based expert systems are not generally designed to include the preferences (e.g., risk attitude) of the decision maker (i.e., the use of the system). These expert systems usually rely implicitly on the preferences of the expert(s).

Disagreement between the experts and the user can result from a conflict of objectives. This can be illustrated, for example, by a disagreement between a customer and an auto mechanic who may have a policy of changing a whole part in a car rather than opening a subsystem and performing an inexpensive intervention. Therefore, there may be a need for a "custom-tailored system" that recognizes the criteria of the expert and accounts specifically for the user's preferences. The diagnosis part, itself involves a choice in the depth of the investigation (e.g., a whole generator versus a brush).

Two experts may therefore recommend different actions, and this may occur either on the basis of different diagnoses, or on the basis of different preference systems, for instance, if a second mechanic is ready to pursue the investigation further in order to reduce the cost of repair to the customer. Indeed, experts themselves, and therefore expert systems, often disagree in their conclusions. It is important for the decision maker to understand the source of these diagreements (e.g., disagreements about facts vs disagreements about value systems) in order to choose an appropriate method of resolution of conflict among divergent experts' opinions (Bonduelle, 1987). In a disagreement about facts, it may be useful to understand first the source of the divergences (e.g., fundamental condition versus severity levels) and the experts' reasonings, given their experience. One solution may be to seek a third independent opinion. In a disagreement about values, the logical solution is to try to obtain from the experts their assessment of facts, then to use the preferences of the decision maker himself. These are, for example, the values of the patient who may prefer to take some risks to minimize pain, or those of the car owner who may prefer to minimize cost as opposed to repair time and effort.

In the first mode of the described hybrid system, the problem, when constructing the diagnosis system, may be to aggregate experts' opinions about facts. In the second mode, the recommendations involve directly preferences. These preferences have to be explicit for the user to be able to choose the heuristics that fit specific circumstances. In the third mode, there is no foreseeable value problem, except, perhaps, hidden ones in the choice of the information to be encoded and stored. For example, an expert who is also a proponent of a technology may have convinced himself that a specific failure mode is so unlikely that it is unnecessary to store information that might be useful should it occur.

Structure of a hybrid system: choice of mode and decision criteria

The general framework proposed by Tse (Paté-Cornell, Lee, Tse, and Schachter 1986) does not involve any assumption about rules: the heuristics can be based on an optimization model of the type proposed above, or on another set of simple quantitative criteria, or on more general information loosely connected to the functions to be performed. The process is the following:

1. A set of symptoms is observed.
2. Determine whether the observed symptoms imply possible faults that need to be corrected.
3. If a fault is identified, determine the corrective actions to be taken.
4. If there are multiple plausible faults, determine either a corrective procedure to be carried out or a verification procedure to discriminate the true fault from the ambiguous ones.
5. If a corrective procedure is carried out, observe to see whether symptoms prevail; if not, the fault is corrected, otherwise a new piece of evidence is available.
6. If a verification procedure is carried out, the newly observed symptoms

Hybrid Systems For Failure Diagnosis

151

provide new evidence.

7. Repeat the process until either the fault is identified and corrected, or symptoms leading to possible fault disappear.

Most *ad hoc* procedures used by experts can be encoded in this framework. Step 3 involves a decision to repair, replace, or leave as is. Step 4 is the one that can be addressed from different angles. An optimization procedure is one possible approach. In the knowledge encoding typical of expert systems, however, the procedure is to replicate the reasoning of an expert when he is faced with a particular situation. A critical step is thus the choice of a mode of operation. This step must be added following directly the observation of symptoms (step 1'). It involves the decision to treat the symptoms either (1) in the optimization mode, or (2) in a simple heuristic mode, or (3) in the information providing mode in which the decision is left to the user. Among the discriminating characteristics that guide the choice of one of these three modes one finds the following: are there constraints to the decision at each inspection step? Is the system too complex to be treated in the optimization mode? Are there partial failures that are incompatible with the PRA approach?

The structure of the knowledge base is therefore such that it can be accessed at different levels: a fundamental level at which basic information is stored (for example, physical characteristics of the components), a heuristic level at which simple rules of operation are proposed for cases where the optimization is undesirable but where some direct guidance can be provided on the basis of encoded rules, and an optimization level for the cases where the decision circumstances allow it and where a PRA-based diagnosis model has been developed. The user, therefore, needs to specify the type of decision and the type of constraint that will determine the mode selection in the decision support system. Several types of decisions can be envisioned, for example, minor repair, or retrofitting, or emergency decisions of different categories. Constraints can thus involve time, costs, or mixed criteria that ensure that neither excessive time nor excessive costs are involved in the intervention.

The general organization of a hybrid system includes a knowledge base, a reasoning module, an explanation module, and an interface module. The knowledge base is extended to include (1) functional and PRA-related information (failure probabilities, characteristics of external events, costs and duration of inspection and repair, etc.), (2) heuristics that can be called for in cases where the optimization is infeasible or undesirable, and (3) other types of data physical, functional, and operational. A full reliability analysis of the system is thus included in the knowledge base. The reasoning module involves first a choice of mode facility, then a reasoning submodule adapted to the chosen mode (choice of decision criterion, optimization capability, heuristic processing, or hierarchical access to general information).

The performances of heuristic and analytical methods --and therefore the circumstances under which a shift of mode might be desirable-- depend upon the characteristics of the physical system under consideration and upon the decision conditions.

A large number of elements in the system can make analytical methods time consuming in times of emergency, unless probabilistic independencies simplify the updating of probabilities and, therefore, the optimal search procedure. If the time constraint becomes incompatible with the optimization requirements, then simpler heuristics may perform better and the choice of one set of heuristics in the second mode may be preferable. By contrast, in cases in which the time problem can be resolved to the satisfaction of the decision maker, the complexity of the system can make the analytical approach more attractive for several reasons. Experts may have difficulties thinking through the different functions to be performed and the interface between the different subsystems. In addition, in cases where external loads may have affected the physical system, the experts' knowledge and experience may be too limited to allow them to assess and process the new failure probabilities of the different parts, conditional on the external load. In other terms, combining intuitively the hazard analysis and the fragility analysis may simply be beyond their capability.

Probabilistic dependencies of element failures can be treated by analytical methods, but they require development of dynamic programs including efficient procedures of probabilistic updating. Heuristic methods can also be useful in the presence of failure dependencies, provided that the search procedures properly account for these dependencies. Similarly, the particular problem of dependencies among element failures caused by the occurrence of external events can be addressed either by improved analytical methods or by heuristics that recognize probabilistic dependencies.

Heuristic methods may perform better for continuous systems and for systems in which partial failures do affect the choice of diagnosis and repair procedures. This is true because analytical methods, in the current state of the art, are not fully adequate to treat partial failures. Also, an exhaustive analysis can be extremely cumbersome when the system must be arbitrarily discretized into a large number of elements.

THE ANALYTICAL MODE: OPTIMIZATION

Minimum repair

The first mode relies on a model-based, problem-oriented approach to the question of diagnosis and repair. First, this approach assumes that the range of problems to be resolved is known at the time of the conception of the information system. This means, for example, that the model will be used for the minimization of the overall cost or time of intervention. It also means that the goal of the

Hybrid Systems For Failure Diagnosis

diagnosis and repair problem is well defined, for example, to achieve "minimum repair" which is defined here as repairing a sufficient number of parts (but not necessarily all of them) so that the system works. Second, this approach is model-based in that it assumes that the system is sufficiently known to be amenable to reliability analysis. This means that the functional relationships between the different subsystems can be described without ambiguity, and that the probability of failures of the different parts can therefore be combined using Bayesian methods of analysis.

Following the study of this reliability model, dynamic programming is used to optimize the overall cost or time function associated to inspection and repair (Bellman, 1957). This method allows identification of the optimal inspection path at the same time as repair decisions in the cases where the goal is to achieve operability through minimum repair.

Reliability analysis for a simple engineering system. Illustrative example

Consider for illustration the simplified case of a feed-water system designed for cooling in an industrial process and involving two functions in series: water supply and pumping. Water supply is assumed to involve two water tanks in parallel. Pumping can be performed either by an electric pump and its generator, or by a turbine pump. These relations can be represented on a functional diagram (see Figure 1).

Assume that the symptom observed is: no water comes out of the system (event noted T). Reliability analysis, performed before the fact, addresses two questions: (1) How can the system fail, i.e., what are the conjunctions of events (or failure modes) leading to T, and (2) what is the probability of T per time unit? In case of emergency, this probabilistic evaluation can also be used to answer the question: given that T has occurred, what is the probability of each of the failure modes? In order to conduct this reliability study, one can use fault tree analysis (Paté-Cornell, 1984). The next step is to use this tree to draw the logical link between the occurrence of T and the failure of the basic elements. The Boolean polynomial used to do this is of the following form:

$$T = T1 * T2 + TP * G + TP * EP = M1 + M2 + M3$$

in which the sign * represents an AND function, the sign + represents an OR function, and the letters represent the failure of the basic elements.
T1: failure of tank 1 (A)
T2: failure of tank 2 (B)
TP: failure of the turbine pump (E)
G: failure of the generator (C)
EP: failure of the electric pump (D)

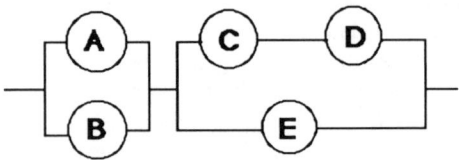

Legend:
A: operation of tank 1
B: operation of tank 2
C: operation of the generator
D: operation of the electric pump
E: operation of the turbine pump

Figure 1: Functional diagram for a hypothetical feed-water system

Finally, Mi (i= 1, 2, or 3) represents a set of events leading to failure. This Boolean (logical) polynomial allows identification of the failure modes M1, M2, and M3. For example, the joint failure of the generator and of the turbine pump means that there is no pumping capability, and therefore no water is delivered even though the reservoirs may be functioning. If no failure mode occurs, the system works even though some of the basic parts may have failed. For example, if tank 1 and the turbine pump work, both the water supply and the pumping will be ensured, even though the other tank and the electric pump system may be out of order.

One can, therefore, define the sets of events that are sufficient to ensure that the failure does not occur. Such a set is called a success path (SPi). Define the negation of an event by the notation N(.), and event S as the negation of event T. The success paths are found by taking the negation of the set of failure modes, i.e., replacing ORs by ANDs, ANDs by ORs, and each event by its negation.

$$S = N(T1)*N(TP) + N(T1)*N(G)*N(EP) + N(T2)*N(TP) + N(T2)*N(G)*N(EP)$$

Minimum repair means that at least one success path is achieved, for example, that tank 2, the generator and the electric pump are all operational.

The probability of failure occurrence for the whole system (T) can then be expressed as a function of the probabilities of occurrence of the basic failures.

$$p(T) = p(M1) + p(M2) + p(M3) - p(M1,M2) - p(M2,M3) - p(M1,M3) + p(M1,M2,M3)$$

$p(M1) = p(T1) \times p(T2|T1)$

in which $p(X|Y)$ represents the conditional probabilty of X given the occurrence of Y.

There are two possible simplifications of this expression. The first one is the "rare event approximation" which applies if the probability of occurrence of two failure modes at the same time is so low that it is negligible (=> $p(Mi,Mj) \approx 0$). The second is the probabilistic independence of failure events which applies either when such independence can be proved or when the approximation is acceptable. For example, if the failure of the two tanks are independent events, one can write the probability of the first failure mode as:

$p(M1) = p(T1) \times p(T2)$

One major source of dependences among failures is the occurrence of a common cause, often an external event that affects several components at the same time. It can be, for example, a surge in the electrical voltage that affects several elements of an electrical system, or an earthquake that affects several subsystems of a nuclear reactor. The probability of the overall failure is computed as a function of the occurrence of this external event E (hazard analysis) and the probabilities of failure of the different components given the event (fragility analysis).

$p(T) = p(T,N(E)) + p(T,E)$
$p(T,E) = p(E) \times p(T|E)$

The probability of T given E is then computed using the same formula as above and replacing each failure probability by the probability that failure occurs conditional on the external event.

Reliability analysis as a diagnosis tool

Consider the following situation: the water arrives to the main system but the pressure is insufficient or nil (symptom). Assume that the objective is to achieve minimum repair at minimum expected cost. The problem is to find an optimal sequence of inspection to identify total or partial failures of basic elements, and to make the decision to fix, replace, or leave as is and continue inspection, every time a failure is discovered. For example, if the turbine pump has been found broken, it can be fixed, or replaced, but it may be quicker, in an emergency situation, to inspect and fix the electric pump system instead. Note that in addition to reliability-based information, other types of data may be needed involving (1) water pressure as a function of the state parameters of the different components and (2) the repair costs. It can be included in a knowledge base for retrieval at the time of the incident.

In this case, where the objective is to ensure that at least one success path is achieved, reliability analysis and estimates of repair times or costs can be efficiently used to find an optimal strategy. This is true provided that the system's model is manageable and that the time of processing of the information is not so large as to be unusable in times of emergency. Dynamic programming is then an effective method to find the optimal strategy. The principle is to construct a recursive relationship in order to find an optimal path in a network starting from the final objective on the basis of all the possible states of the system. Define a network in which the nodes are the functions to be performed, for example, the operation of the basic elements. Let T be the symptom, (here: the system does not work) and U be the set of all nodes (all possible failures).

At any given stage of the inspection process, define:
X0: set of nodes known to NOT be working
X1: set of nodes known to be working
X: set of all previously inspected nodes

Obviously, the intersection of X0 and X1 is the empty set, and the union of X0 and X1 is equal to X.

Define the function MC(X0,X1) as the expected minimum cost to make the system work, given T, X0 and X1 at the particular stage being considered. When all nodes have been inspected (X=U), MC(X0,X1) is well defined: it is the minimum cost to repair identified failure modes. When at least one success path has been achieved (X1=SPi), the cost is zero: MC(X0, SPi)=0.

For each node x, define the relevant value function (cost, time, or utility function of either or both):
h(x): cost of inspecting node x
r(x): cost of repairing node x
p(x|X0,X1): probability that x is working given that the elements of X0 are not working and that the elements of X1 are working.
q(x|X0,X1) = 1 - p(x|X0,X1)
Q(X0,X1): probability that the system does not work, given that the elements of X0 are not working and that the elements of X1 are working. Note the this probability is equal to zero (1) if X0 is the empty set and X1 is the set U, or (2) if X1 contains at least one success path.

In addition, define:
ck: fixed cost of checking if the system works

At any given step of the inspection process characterized by the sets of events X0 and X1, define the following recursive relationship:

Hybrid Systems For Failure Diagnosis 157

$$MC(X0, X1) = \text{Min} \begin{cases} \text{Min } \{ [Q(X0 \setminus \{x\}, X1 \cup \{x\}) / Q(X0, X1)] \\ \qquad MC(X0 \setminus \{x\}, X1 \cup \{x\}) + r(x) + ck \} \\ \qquad\qquad\qquad \text{for all } x \text{ element of } X0 \\ \\ \text{Min } \{ h(x) + p(x \mid X0, X1) \; MC(X0, X1 \cup \{x\}) \\ \qquad + q(x \mid X0, X1) \; MC(X0 \cup \{x\}, X1) \} \\ \qquad\qquad\qquad \text{for all } x \text{ element of } U \setminus X \end{cases}$$

in which the sign U means union of sets, and the sign \ means subtraction of sets.

This recursive relationship means that the strategy is to choose at each step the least costly of two options:
1. Repair x previously identified as not working. Remove x from X0. Add it to X1. Check if the system works. If it does, stop.
2. Inspect a new x. If it works, add it to X1. If it does not, add it to X0. Reconsider options 1 and 2 (i.e., move to next stage following the updating of the information).

In case where the first option is chosen, the system may still not work and the search will have to be continued. The probability that the system does not work (event T) after the repair of x can be computed as follows. Define for simplicity:
T after = T after repair of x
X0\x after = X0\{x} not working after repair of x
X1 U {x} after = X1 U {x} working after repair of x
T before = T before repair of x
X0 before = X0 not working before repair of x
X1 before = X1 working before repair of x

p(T after, X0\{x} after, X1 U {x} after | T before, X0 before, X1 before)
$\quad = \dfrac{p(T \text{ before, T after, X0 before, X0}\setminus\{x\} \text{ after, X1 before, X1 U }\{x\}\text{ after)}}{p(T \text{ before, X0 before, X1 before)}}$

Note that:
T after AND T before = T after,
   since the set of scenarios leading to T after repair of x are included in the set of scenarios leading to T before repair. Therefore "T after" represents the intersection of the two.
X0 before AND X0\x after = X0\x after,
   since x is known to be in a failed state before repair and consequently fixed; therefore the two events X0 before and X0\{x} after are logically equal.
X1 before AND X1 U {x} after = X1 U {x} after, for the same reason.

Therefore, the probability that the system does not work after repair of x given that it does not work before is the following:

p(T after, X0\{x} after, X1 U {x} after I T before, X0 before, X1 before)
   = p(T after, X0\{x} after, X1 U {x} after)
   p(T before, X0 before, X1 before)
   = Q(X0 \ {x}, X1 U {x}) / Q(X0, X1)

Bayesian updating of the information is needed to modify the probabilities that uninspected parts are in a failed state given the information acquired in the last inspection step. This might be the most time consuming part of the information processing. An efficient method for doing so has been developed on the basis of influence diagrams (Schachter, in Paté-Cornell, Lee, Tse, and Schachter, 1986). This procedure is based on a structure called "fault influence diagram". Using two classical operations of influence diagrams (arc reversal, and arc removal) to process the fault influence diagram, the method allows computation of the probability that a particular yet-uninspected element is in a failure state conditional on the top event (failure of the entire system) and the results of previous inspections (set of inspected elements that work and set of those that don't). The procedure involves a pivotal decomposition algorithm and a mixed algorithm to treat the cases of independent (as opposed to intermediate) events (ibid.).

The dynamic program begins with the stage in which the states of the nodes in the system are known, i.e., X0 U X1 = U. Using the recursive relationship, we successively determine the MC value as well as the course of action when the state of one less node is known. The process leads to the MC value as well as the first course of action when the states of all the nodes are unknown. This constitutes the solution to the inspection and minimal repair problem.

Potential problems arise either when the necessary information (reliability model) is not available or when the optimization is so cumbersome as to be infeasible in reasonable time when needed. First, there may be too many elements in the system and the time of information processing becomes too great for emergency situations. Secondly, there may be too many probabilistic dependencies among failures and the updating the probabilistic information at each inspection step may take so long as to make the process impractical.

Altogether, the method of optimization by dynamic programming assumes that the reliability model is feasible, that the problems are well defined, and that the basic failures are complete. When these features exist, and in particular when the reliability study of the system has been completed, it is extremely attractive to convert an existing risk assessment program into a diagnosis tool. Among other things, this method allows the consideration of rare failure modes that the experts may never have encountered before and provides for an optimal strategy.

HEURISTICS AND ADDITIONAL INFORMATION

In the second mode, heuristics can be powerful ways of obtaining suboptimal but efficient guidance in the diagnosis and repair procedure. Some of these heuristics can be provided by the experts under the form of rules. Others can be probability-based and derived from the PRA model itself. There is a rich operations research literature on the use of heuristics in solving complex problems. The most common ones are based on marginal analysis, where the best immediate solution is used, and hence the term "greedy heuristics." Examples of such methods are numerous: Fisher and Wolsey (1982) used greedy heuristics to solve the continuous version of the set covering problem and developed worst-case bounds of the heuristics; Rolfe (1971) and Dyer and Proll (1977) have studied the convexity of M/D/c and M/M/c queueing systems and consequently shown that the greedy heuristic to allocate a fixed number of servers to a multiple facility system would indeed be optimal for such systems; Cohen et al. (1989) developed a greedy algorithm to an inventory stocking problem and found that the algorithm's solution can be different from the optimal one by 0 to 70%; Smith et al. (1980) also used a greedy heuristic to solve a repair kit stocking problem and proved that the heuristic is indeed optimal. The development of efficient heuristics and the testing and evaluation of them are topics of continual research.

Consider the general form of technical systems: series of functions involving redundancies, and therefore elements in parallel. One success path includes the operation of one of each of these redundancies in each of the subsystems in series. One failure mode (or min-cut set) involves the failure of one of the subsystems in series, and therefore of all its redundancies. In addition to the probability of the top event (system's failure) one can therefore store in the knowledge base PRA-related information such as the nature and the probability of each of the failure modes, the probabilities of failure of the individual components (and conditional probabilities when needed to account for dependences), the relative contribution of each failure mode Mi to the overall system's fragility (=$p(Mi|T)$), and the expected costs (inspection and repair) of achieving each of the success paths. From this information, a certain number of heuristics can be generated.

To illustrate the use of these probability-based heuristics, return to the example of the physical system described in Figure 1 and assume that the optimization option is not available. The results obtained in the heuristic mode generally differ from the optimization results and vary according to the rules used. For example, if the rule is to inspect parts by increasing order of inspection and repair costs, the procedure is: inspect first C, D, or E (indifferently), then A or B. Obviously, this procedure does not take into account the probabilistic contribution of each part to the system fragility and, therefore, does not involve any updating of information. Another possible heuristic is to inspect parts by increasing order of failure probability. The corresponding solution for the example is to inspect C first, then A or E indifferently, then B or D. This procedure does not take into account the cost of inspection and repair. A third possible rule is to inspect the parts by increasing order of ratios of the costs of inspection and repair to failure probabilities ($[h(i)+r(i)]/p(i)$). The procedure, for the proposed example, is to inspect first C, then, successively, E, A, D, and B. For all three heuristics, it is assumed that a failed part is systematically fixed or replaced. More sophisticated rules are required if this assumption is not made.

The decision maker (system's user) thus has the option to select the heuristic mode at some penalty cost when immediate response is needed. These procedures are not optimal in that they rely on prior expected values and do not account for updating of the information as the diagnosis progresses. They may be efficient enough, however, either when optimization is unnecessary, or in situations of emergency when time is critical. This is true, in particular, in cases in which the total time required for the full optimization procedure would be so long as to cancel the benefits of the optimization over those of the simple heuristics. Indeed, a complete optimization procedure includes the time (or the costs) of information processing in the objective function.

The user of the diagnosis system may also face situations in which, for example, the physical system has been subjected to external loads such as heat, vibrations, or moister that modified the values of the PRA-related data. First, the knowledge base used in the first mode can be enriched to include conditional probabilities of failure of the different elements given the level of the external loads. In this case, the optimization can be performed on the basis of these modified data. The physical system may also be subjected to loads still unknown or unexpected, the actual loads may be higher than previously anticipated, and for different reasons, additional information is needed to make inspection and repair decisions. The data relevant to each subsystem, their interfaces, and the nature of their environments, are more loosely related to failures *per se* and include: material nature and characteristics, types of connections and interfaces, resistance to heat, vibrations, or any anticipated external loads, possible techniques of repair, previous experience (trend analysis, types and circumstances of past failure, etc.), and all other informations that may be useful in diagnosis and repair. This information can be stored in the knowledge base and accessed by the user in the third mode of the hybrid system's operation when heuristics or optimization procedures are inadequate or irrelevant.

CONCLUSION

Considering the whole range of decision circumstances and introducing the flexibility of options in a computer-aided fault diagnosis system allows more efficiency when optimization is desired, more simplicity when rules and heuristics suffice, and access to a richer data base when neither apply.

The range of diagnosis and repair problems for which one can construct a computer-based decision support system increases constantly. This is due on the one hand to the progress of computers and the development of AI methods, and on the other hand to the progress in the development of risk analysis and operations research methods. A key advantage of the hybrid system described above is its flexibility.

The first mode --optimization of the inspection and repair procedure-- is designed for situations in which the failure characteristics can be handled by the classical reliability models, and for which the optimization is worth the effort and the resources involved. It may be the case, for example, for a computer network whose structure is well known, the failures are well documented, and time is critical.

The second mode --generation of recommendations based on heuristics-- is desirable for cases where the failures, although predictable, are not adequately described by PRA models, for example, some types of partial failures or continuous phenomena; or when optimization is unnecessary; or when optimization, although theoretically feasible, is undesirable due to system's complexities. In this case the recommendations are based on experts' knowledge in the classical expert systems' format.

The third mode is designed to provide information to be interpreted in the light of the current circumstances rather than (or in addition to) explicit recommendations for action. This option can be useful when the problem is too complex to be treated by the other two approaches, for instance, when the failures are ill-defined, the physical system is poorly known, and the decision circumstances have not been anticipated at the time where the diagnosis system was designed. In this kind of situation, an extended data base is necessary. It may involve fundamental information as well as heuristics and reliability-related data.

REFERENCES

American Association for Artificial Intelligence, 1985. "Uncertainty and Probability in Artificial Intelligence." Proceedings of the UCLA Workshop of August 14-16, 1985, Los Angeles, California.
Bellman R., 1957. Dynamic Programming Princeton University Press. Princeton, New Jersey.
Bellman R. and Dreyfus S., 1962. Applied Dynamic Programming. Princeton University Press, Princeton, New Jersey.
Bonduelle, Y., 1987., "Aggregating Expert Opinions by Resolving Sources of Disagreement", Doctoral Thesis, Department of Engineering-Economic Systems, Stanford University, Stanford, California.

Cohen, M.A., Kleindorfer, P.R., and Lee, H.L., 1989. "Near-Optimal Service-Constrained Stocking Policies for Spare Parts." Operations Research, Vol. 37, No. 1, pp. 104-117.

Dyer, M.E. and Proll, L.G., 1977. "On the Validity of Marginal Analysis for Allocating Servers in M/M/c Queues." Management Science, Vol. 23, pp. 1019-1022.

Electric Power Research Institute, 1985. Workshop on Artificial Intelligence Application to Nuclear Power Plants. EPRI, Palo Alto, California. Electric Power Reearch Institute, 1987. Artficial Intelligence/Expert Systems Research and Development. EPRI, Palo Alto, California.

Fisher, M.L. and Wolsey, L.A., 1982. "On the Greedy Heuristic for Continuous Covering and Packing Problems." Siam J. Alg. Disc. Meth., Vol. 3, pp. 584-591.

Genesereth, M. R., 1978. Automated Consultation for Complex Computer Systems. Doctoral Dissertation. Harvard University. September.

Gluss B., 1959. An Optimal Policy for Detecting a Fault in a Complex System. Operations Research, Vol. 7, pp. 468-477.

Henley E. and H. Kumamoto, 1981. Reliability Engineering and Risk Assessment. Prentice Hall, Inc. Englewood Cliffs, New Jersey.

Henrion M., 1987. Uncertainty in Artificial Intelligence: is Probability Epistemologically and Heuristically Adequate? in Expert Systems and Expert Judgment, J. Mumpower (Ed.), NATO ISI Series, Springer-Verlag.

Holtzman S., 1985. "Intelligent Decision Systems." Doctoral Thesis, Department of Engineering-Economic Systems, Stanford University, Stanford, California.

Holtzman S. 1989. Intelligent Decision Systems. Addison-Wesley Pub. Reading, Mass.

Howard R.A., 1968. "The Foundations of Decision Analysis." IEEE Transactions on Systems, Science, and Cybernetics. SSC-4, 211-219.

Keeney R. and H. Raiffa, 1976. Decisions with Multiple Objectives: Preferences and Value Trade-Offs. New York: John Wiley & Sons, Inc.

Lewis, H., 1984. Probabilistic Risk Assessment: Merits and Limitation, in Proceedings of the 5th International Meeting on Thermal Nuclear Reactor Safety, Nuclear Research Center, Karlsruhe, Federal Republic of Germany, 1984.

Nilsson N., 1980. Principle of Artificial Intelligence. Tioga Publishing Company. Palo Alto, California.

Paté-Cornell M.E., 1984. Fault Trees vs Event Trees in Reliability Analysis. Risk Analysis, Vol.4, No. 3, pp. 177-185.

Paté-Cornell, M.E., H. Lee, E. Tse, and R. Schachter, 1986. Fault Diagnosis and Repair in Complex Systems: Heuristics vs Optimization. Research Report to IBM. Department of Industrial Engineering, Stanford University, Stanford, California.

Paté-Cornell, M.E., H. Lee, and G. Tagaras, 1987. Warnings of Malfunction: The Decision to Inspect Production Systems on Schedule or on Demand. Management Science, Vol. 33, No. 10, pp. 1277-1290.

Raiffa H., 1968. Decision Analysis. Addison-Wesley Pub. Reading, Mass.
Rolfe, A.J., 1971. "A Note on Marginal Allocation in Multiple-Server Service Systems." Management Science, Vol. 17, pp. 656-658.
Rose, F., 1988. "Thinking Machine and Electronic Clone of Skilled Engineer is Very Hard to Create", The Wall Street Journal, August 12, 1988.
Savage L., 1972. The Foundation of Statistics. Dover Publishers. New York.
Schachter R. D. 1986. Evaluating Influence Diagrams. Operations Research, Vol. 34, pp. 871-882, 1986.
Shafer G., 1976. A Mathematical Theory of Evidence. Princeton University Press. Princeton, New Jersey.
Smith, S.A., J.C. Chambers, and Shlifer, E., 1980. "Optimal Inventories Based on Job Completion Rate." Management Science, Vol. 26, pp. 849-852.
Ted Shortliffe and Bruce Buchanan, 1975. "A Model of Inexact Reasoning in Medicine." Mathematical Biosciences, 23, 351-379.

APPENDIX:
NUMERICAL ILLUSTRATION OF THE OPTIMIZATION METHOD

In this illustration, the functional diagram is the one described in Figure 1 and the corresponding figures are as follows:

Node i	p(i)	h(i)	r(i)
A	0.2	0.3	1.5
B	0.05	0.3	1.5
C	0.3	0.5	1.0
D	0.05	0.5	1.0
E	0.2	0.5	1.0

p(i): failure probability of element i
h(i): inspection cost of element i
r(i): repair cost of element i
ck: cost of checking if the system works = 0.1

All failures are assumed to be probabilistically independent. The set U is equal to {A,B,C,D,E} and thus includes five elements.

The use of dynamic programming requires finding the minimum costs (MC's) recursively. We start with the minimum costs when the states of all the nodes are known, i.e., X0 U X1 = U, which can be trivially found. From the equation giving the recursive relationship, we can then find the minimum costs when all but one nodes are known. Next, we find the minimum costs when the states of all but two nodes are known. Proceeding this way, we find the minimum cost when the states of all nodes are unknown, i.e., X0 U X1 = ø. At this point, there is no information about the state of the individual nodes of the system and yet the system is not working. The solution then gives the optimal first step for fault diagnosis and repair. To illustrate the dynamic programming method, the calculation corresponding to one step in the recursive computation is illustrated below.

Assume that one is at a step where X0 equals {A} and X1 equals {D}. At this point, the system is not working (event T) and it was found that A has failed whereas D is in operating condition. The decision then is either to repair A immediately or to keep inspecting the other nodes. At this step, the minimum costs for cases when X0 U X1 contains three elements or more have already been computed.

(i) Note that $Q(\{A\},\{D\})$ $= p(B) + [1-p(B)]\, p(E)\, p(C)$
$= 0.107$

$Q(ø,\{A, D\})$ $= p(E)\, p(C)$
$= 0.06$

If the decision is to repair A right away, there is a probability $0.06/0.107 = 0.56$ that the system will still not function after the repair. Hence, the expected cost for the decision to repair A is $1.5 + 0.1 + 0.56\, MC(ø, \{A,D\})$ where $MC(ø,\{A,D\})=1.6$ as found in a previous step. This cost is computed to be 2.5.

(ii) The probability that B has failed, given that X0={A} and X1={D} and the system has failed, is $0.05/0.107 = 0.47$. Hence, the expected cost to inspect node B now is $0.3 + 0.47\, MC(\{AB\},\{D\}) + 0.53\, MC(\{A\},\{BD\})$ where $MC(\{AB\},\{D\}) = 1.82$ and $MC(\{A\},\{BD\}) = 1.6$ have been computed in a previous step. This gives a cost of 2.00.

(iii) Similarly, the conditional probability that C has failed is:
$\{ p(B)\, p(C) + [1-p(B)]\, p(C)\, p(E) \} / 0.107 = 0.67$

The expected cost to inspect node C is:
$0.5 + 0.67\, MC(\{AC\},\{D\}) + 0.33\, MC(\{A\},\{CD\}) = 2.07$
where $MC(\{AC\},\{D\})=1.55$ and $MC(\{A\},\{CD\})= 1.6$ as computed in a previous step.

(iv) The conditional probability that E has failed is:
$\{ p(B)\, p(E) + [1-p(B)]\, p(C)\, p(E) \} / 0.107 = 0.63$

The expected cost to inspect node E is:
0.5 + 0.63 MC({AE},{D}) + 0.37 MC({A},{DE}) = 2.02
where MC({AE},{D}) = 1.48 and MC({A},{DE}) = 1.6 as computed in a previous step. Hence MC({A},{D}) = 2.00 and the optimal solution at this point is to inspect node B.

The overall solution from solving the dynamic program is the following:
(i) Inspect node E
(ii) If X0 = {E}, X1 = ø, then fix E. This leads to:
 X0 = ø , X1 = {E}.
(iii) Inspect node A or B.
Here, we must have: X0 = {A}, X1 = {E} or X0 = {B}, X1 = {E} depending on whether A or B is inspected.
(iv) Fix A (or B). The system must then work.

ACKNOWLEDGEMENT

The original work presented in this paper was funded in part by a research grant (IBM 551316) from IBM Corporation.

III. IMPRECISE REASONING

169

This section is comprised of two papers. The first paper, Post and Bell's ("Default Reasoning Through Linear Programming"), presents the Least Exception Logic (LEL) model for default reasoning. Some form of non-monotonic reasoning is often required to provide a knowledge representation which is rich enough for real world applications. One convenient non-monotonic reasoning paradigm expresses knowledge as sets of (1) facts, (2) inviolable rules, and (3) default rules for which there may be exceptions. Traditional first order logic representations incorporate (1) and (2). In addition, LEL allows for the explicit expression of possible exceptions to rules and thereby facilitates modeling of default rules.

Statements of (1), (2), and (3) are viewed as constraints on truth values. These are constraints of a 0-1 integer linear program (ILP). By choosing an appropriate ILP objective function, the optimal ILP solution makes the "least" possible use of exceptions. LEL is a heuristic approach in two senses: choice of the objective function uses heuristic weights, and the LEL solution procedure examines a sequence of ILP's with fewer constraints than are implied by (1), (2), and (3). In part because of these heuristic compromises, LEL has found useful practical applications. Post and Bell believe that the following three features of LEL might well be considered by developers of future knowledge representations for practical applications: explicit representation of exceptions to rules, computational mechanisms for choosing one assignment of truth values over another, and recognition that completeness may have to be sacrificed for a more computationally feasible heuristic approach.

The second paper ("The Problem of Determining Membership Values in Fuzzy Sets in Real World Situations" by Triantaphyllou et al.) notes that fuzzy set theory plays a critical role in many AI developments. One of the fundamental concepts in fuzzy set theory is the one of membership values. Membership values are used to determine the degree of membership of the elements of a fuzzy set. An appealing procedure for deriving information about membership values is to use a matrix of pairwise comparisons. A number of OR approaches that are based on eigenvalue theory and mathematical programming have been proposed to manipulate these pairwise comparison matrices and estimate membership values. The findings of this paper reveal that although some methods appear to be more effective than others, their performance is still not very good. In this way, the need for additional OR and AI research on the membership value problem becomes apparent.

Default Reasoning Through Integer Linear Programming

Stephen D. Post
Applied Information Technology
6805 Rosemont Drive
McLean, Virginia 22101

Colin E. Bell
Department of Management Sciences
University of Iowa
Iowa City, IA 52242

ABSTRACT

Least Exception Logic (LEL) is a model for default reasoning that is based upon integer linear programming (ILP) and the first order predicate calculus. The basic LEL model was presented by Post in [1987b and 1988]. This paper summarizes the current research on LEL and presents a new example of using LEL for hardware diagnosis.

INTRODUCTION

It is well established that some form of default reasoning is necessary for real-world problem domains. Classical logic is inadequate for this purpose because it requires complete and certain information about the problem domain and the relationships within that domain that support inferences, whereas real-world problem domains are pervaded by incomplete and uncertain information. Accordingly, practical reasoning about real-world problems, such as is routinely done by people, calls for methods that can bridge gaps in information and make likely, though uncertain, choices to arrive at good, overall solutions that are consistent with what is known.

Default reasoning is an active field of research that includes a wide variety of approaches, each with differing terminology, focus, and methods. No clear theory of default reasoning has arisen, and the principles and requirements remain open to argument.

Default reasoning can be roughly categorized under logic-based approaches and measure-based approaches. The common principle among the logic-based approaches is to accept a tentative proposition based on some supporting (but inconclusive) evidence and a (perhaps temporary) lack of contrary evidence. Non-monotonic logic [McDermott and Doyle 1980], default logic [Reiter 1980], and circumscription [see Genesereth and Nilsson 1987 and McCarthy 1980] are among the best known variations on this theme. Most logic-based approaches to default reasoning are nonmonotonic and require some form of truth (or reason) maintenance. That is, the inferencing mechanism must be able to retract individual conclusions and readjust the overall solution accordingly.

The reason maintenance system described in chapter 15 of Charniak et al [1987] and based on the work of McAllester [1982] is typical. It maintains a set of truth values for propositions consistent with a set of current assumptions and logical implications. The Least Exception Logic (LEL) system described in this paper is a heuristic reasoning device which bears some similarity to this reason maintenance system. Like those in Charniak et al [1987], well-formed formulas in first order predicate calculus are viewed as constraints on truth values of atomic formulas. The system seeks an assignment of truth values to atomic formulas consistent with the expressed constraints.

The handling of exceptions is one of the thorny issues in default reasoning. Brachman [1985] explains some of the difficulties. Rules used in Charniak et al [1987] do not allow exceptions. This reason maintenance system accepts as inputs sets of rules and facts (some of which may be considered to be "current assumptions"). The computational process yields a partial assignment of truth values to propositions consistent with the entire set of constraints (rules and facts). Any given proposition is assigned the value "true" when a value of "false" is demonstrated to be inconsistent with the constraints; likewise a proposition is labeled "false" if a value of "true" is demonstrated to be inconsistent with the constraints. It is possible that truth values of some propositions would be left undetermined by this process. In the course of reasoning with such a system, it may be necessary to analyze the impact of additional assumptions. This increases the computational burden of the reasoning process. For example, the two rules: (1) a => b and (2) a => $\neg$b (together with no additional facts) yield no assignment of truth values to a and b. However, if a is assumed true then an inconsistency is revealed; no such inconsistency results when a is assumed false. Thus, with this additional reasoning, it is possible to assign truth value "false" to a.

Another mechanism for reaching the same conclusion in this simple example is to view the rules (1) and (2) as linear constraints in a (binary) integer linear program (ILP). Throughout this paper we build ILP representations using Boolean (i.e. 0,1) variables to represent the truth values assigned to atomic formulas. We use the convention of identical notation for an atomic formula and the decision variable which represents it in the corresponding ILP. The value 1 for an ILP decision variable corresponds to an assignment of "true" to the atomic formula; 0 corresponds to "false". Using a process to be explained later, rule (1) is translated into the ILP constraint: $(1-a) + b \geq 1$ while rule (2) is translated into $(1-a) + (1-b) \geq 1$. These constraints together with the requirements that a and b are 0-1 variables yield only two feasible solutions: a=0, b=0 and a=0, b=1. The LEL approach will use an integer programming algorithm to obtain a consistent assignment of truth values to propositions. The LEL software can solve integer programming problems with many variables and constraints because of the problems' special structure. To this end, constraint propagation in a spirit similar to that shown by Charniak et al [1987] can be used.

Default Reasoning Through Integer Linear Programming 173

The common principle among the measure-based approaches to default reasoning is to use numerical measures that are somehow related to probability. These measures are applied to the database and propagated through the reasoning process. Bayesian analysis, Dempster-Shafer, Fuzzy set theory, and confidence factors are among the best known variations on this theme [Kanal and Lemmer 1986]. Heuristic approaches are employed when measure-based approaches are used. For example, MYCIN [see Shortliffe 1976] employs a somewhat arbitrary rule that conclusions whose associated certainty factor is sufficiently close to 0 can be ignored and not used in further reasoning. Although such heuristic approaches may be arbitrary and when related to probability may require independence assumptions which are difficult to justify, they often appear to work well in practice.

Although artificial neural networks [see Rummelhart and McClelland 1986] are not commonly included among default reasoning paradigms, they actually are a measure-based approach, and there is considerable similarity between neural nets and LEL.

The development of LEL presented in the next section relates it to logic-based approaches to default reasoning. However, in a subsequent section we discuss the interpretation of probabilities with respect to LEL. As is often the case with measure-based approaches, a probabilistic interpretation requires additional and possibly unrealistic assumptions. This probabilistic interpretation is used later in a fault-diagnosis example.

The next section provides details of LEL. It is important to keep in mind that LEL is a heuristic approach to default reasoning. We start from the assumption that a full-blown reason maintenance system such as McAllester's is too expensive computationally. There are two principal ways in which we have lessened the computational burden: (1) rules can be stated to allow exceptions explicitly, thus relaxing some of the problem's constraints and making a feasible solution easier to find, and (2) search is restricted to find a solution which is feasible with respect to only those constraints which have been discovered. Undiscovered constraints are ignored. With the explicit representation of exceptions to rules there are often multiple feasible solutions (assignments of truth values to propositions). In LEL a simple (linear) measure of the use of exceptions can be associated with any feasible solution. With the use of ILP to manipulate constraints, we can simultaneously seek a feasible solution which minimizes this measure of the use of exceptions. Each exception is given a positive weight and we seek to minimize the sum of these weights over all exceptions used in arriving at a solution. Thus, LEL takes advantage of integer programming technology both for reasoning about the feasibility of solutions and for finding the solution with the "least" use of exceptions.

174 Default Reasoning Through Integer Linear Programming

LEAST EXCEPTION LOGIC

In LEL, an ILP performs logical inference using default rules, and selects which defaults to accept when the default rules support multiple extensions (i.e., alternative solutions). The objective function is to minimize the total weight of exceptions that are made to the default rules. The approach handles symbolic information by stating the constraints as formulas in the first order predicate calculus. Each predicate calculus formula that is instantiated through unification (in this case, substitution of constant symbols for universal variables) becomes a constraint that augments the ILP, and its possible exception is included in the objective function. The solution process interleaves constraint instantiation with solution of the augmented ILP.

An expert system shell based on LEL has been developed and used on a moderately large problem in natural language understanding, and on smaller problems in diagnosis, and robotics. One natural language test case used 300 rules that instantiated 2,000 ILP constraints through substitution for universal variables in well-formed formulas. The resulting constraints included 4,000 binary decision variables representing atomic formulas. One hundred of these binary decision variables represented exceptions. The total process took 12 minutes to execute on a Macintosh II.

Representation

In LEL, knowledge is represented as arbitrary, well-formed formulas in the first order predicate calculus. The formulas are automatically translated to conjunctive normal form for use by the inferencing mechanism. In conjunctive normal form, Skolem constants (and/or functions) are used to replace existential quantifiers and all remaining variables are universally quantified. In conjunctive normal form, a Horn clause contains at most one positive literal. Some approaches to default reasoning are restricted to Horn clauses, but this restriction is not required in LEL. Also, LEL does not require modal operators or other extensions to predicate calculus, and each formula is sound.

As described so far, LEL uses the same representation as a resolution refutation theorem prover, a classical inference procedure for first order predicate calculus described in Nilsson [1980] and Kowalski [1979]. The difference is that LEL uses a convention to represent default knowledge that, while valid in the first order predicate calculus, would not support a resolution theorem proving approach. However, it does support inference through an ILP.

The convention used in LEL to represent default knowledge that could be subject to exception is to explicitly include the possibility of exception in the clause. This is done by reifying the exception (i.e., regarding the exception to be an object) and including it in the clause. For example, the classic default rule that birds fly can be stated to explicitly allow the possibility of exception, as follows:

Default Reasoning Through Integer Linear Programming

$\forall X$ bird(X) $\Rightarrow$ fly(X) $\vee$ $\exists Y$ (exception(Y, 2)).

We use upper case letters (often X, Y, or Z) and words starting with upper case letters for logical variables; constants and predicate names start with a lower case letter. This formula can be read as, "If X is a bird, then X can fly or an exception exists." This causes fly(X) to be the default conclusion, since it is the alternative to making an exception. The second argument, 2, of "exception" is a measure of the importance of making such an exception. It allows exceptions to be independently measured against a common scale. Its interpretation and uses are explored later.

In conjunctive normal form this clause becomes:

¬bird(X) $\vee$ fly(X) $\vee$ exception(non_flying_bird(X),2).

where non_flying_bird is a Skolem function which appears nowhere else in the knowledge base. Since exception(non_flying_bird(X),2) could never be unified with any other literal, a successful resolution refutation could not make use of this clause. However, this clause does represent a family of constraints on truth values of atomic formulae (ground literals). If the objects (constants) appearing in the knowledge base are {tweety, clyde, dodo, g_bush, ...} then the clause above is shorthand for the family of constraints:

¬bird(tweety) $\vee$ fly(tweety) $\vee$ exception(non_flying_bird(tweety),2)
¬bird(clyde) $\vee$ fly(clyde) $\vee$ exception(non_flying_bird(clyde),2)
¬bird(dodo) $\vee$ fly(dodo) $\vee$ exception(non_flying_bird(dodo),2)
¬bird(g_bush) $\vee$ fly(g_bush) $\vee$ exception(non_flying_bird(g_bush),2)
....

These, in turn, can be translated into ILP constraints:

(1 - bird(tweety)) + fly(tweety) + exception(non_flying_bird(tweety),2) $\geq$ 1
(1 - bird(clyde)) + fly(clyde) + exception(non_flying_bird(clyde),2) $\geq$ 1
etc.

In many instances we will leave ILP constraints in such a non-standard form. Of course, they are easily converted to a standard form with a sum of multiples (always 0, 1, or -1) of decision variables on the left of either a "$\leq$" or "$\geq$" inequality and a non-negative integer on the right.

Clauses stated under this convention are intrinsically sound since they can always be satisfied by making the exception. (Any ILP constraint involving an exception is satisfied whenever its "exception" decision variable is set to 1.) This also means that these clauses will not support useful logical deduction unless the inferencing process is prevented from making arbitrary exceptions.

Inferencing

LEL reasons by interleaving unification and ILP. The approach diverges from logical inference (i.e., resolution) in that it decouples unification from solution, and performs solution as constrained optimization rather than through piecewise logical inference.

Unification. LEL performs unification separately from solution. The first order predicate calculus clauses are instantiated through the same series of unifications that would occur under resolution. However, each unification merely substitutes for universal variables in the clause. Unification is not accompanied by the aspect of Boolean solution that is present in resolution. For example, the clause

$\neg$bird(X) $\vee$ fly(X) $\vee$ exception(non_flying_bird(X), 2).

can be unified with bird(tweety) to instantiate the following propositional clause:

$\neg$bird(tweety) $\vee$ fly(tweety) $\vee$ exception(non_flying_bird(tweety), 2).

Solution. The ground clauses that have been instantiated through unification (or given) define an ILP in Boolean (i.e., $\{0,1\}$) variables. The ground clauses comprise the system of constraints. The system of constraints must be satisfied by assigning a Boolean value to each proposition (i.e., ground atom) such that at least one ground literal per clause is true. The exception relations are the only literals to appear with non-zero coefficients in the ILP's objective function, which is to minimize the sum of the weights of exceptions set to "true" (corresponding ILP decision variable set to 1).

The mapping from ground clause to linear constraint is straightforward.

$\neg P$ becomes $(1 - P)$

$\underset{i=1}{\overset{n}{OR}} P_i$ becomes $\sum_{i=1}^{n} P_i \geq 1$

For example,

$\neg$bird(tweety) $\vee$ fly(tweety) $\vee$ exception(non_flying_bird(tweety), 2).

becomes:

$(1 - \text{bird(tweety)}) + \text{fly(tweety)} + \text{exception(non\_flying\_bird(tweety), 2)} \geq 1$

Default Reasoning Through Integer Linear Programming 177

The objective function is built from the sum of the products of the exceptions and their weights. For example,

Minimize C = 2 • exception(non_flying_bird(tweety), 2) + ...

An ILP formed in this manner is complete in terms of logical (Boolean) inference. This is as expected since both integer linear programming and Boolean satisfiability are in the class of NP-complete problems (see Garfinkel and Nemhauser [1972] and Hooker [1986]). For example, modus ponens and contrapositive inference rules are supported in the "if tweety is a bird then tweety flies" constraint above, (assuming the exception is not allowed). Modus ponens is supported because if bird(tweety) equals 1 then fly(tweety) must equal 1. Similarly, contrapositive inference is supported since if fly(tweety) equals 0 then bird(tweety) must equal 0.

This ILP formulation has a typical geometric interpretation. That is, the problem is constrained to (the vertices of) a unit hypercube with dimensionality equal to the number of ground literals, including the exceptions. Each constraint (i.e., clause) represents a hyperplane that slices off an infeasible region from the hypercube to leave a convex polytope. The feasible solutions are the integer vertices of the polytope, and the best solution is selected by the objective function.

However, we prefer a nonstandard interpretation of the ILP in which the exceptions are considered to be artificial variables. This interpretation is nonstandard because exceptions might remain in the final solution, which is precluded in linear programming. In standard linear programming, the constraints are held to be inviolable, while feasibility is to be determined. In LEL feasibility is held to be inviolable, while the enforcement of some questionable (i.e., default) constraints is to be determined.

Control. Control is a critical issue since both the unification and ILP phase can be combinatorial. Control is implemented in the LEL shell as a module that uses the blackboard paradigm (see Nii [1986a] and [1986b]) to task and monitor the unification and ILP modules. However, this paper focuses on the basic LEL model and covers control issues only briefly.

The overall control strategy is to interleave unification with solution of the ILP until a solution is reached. The unification phase generally unifies upon propositions whose values have been set by the ILP phase, and the ILP phase readjusts its solution as necessary for each new constraint and possible exception spawned by the unification phase. This process is illustrated with the following simple example.

Consider a system with one default rule and two sound rules, where X and Y are universal variables:

$a(X) \Rightarrow b(X) \lor \text{exception}(f(X), 3)$.
$b(Y) \Rightarrow c(Y)$.
$a(\text{tom})$.

Since only one of these three clauses is fully instantiated, the initial ILP has only one decision variable, a(tom), and only one constraint, a(tom) = 1. Since a(tom) is not an "exception" literal, it appears in the objective function with coefficient 0, thus the objective function is:

Minimize $z = 0$.

The optimal solution is: a(tom) = 1. The set of ground literals assigned the value "true" is {a(tom)} and the set of ground literals assigned the value "false" is empty.

The unification phase uses this assignment of truth values to unify a(tom) with the first formula to spawn a new ground formula:

$a(\text{tom}) \Rightarrow b(\text{tom}) \lor \text{exception}(f(\text{tom}), 3)$.

Since this formula is ground, it translates into the ILP constraint:

$(1 - a(\text{tom})) + b(\text{tom}) + \text{exception}(f(\text{tom}), 3) \geq 1$

The ILP is then augmented by this constraint and the term
3 • exception(f(tom), 3) is added to the (previously empty) objective function to yield the new ILP:

Minimize $z = 3 \cdot \text{exception}(f(\text{tom}), 3)$
Subject to: $a(\text{tom}) = 1$
             $(1 - a(\text{tom})) + b(\text{tom}) + \text{exception}(f(\text{tom}), 3) \geq 1$

Then the ILP phase solves the ILP as follows:

$a(\text{tom}) = 1$
$b(\text{tom}) = 1$
$\text{exception}(f(\text{tom}), 3) = 0$
$z = 0$

The set of ground literals assigned the value "true" in this solution is {a(tom), b(tom)} and the set of ground literals assigned the value "false" is {exception(f(tom),3)}. The unification phase uses this assignment of truth values to unify b(tom) with the second formula to spawn a new ground formula:

$\neg b(\text{tom}) \lor c(\text{tom})$.

Default Reasoning Through Integer Linear Programming 179

Since this formula is ground, it translates into the ILP constraint:

(1 - b(tom)) + c(tom) ≥ 1.

The ILP is then augmented by this constraint, the objective function is left unchanged (since this new constraint contains no "exception" literal) resulting in the new ILP:

Minimize   z = 3 • exception(f(tom), 3)
Subject to: a(tom) = 1
   (1 - a(tom)) + b(tom) + exception(f(tom), 3) ≥ 1
   (1 - b(tom)) + c(tom) ≥ 1.

The ILP phase solves the ILP as follows:

a(tom) = 1
b(tom) = 1
c(tom) = 1
exception(f(tom), 3) = 0
z = 0

Use of the truth values in this solution for unification fails to develop any new constraints. Thus the process terminates.

In this simple example where the universe of constant symbols appearing in the original three clauses is {tom}, it is clear that this last ILP incorporates all possible ground constraints and determines the assignment of truth values which minimizes the weighted sum of exceptions (by avoiding exceptions altogether). It should be emphasized that LEL is a heuristic approach to default reasoning and that there is no guarantee that such a global optimum will be found in more complicated cases.

The control module has several strategies by which to focus the search. Both data driven and goal driven control is supported. Data driven control is implemented by unifying upon the input data and subsequent conclusions. The example given above was data driven; it uses "forward chaining".

Goal driven control is implemented through a refutation proof. The negation of the goal is entered as a default rule (with a very low weight). When an exception is made to the rule, the goal has been proved since its refutation was found to be inconsistent. A goal driven example is given by Post [1987b]; goal driven control will not be discussed further in this paper.

As seen in the above example, an ILP is generated from its predecessors by adding constraints and adding terms to the objective function. In general, an ILP with an augmented set of constraints requires the full machinery of ILP to find a new solution. However, there is one important case where the iteration process is particularly easy. Thus, careful selection of which unifications to perform can drastically limit the search required in the ILP. In particular, the LEL shell is programmed to prefer constraints in which all but one of the instantiated propositions (excluding "exceptions") has a known truth value. This allows a value for the single new proposition to be immediately determined without search; i.e., set any exceptions to 0 and set the new proposition as required to satisfy the new constraint. Detecting and choosing to unify on such constraints is analogous to constraint propagation in the approach to reason maintenance of Charniak et al [1987].

In fact, the tractability of an LEL problem formulation largely depends on exploiting this type of constraint. This is done by defining rules that have a single consequent or conjunction of consequents, or where the consequents are disjoined with an exception. When the resultant constraints are instantiated, ILP decision variables representing the antecedents will have known values (in the currently best solution), new exception decision variables can be set to 0, so that ILP decision variables representing the individual consequents can all be set to 1. The ILP problem would become intractable for practical-sized problems unless most rules were of this form. This approach has been used to solve ILPs with thousands of variables and constraints within minutes on a microcomputer. The ILP given below in the electronic circuit example has 89 binary variables. Its special structure allows it to be solved easily. ILP's with 89 variables and without special structure can be much more difficult to solve.

When more extensive search is required, the control module can also assist the ILP search algorithm. When the algorithm encounters a large number of branches with the same objective value, the control module can halt the search and task the unification module to further constrain the critical variables so that choices can be guided.

MEASURE-BASED INTERPRETATION OF EXCEPTION WEIGHTS

In this section the symbols A, B, C, and F stand for probabilistic events (or simple propositions); the only logical variable appearing in this section is denoted by X. In standard logic, the formula: $A \wedge B \Rightarrow C$ means that there is no case in which A and B are true and C is false. This is an absolute statement that allows for no uncertainty. This is shown in the following Venn diagram:

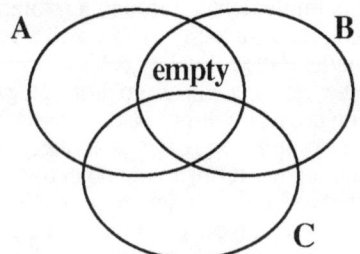

Default Reasoning Through Integer Linear Programming   181

The closest analog to A ∧ B ⇒ C in a probabilistic model would involve the assignment of a high probability to P(C | A,B). An example where P(C | A,B) = 0.9 is shown in the following Venn diagram:

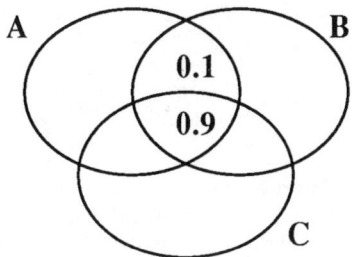

Use of such conditional probabilities allows for the expression of rules which do not hold with absolute certainty. This modeling flexibility comes at the cost of estimating a possibly large number of conditional probabilities. Use of Bayes rule for updating probability assignments is limited in practice by the difficulty of estimating all necessary conditional probability inputs.

As we have seen, LEL shares the feature of the above probabilistic model that it need not totally exclude the case where A and B are true and C is false. Instead, that case is the exception and can be allowed at some cost. The corresponding LEL clause might be:

A ∧ B ⇒ C ∨ exception(e1,2).

which in conjunctive normal form becomes:

¬A ∨ ¬B ∨ C ∨ exception(e1,2).

The corresponding ILP constraint is:

(1-A) + (1-B) + C + exception(e1,2) ≥ 1.

This is shown in the following Venn diagram:

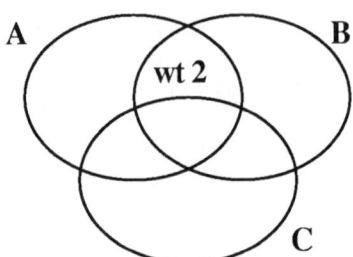

In the remainder of this section we examine the legitimacy of assigning probabilistic interpretations to exception weights in LEL. For example, we might want an exception weight to be higher if we know that a particular exception is extremely unlikely to occur. Thus we explore the possible relationship between exception weights and the assessment of probabilities of exceptions. As one might expect, at best this association requires some care; at worst associating a probabilistic interpretation to exception weights might be unjustifiable. There are two main concerns: (1) how should the user convert his assessed probability such as P(¬C | A,B) into an exception weight? and (2) is such a relationship consistent with the user's assessment of probabilities. Leaving the issue of mapping probabilities into exception weights for now, we concentrate on (2).

One difficulty with a probabilistic interpretation of exception weights arises from the fact that a single clause in conjunctive normal form can represent multiple rules. This presents a problem only when the user wishes to make such a probabilistic interpretation and does not detract from the value of LEL in other cases. The clause above:

¬A ∨ ¬B ∨ C ∨ exception(e1,2).

(and its associated ILP constraint) can represent any of the following three rules:

A ∧ B ⇒ C ∨ exception(e1,2).
A ∧ ¬C ⇒ ¬B ∨ exception(e1,2).
B ∧ ¬C ⇒ ¬A ∨ exception(e1,2).

If we are to relate exception weights to probabilities then these three rules require us to map each of the three probabilities: P(¬C | A,B), P(B | A,¬C), and P(A | B,¬C) into the same exception weight, 2. If our chosen mapping is 1 to 1, then the three probabilities must be equal. These probabilities are equal if and only if P(A ∩ B) = P(A ∩ ¬C) = P(B ∩ ¬C). Examples which violate this condition are easily constructed.

A simpler example is the following: Consider an experiment involving a single draw from a bag containing three white marbles, five black marbles, one white block, and one black block. If each of the 10 objects is equally likely to be drawn, then P(¬marble|white) = 0.25, while P(white|¬marble) = 0.5. Therefore, a probabilistic interpretation of exception weights related to the clause

¬white(X) ∨ marble(X) ∨ exception(f(X), 1).

should not be used since this clause corresponds to each of the following two rules:

white(X) ⇒ marble(X) ∨ exception(f(X), 1).
¬marble(X) ⇒ ¬white(x) ∨ exception(f(X), 1).

and a probabilistic interpretation requires P(¬marble | white) = P(white | ¬marble), a relationship which does not hold.

We have seen that associating a probabilistic interpretation with an exception weight can be realistic only when certain conditional probabilities are all equal. This may or may not be the case in examples of practical interest. Assuming that we are willing to live with the assumption that such conditional probabilities are equal, we now turn attention to one possible way of assigning a probabilistic interpretation to exception weights. The example in the next section concerns system reliability. In this example, an exception to a rule occurs when a component of an electronic system fails. If failures of distinct components are considered to be independent events, the exception weights have a natural interpretation as the negative log odds that the component will fail and an exception must be made to the rule involving that component. If F is an event with probability P(F), then the negative log odds of F is

NLO(F) = -log(P(F)) + log(1-P(F)).

For our purposes, any logarithmic base can be chosen provided it is used exclusively. Assume that an electronic device consists of components 1,2,...,n and events F_1, F_2, ... , F_n denote failures of the respective components. Assume also that these failure events are mutually independent. Let S be any subset of {1,2,...,n} and let E_S be the event that exactly those components in S fail. Then it can be easily shown that

log(P(E_S)) = c + ΣNLO(F_i)

where c is a constant and the summation is taken over all i in S. If NLO(F_i) is the exception weight in a rule which must hold unless component i fails, then the optimal solution to LEL's final ILP will correspond to the subset S of failed components which has the highest probability over all such subsets consistent with knowledge expressed in the program. Thus, LEL when used for fault diagnosis will return the most likely set of causes of the fault.

We have seen that great care must be taken when attempting to assign a probabilistic interpretation to exception weights. In the system reliability example it was necessary to assume independence of certain events. The need to assume independence is shared, in one form or another, by all other measure-based approaches (except for Bayesian approaches discussed above). However, the assumption of independence is unjustified in many cases, so further research is required to allow the weights to change in light of new evidence, perhaps through offsetting, negative weights or a nonlinear objective function.

LEL has been found in practice thus far to be fairly insensitive to exception weights. This is perhaps because the exceptions made have been somewhat sparse, and wrong choices tend to be eliminated during solution because they cause conflicts with other rules. Moreover, knowledge refinement is generally accomplished by more carefully specifying the logic rather than by adjusting the weights.

AN EXAMPLE

LEL appears to have good applicability for system diagnosis, because the relationships between types of components can be stated in a general form, i.e., with symbolic variables, and the exception weights can correspond to the negative log odds of component failure, in keeping with the general practice to consider component failures as independent events. In this approach, the constraint network that corresponds to a given system will be automatically constructed from the symbolic rules by the unification module. As pointed out in the last section, minimizing the ILP objective function yields the feasible set S of failed components whose probability is maximized. The following experiment illustrates the approach.

The circuit diagram shown in Figure 1 shows four components, each with a negative log odds of failure. The wire, buzzer, and light failures correspond to a broken circuit and failure to function. The buzzer arm failure corresponds to the buzzer arm failing to buzz even though current may be flowing.

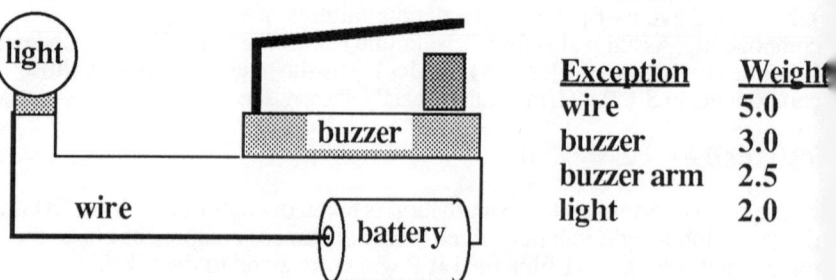

Exception	Weight
wire	5.0
buzzer	3.0
buzzer arm	2.5
light	2.0

Figure 1. A simple electric circuit

Default Reasoning Through Integer Linear Programming 185

To handle this problem, the LEL shell was given: (1) a classification hierarchy, (2) Rules 1-10 that defined component behavior by type, and (3) Rule 11 that specified a particular set of components with a specific configuration. The classification hierarchy specified: (a) that a battery, switch, or device was a component and (b) that a wire, buzzer, or lightbulb was a device. Rules 1-10 encoded the following information:

- Each device is a closed circuit
  or there is an exception with the corresponding cost.

- If there is a complete circuit through a battery, then current flows

- For each lightbulb,
  the lightbulb shines exactly when the current flows

- For each buzzer,
  if the current flows then the buzzer will buzz or the arm is stuck.

The listing below provides brief descriptions of Rules 1-10 together with clauses of each rule in conjunctive normal form.

Rule 1:   "+ and - are opposites"
opposite(+, -).
opposite(-, +).

Rule 2:   "Electrical connectivity is commutative"
¬connect(X,Y) ∨ connect (Y,X).

Rule 3:   "Define the circuit reliability for different classes of devices"
¬device(D) ∨ ¬wire(D) ∨ circuit-reliability(D,5.0).
¬device(D) ∨ ¬switch(D) ∨ circuit-reliability(D,4.0).
¬device(D) ∨ ¬buzzer(D) ∨ circuit-reliability(D,3.0).
¬device(D) ∨ ¬lightbulb(D) ∨ circuit-reliability(D,2.0).

Rule 4:   "There is a connection across a device, or an exception exists"
¬device(D) ∨ ¬terminals(D,T1,T2) ∨ ¬circuit-reliability(D,R) ∨
    connect(T1,T2) ∨ exception(broken-circuit(D),R).

Rule 5:   "If a broken-circuit exception has been made, no current flows"
¬exception(broken-circuit(D),X) ∨ ¬current_flow(D).

Rule 6:   "Electrical potential propagates over the circuit"
¬connect(X,Y) ∨ ¬potential(X,P) ∨ potential(Y,P).

Rule 7:   "If positive and negative potential has propagated to a device, current flows"
¬terminals(X,T1,T2) ∨ ¬potential(X,P1) ∨ ¬potential(X,P2) ∨
    ¬opposite(P1,P2) ∨ current_flow(X).

Rule 8:   "A battery provides positive and negative potential"
¬battery(B) ∨ ¬terminals(B,Plus,Minus) ∨ potential(Plus,+).
¬battery(B) ∨ ¬terminals(B,Plus,Minus) ∨ potential(Minus, -).

Rule 9:   "In a lightbulb, the light shines if and only if the current flows"
¬lightbulb(L) ∨ ¬shine(L) ∨ current_flow(L).
¬lightbulb(L) ∨ ¬current_flow(L) ∨ shine(L).

Rule 10: "Current flows through a buzzer if and only if (the buzzer buzzes or is stuck)"
¬buzzer(B) ∨ ¬current_flow(B) ∨ buzz(B) ∨
    exception(buzzer-stuck(B),2.5).
¬buzzer(B) ∨ ¬buzz(B) ∨ current_flow(B).
¬buzzer(B) ∨ ¬exception(buzzer-stuck(B),2.5) ∨ current_flow(B).

Finally, the 12 facts of Rule 11 define the specific circuit for our example. Any number of other circuits could have been defined from the same types of components that were covered by the rules.

Rule 11:

battery(bat).
terminals(bat,b+,b-).
wire(w1).
terminals(w1,t1,t2).
lightbulb(lb).
terminals(lb,t3,t4).
buzzer(b).
terminals(b,t5,t6).
connect(b+,t1).
connect(t2,t3).
connect(t4,t5).
connect(t6,b-).

Default Reasoning Through Integer Linear Programming 187

Executing this hardware diagnosis example produced the following behavior. First, the system concluded that the lightbulb would shine and the buzzer would buzz, with no exceptions. Then, the system was given further information that the buzzer was not buzzing, so it nonmonotonically concluded that the lightbulb was burned out (r=2.0), since the set containing this failure alone was the most probable set of failures that would prevent the buzzer from buzzing. Then, the system was given further information that the lightbulb was shining, so it nonmonotonically concluded that the lightbulb was not burned out, but the buzzer arm was stuck (r=2.5). Now the set of failures including only buzzer arm failure was the most probable consistent with the additional knowledge that the lightbulb had not failed.

This inferencing process took the 10 rules and the circuit specification to build a system of 89 constraints with 89 variables. Eight of the 16 possible extensions (distinct subsets of failed components) were opened, and three were fully explored. The experiment reinforced a lesson from other examples that a (seemingly) large number of constraints are spawned from simple examples, and yet, because of its special structure, the ILPs can be readily solved.

The final ILP is included below to illustrate the example. Note that the LEL shell automatically enforces classification relationships (for efficiency), so those constraints do not show in the ILP (although they could if the shell was so designed).

Minimize z = 2.5 • exception(buzzer-stuck(b),2.5) +
5.0 • exception(broken-circuit(w1),5.0) +
2.0 • exception(broken-circuit(lb),2.0) +
3.0 • exception(broken-circuit(b),3.0)

Subject to:

opposite(-,+) ≥ 1
opposite(+,-) ≥ 1
connect(t6,b-) ≥ 1
connect(t4,t5) ≥ 1
connect(t2,t3) ≥ 1
connect(b+,t1) ≥ 1
terminals(b,t5,t6) ≥ 1
buzzer(b) ≥ 1
terminals(lb,t3,t4) ≥ 1
lightbulb(lb) ≥ 1
terminals(w1,t1,t2) ≥ 1
wire(w1) ≥ 1
terminals(bat,b+,b-) ≥ 1
battery(bat) ≥ 1
- terminals(bat,b+,b-) - battery(bat) + potential(b-,-) ≥ -1
- terminals(bat,b+,b-) - battery(bat) + potential(b+,+) ≥ -1
- current_flow(lb) + shine(lb) - lightbulb(lb) ≥ -1

- shine(lb) + current_flow(lb) - lightbulb(lb) ≥ -1
 buzz(b) - current_flow(b) - buzzer(b) + exception(buzzer-stuck(b),2.5) ≥ -1
 current_flow(b) - buzzer(b) - exception(buzzer-stuck(b),2.5) ≥ -1
- buzz(b) + current_flow(b) - buzzer(b) ≥ -1
- connect(b+,t1) + connect(t1,b+) ≥ 0
- connect(b+,t1) - potential(b+,+) + potential(t1,+) ≥ -1
- connect(t2,t3) + connect(t3,t2) ≥ 0
- connect(t4,t5) + connect(t5,t4) ≥ 0
- connect(t6,b-) + connect(b-,t6) ≥ 0
- wire(w1) + circuit-reliability(w1,5.0) - device(w1) ≥ -1
- switch(w1) + circuit-reliability(w1,4.0) - device(w1) ≥ -1
- buzzer(w1) + circuit-reliability(w1,3.0) - device(w1) ≥ -1
- lightbulb(w1) + circuit-reliability(w1,2.0) - device(w1) ≥ -1
- wire(lb) + circuit-reliability(lb,5.0) - device(lb) ≥ -1
- switch(lb) + circuit-reliability(lb,4.0) - device(lb) ≥ -1
- buzzer(lb) + circuit-reliability(lb,3.0) - device(lb) ≥ -1
- lightbulb(lb) + circuit-reliability(lb,2.0) - device(lb) ≥ -1
- wire(b) + circuit-reliability(b,5.0) - device(b) ≥ -1
- switch(b) + circuit-reliability(b,4.0) - device(b) ≥ -1
- buzzer(b) + circuit-reliability(b,3.0) - device(b) ≥ -1
- lightbulb(b) + circuit-reliability(b,2.0) - device(b) ≥ -1
- connect(b-,t6) - potential(b-,-) + potential(t6,-) ≥ -1
- connect(t1,b+) - potential(t1,+) + potential(b+,+) ≥ -1
- connect(t1,b+) + connect(b+,t1) ≥ 0
- connect(t3,t2) + connect(t2,t3) ≥ 0
- connect(t5,t4) + connect(t4,t5) ≥ 0
- connect(b-,t6) + connect(t6,b-) ≥ 0
- device(w1) - terminals(w1,t1,t2) - circuit-reliability(w1,5.0) + connect(t1,t2)
    + exception(broken-circuit(w1),5.0) ≥ -2
- device(lb) - terminals(lb,t3,t4) - circuit-reliability(lb,2.0) + connect(t3,t4) +
    exception(broken-circuit(lb),2.0) ≥ -2
- device(b) - terminals(b,t5,t6) - circuit-reliability(b,3.0) + connect(t5,t6) +
    exception(broken-circuit(b),3.0) ≥ -2
- connect(t6,b-) - potential(t6,-) + potential(b-,-) ≥ -1
- connect(t1,t2) + connect(t2,t1) ≥ 0
- connect(t1,t2) - potential(t1,+) + potential(t2,+) ≥ -1
- connect(t3,t4) + connect(t4,t3) ≥ 0
- connect(t5,t6) + connect(t6,t5) ≥ 0
- connect(t2,t3) - potential(t2,+) + potential(t3,+) ≥ -1
- connect(t2,t1) - potential(t2,+) + potential(t1,+) ≥ -1
- connect(t2,t1) + connect(t1,t2) ≥ 0
- connect(t4,t3) + connect(t3,t4) ≥ 0
- connect(t6,t5) + connect(t5,t6) ≥ 0
- connect(t6,t5) - potential(t6,-) + potential(t5,-) ≥ -1
- connect(t3,t2) - potential(t3,+) + potential(t2,+) ≥ -1
- connect(t3,t4) - potential(t3,+) + potential(t4,+) ≥ -1
- connect(t5,t4) - potential(t5,-) + potential(t4,-) ≥ -1

Default Reasoning Through Integer Linear Programming 189

- connect(t5,t6) - potential(t5,-) + potential(t6,-) ≥ -1
- connect(t4,t5) - potential(t4,+) + potential(t5,+) ≥ -1
- connect(t4,t3) - potential(t4,+) + potential(t3,+) ≥ -1
- connect(t4,t5) - potential(t4,-) + potential(t5,-) ≥ -1
- connect(t4,t3) - potential(t4,-) + potential(t3,-) ≥ -1
- connect(t5,t4) - potential(t5,+) + potential(t4,+) ≥ -1
- connect(t5,t6) - potential(t5,+) + potential(t6,+) ≥ -1
- terminals(b,t5,t6) - potential(t5,-) - potential(t5,+) - opposite(-,+) + current_flow(b) ≥ -3
- terminals(b,t5,t6) - potential(t5,+) - potential(t5,-) - opposite(+,-) + current_flow(b) ≥ -3
- connect(t3,t2) - potential(t3,-) + potential(t2,-) ≥ -1
- connect(t3,t4) - potential(t3,-) + potential(t4,-) ≥ -1
- terminals(lb,t3,t4) - potential(t3,+) - potential(t3,-) - opposite(+,-) + current_flow(lb) ≥ -3
- terminals(lb,t3,t4) - potential(t3,-) - potential(t3,+) - opposite(-,+) + current_flow(lb) ≥ -3
- connect(t6,b-) - potential(t6,+) + potential(b-,+) ≥ -1
- connect(t6,t5) - potential(t6,+) + potential(t5,+) ≥ -1
- connect(t2,t3) - potential(t2,-) + potential(t3,-) ≥ -1
- connect(t2,t1) - potential(t2,-) + potential(t1,-) ≥ -1
- connect(b-,t6) - potential(b-,+) + potential(t6,+) ≥ -1
- connect(t1,b+) - potential(t1,-) + potential(b+,-) ≥ -1
- connect(t1,t2) - potential(t1,-) + potential(t2,-) ≥ -1
- terminals(w1,t1,t2) - potential(t1,+) - potential(t1,-) - opposite(+,-) + current_flow(w1) ≥ -3
- terminals(w1,t1,t2) - potential(t1,-) - potential(t1,+) - opposite(-,+) + current_flow(w1) ≥ -3
- connect(b+,t1) - potential(b+,-) + potential(t1,-) ≥ -1
- terminals(bat,b+,b-) - potential(b+,+) - potential(b+,-) - opposite(+,-) + current_flow(bat) ≥ -3
- terminals(bat,b+,b-) - potential(b+,-) - potential(b+,+) - opposite(-,+) + current_flow(bat) ≥ -3
- buzz(b) ≥ 0
- current_flow(lb) - exception(broken-circuit(lb),2.0) ≥ -1
- shine(lb) ≥ 1.

EVALUATION OF LEL AND COMPARISON WITH OTHER METHODS

The next two subsections briefly detail some advantages of LEL when compared with leading logic-based and measure-based approaches. The third subsection summarizes possible shortcomings of LEL. The next section offers additional conclusions.

Logic-based Approaches

The nonmonotonic logic of McDermott and Doyle, and the default logic of Reiter both use the idea that default conclusions can be believed when they are consistent. For example, "If X is a bird and it is consistent to believe that X can fly then one may believe that X can fly". Consistency is enforced by use of a modal operator or by specialized rules of inference. LEL uses the same idea, but no modal operator is required and consistency is enforced by the simultaneous solution of the system of constraints. Although the consistency checking in LEL covers only the instantiated rules in the ILP, this allows a relatively practical implementation, and inconsistencies that show up later can be accommodated by re-solving the ILP.

Circumscription is an augmentation of ordinary first order predicate calculus that provides a formalism for drawing conclusions based on the assumption that all information that would be needed to derive exceptions is present. Circumscription is based on the idea of a minimal model that contains exactly as much information as necessary to make sense of the knowledge in the theory. Default reasoning is done by rules such as, "If X is a bird and X is not abnormal (in the sense of this rule) then X can fly." Since the minimal model will not allow the abnormality (or any circumscribed relation) to exist unless it is necessary, default conclusions are enforced. This is similar to the convention used in LEL, with abnormalities being analogous to exceptions in LEL. The minimal model is analogous to a solution with the least exceptions. However, LEL seeks a minimal model through global optimization rather than by augmenting a set of beliefs.

LEL compares favorably with these logic-based methods because consistency checking in the ILP is more practical than consistency checking in the first order predicate calculus, and LEL can choose the best combination of default rules to apply when various feasible combinations are in contention.

Consider the sentence, "The astronomer married a star." (taken from Waltz and Pollack [1985]). One default rule might select the celestial object sense of "star" based on the presence of "astronomer". Another default rule might select the performer sense of "star" as the object of "married", based on animacy. (The latter must be a default rule because one can also be "married to his job," etc.) These defaults are both feasible and are in contention, but the performer sense of "star" would seem to be preferred.

In general, a large problem would involve a large set of defaults, it would be inconsistent to accept the entire set, and many different subsets would be consistent. How can the consistent subsets be ordered? Of the logic-based methods, only circumscription addresses this problem and it supports only a partial ordering based on complete inclusion of one subset within another, and with all exceptions being of equal measure. LEL supports an ordering of the subsets based on the total weight of exceptions.

Measure-based Approaches

All of the measure-based approaches (e.g., Bayesian, Dempster-Shafer, fuzzy set theory, confidence factors, etc.) assign probability-like measures to individual propositions, and attempt to aggregate and combine them into a global certainty measure for the final conclusion. These approaches are subject to deteriorating confidence in the accuracy of certainty factors as uncertain conclusions build upon uncertain conclusions. The conclusions are somewhat like an averaged overlay of the different sets of possibilities. In effect, the approaches use certainty measures to avoid making (and keeping track of) choices. With such approaches, it might well be unrealistic to ascribe any confidence to certainty measures of final conclusions even though such measures at intermediate stages of the analysis might have played a reasonable role in limiting the search.

Solutions in LEL are also built from a number of uncertain conclusions, but LEL considers only one consistent extension at a time and does not mix together certainty from different extensions. Although each proposition is given a zero or one value in each extension, the different extensions are given weights, and multiple extensions are explored. LEL can find a single consistent extension which is optimal with respect to a well-formulated objective function (minimization of total exception weight).

One advantage of LEL is that the simultaneous solution process resolves interdependencies to the effect that the final solution can be much more likely than would be indicated by the individual certainties of the information that went into the decision. For example, a system for grammatical category disambiguation in DeRose[1988] used mathematical optimization over sentences taken as a whole to determine, for each of the words, the one most likely grammatical category (e.g., word 1 is a noun, word 2 is a verb, etc.). The search was based on the probability distribution for the word category based on the two surrounding words, and on the probability distribution for each word independent of its surroundings. The overall accuracy was 96% over a large corpus of text. The 96% accuracy is much higher than would be calculated by the combination of probabilities (although this was not pointed out in the paper), and can be attributed to the global optimization process that committed to choices and fitted them together.

As mentioned previously, the LEL approach to uncertainty is essentially the same as that of artificial neural networks. The arc weights of neural networks are analogous to exception weights, both approaches reason by minimizing, and neural networks measure their solutions by totaling the weights of activated arcs (e.g., accepted exceptions). In fact, a neural network has been used as a technique for solving the ILP phase of LEL, and the classic neural net model (once the weights are established) can be easily mapped into LEL [Post 1987a]. However, it is not clear how neural network models can handle unification, whereas LEL does, and neural networks performed poorly in comparison to other search algorithms.

Possible Shortcomings of LEL

We have emphasized throughout that LEL is a heuristic approach to default reasoning. Its behavior has been noteworthy in several practical examples. Of course, any method that depends on ILP for part of its computational process is limited in the size of problem that it can accomodate. Although special structure of ILP problems in LEL can be exploited, there remains the possibility that some significant practical problems simply cannot be solved.

Aside from this computational efficiency issue, there remain questions of whether or when an LEL solution is globally optimal (i.e. consistent with all possible constraints). A given set of clauses in conjunctive normal form determines a universe of constants appearing in the set of clauses and a typically huge set of ILP constraints formed by substituting in turn each constant in this universe for each variable in each clause. A consistent solution is an assignment of truth values to all possible atomic formulas (assignment of 0 or 1 to all decision variables in the corresponding ILP) which satisfies this huge set of ILP constraints. LEL, as implemented, builds only the portion of the constraint set which arises as its "forward reasoning" process unfolds. The LEL solution is consistent with all constraints discovered in this process. Theoretical questions which require further investigation are: (1) Under what additional conditions on the structure of clauses will the LEL solution be globally optimal? (2) When will it be straightforward to extend an LEL solution to a globally optimal solution? (3) Are there pathological cases where an LEL solution has little value and can descriptive properties of such cases be easily identified? These are among the questions currently being investigated.

We have seen that extra care must be taken in using probabilistic interpretations of exception weights. An "exception" literal with associated weight appears in the conjunctive normal form representation of a clause. We saw an example where such a clause corresponded to any of three rules. Exception weights should have the property that the cost of an exception is the same for any rule which could give rise to the same clause. Strictly speaking, exception weights should be associated with clauses in conjunctive normal form rather than with individual rules which were used to generate such clauses.

CONCLUSIONS

This volume concerns the integration of operations research and artificial intelligence. Much previous effort toward such integration has involved the use of a rule-based expert system to decide on which specialized operations research procedure to call. LEL is different in that an operations research paradigm, ILP is at the heart of an artificial intelligence inference mechanism.

LEL can be viewed as the formulation of an ILP in an abstract, shorthand form. Each constraint (i.e., clause) that includes universal variables stands for the universal instantiation of the constraint over all substitutions for the

Default Reasoning Through Integer Linear Programming 193

universal variables. Inclusion of universal variables in constraints is useful because constraint instantiation is exponential in the number of universal variables. Therefore, listing all constraints prior to solution can be impractical. In most examples, including every possible constraint would be wasteful. A high percentage of all possible atomic formulas would be assigned truth value "false"; a high percentage of all possible rules would contain one of these false atomic formulas in their premise (e.g., If tweety is a fish then tweety has scales) and the corresponding constraint would be immediately satisfied.

Given an ILP stated in terms of quantified variables, the problem is to instantiate enough of the constraints to lead to a correct solution. Since it is not clear how one could be assured that all relevant constraints were instantiated, it appears that optimality will typically be approached rather than guaranteed. On the positive side, abstract constraints are a powerful representational device. An abstract constraint containing universally quantified variables compactly symbolizes a large number of new ILP constraints that can be instantiated on a contingency basis to further constrain a difficult search.

Exceptions in LEL are analogous to artificial variables in linear programming. With default reasoning, default knowledge is often inconsistent and making exceptions is required to achieve consistency. LEL treats feasibility as a continuum, so the goal is to find the best solution within an expanded space which includes the exception variables. In most linear programming (LP) applications, artificial variables are used only in an all-or-nothing process of determining whether or not a feasible solution exists. Thus there is a difference in spirit between the use of artificial variables in LP and exception decision variables in LEL.

From an AI perspective, the most novel technique in LEL is that basic inferencing goes beyond logic-based satisfaction of constraints through piecewise inference. It optimizes the solution by adding an objective function and using simultaneous solution. Uncertainty is handled not by propagating certainty factors as in other measure-based approaches, but by seeking a least exception solution. In instances where it is reasonable to use a probabilistic interpretation, LEL finds the most likely assignment of truth values consistent with an expressed set of constraints.

Finally, the combination of nonmonotonicity, truth maintenance, and uncertainty in an implemented system is novel. LEL is divergent from other approaches to default reasoning; it is a well-defined and consistent paradigm with a combination of attractive features. LEL has been implemented and has shown adequate performance to date on moderate-sized problems. However, LEL should benefit from additional research on dependencies between exceptions and on theoretical properties of its behavior under various restrictive assumptions on the structure of its rules (e.g. if all clauses were required to contain at most one positive non-exception literal).

Acknowledgement

Colin Bell was supported by a University of Iowa College of Business Administration Faculty Research Award.

REFERENCES

BRACHMAN, R. 1985. "I Lied about the Trees" or, Defaults and Definitions in Knowledge Representation. *AI Magazine* **6** (3), 80-93.
CHARNIAK, E., C. K. REISBECK, D. V. MCDERMOTT, AND J. MEEHAN. 1987. *Artificial Intelligence Programming*, 2nd edition. Lawrence Erlbaum, Hillsdale, N. J.
DEROSE, S. J. 1988. Grammatical Category Disambiguation by Statistical Optimization. *Computational Linguistics* **14** (1).
GARFINKEL, R. S. AND G. L. NEMHAUSER. 1972 *Integer Programming*. John Wiley & Sons, N. Y.
GENESERETH, M. R. AND N. J. NILSSON. 1987 *Logical Foundations of Artificial Intelligence*.. Morgan Kaufmann Publishers, Inc., Los Altos, CA.
HOOKER, J. N. 1986. Generalized Resolution and Cutting Planes. *Working Paper 53-85-86*. Graduate School of Industrial Administration, Carnegie-Mellon University, PA.
KANAL, L. N. AND J. F. LEMMER (editors). 1986. *Uncertainty in Artificial Intelligence*. North-Holland, N. Y.
KOWALSKI, R. 1979. *Logic For Problem Solving*. North Holland, N. Y.
MCALLESTER, D. A. 1982. Reasoning Utility Package User's Manual. Technical Report 667, AI Laboratory, MIT, Cambridge, MA.
MCCARTHY, J. 1980. Circumscription - A Form of Non-Monotonic Reasoning. *Artificial Intelligence* **13**(1-2).
MCDERMOTT, D., AND J. DOYLE. 1980. Non-monotonic Logic I. *Artificial Intelligence* **13**.
NII, P. 1986a. The Blackboard Model of Problem Solving. *AI Magazine* **7**(2).
NII, P. 1986b. Blackboard Systems Part Two: Blackboard Application Systems. *AI Magazine* **7** (3).
NILSSON, N. 1980. *Principles of Artificial Intelligence*. Tioga Publishing Company, Palo Alto, CA.
POST, S. D. 1987a. A Bipartite Connectionist Model to Represent N-ary Boolean and Linear Constraints. *Proceedings of IEEE First Annual International Conference on Neural Networks*. San Diego, CA.
POST, S. D. 1987b. Nonmonotonic Reasoning by Minimizing Contradiction in a Hedged Predicate Calculus. *Proceedings of Expert Systems in Government Symposium*. IEEE Computer Society Press, Los Angeles, CA.

POST, S. D. 1988. A System for Reasoning by Minimizing Exceptions. *Proceedings of the Third IEEE International Symposium on Intelligent Control Symposium.* IEEE Computer Society Press, Los Angeles, CA.
REITER, R. 1980. A Logic for Default Reasoning. *Artificial Intelligence* **13**(1-2).
RUMELHART, E. R. AND J. E. MCCLELLAND. 1986. *Parallel Distributed Processing.* The MIT Press, Cambridge, MA.
SHORTLIFFE, E. H. 1976. *Computer-based Medical Consultations: MYCIN.* American Elsevier, N. Y.
WALTZ, D.L. AND J. B. POLLACK. 1985. Massively Parallel Parsing: A Strongly Interactive Model of Natural Language Interpretation. *Cognitive Science.* **9** (1), 51-74.

The Problem of Determining Membership Values in Fuzzy Sets in Real World Situations

Evangelos Triantaphyllou
Department of Industrial and Management Systems Engineering
The Pennsylvania State University
University Park, Pennsylvania 16802
e-address: E3T@PSUVM.BITNET

Panos M. Pardalos
Department of Computer Science
The Pennsylvania State University
University Park, Pennsylvania 16802
e-address: PARDALOS@SHIRE.CS.PSU.EDU

Stuart H. Mann
School of Hotel, Restaurant, and Institutional Management
The Pennsylvania State University
University Park, Pennsylvania 16802.

ABSTRACT

One of the fundamental concepts in fuzzy set theory is the one of membership values. An appealing procedure for deriving information about membership values, is to use a matrix of pairwise comparisons [18], [19], [20]. A number of OR approaches that are based on eigenvalue theory and mathematical programming have been proposed to manipulate the previous matrices and estimate membership values. The findings of this paper reveal that although some methods appear to be more effective than others, still their performance is dramatically poor.

1. INTRODUCTION

The more than 1,800 references in [3], [4], and [42] describe the importance of fuzzy set theory in engineering and scientific problems. Fuzzy sets are particularly critical in many decision making situations (see, for example, [1], [7], [8], [9], [10], [15], [16], [21], [22], [25], [26], [27]). Recently, an increasingly large number of AI researchers have been faced by the problem that either their data or their background knowledge is fuzzy. This is a pervasive problem in AI. It is particularly critical to people building expert systems, for the knowledge they are dealing with is almost always riddled with vague concepts and judgmental rules [17]. The impact of fuzzy set theory on AI is best illustrated in [3], [4], and [34] - [43]. The understanding of fuzzy sets is of crucial importance to the successful development and operation of many expert systems (e.g. [11], [12], [16], [41]).

The keystone of any new real life application of fuzzy set theory is the successful estimation of the membership values of the elements in a fuzzy set. Saaty [18], [19], [20] proposed the use of a reciprocal matrix with entries that reflect the decision maker's estimates of the relative importances of the elements of a fuzzy set. In this way, membership values are derived from judgments of human experts about the dominance relations among the elements of a fuzzy set.

There are two main approaches in this decision process. The first approach considers the input data as continuous functions (see for example [26]). The second one, uses discrete data. Although there is no evidence as to why the input data have to be continuous or discrete, the discrete data are much easier to obtain. Saaty [20] claims that decision makers can derive effectively the required data by making a number of pairwise comparisons. In this paper we consider the case of dealing with these pairwise comparisons.

The use of the above reciprocal matrices has captured the interest of many researchers (see, for example, [2], [6], [7], [8], [20], [28], [31]). This is mainly due to the nice mathematical properties of the reciprocal matrices and the fact that the input data are rather easy to be obtained. In this paper we review some of the methods that use the above matrices as input and derive membership values. These methods use Operations Research techniques, namely, eigenvalue theory and mathematical programming. Then, the assumption that in reality membership values take on continuous values is made. This assumption is made in order to capture the majority of the real world cases. Using this assumption a forward error analysis reveals that the tested methods yield dramatically high failure rates.

2. LITERATURE REVIEW

2.1. Reciprocal Matrices with Pairwise Comparisons

Let A_1, A_2,...,A_n be the members of a fuzzy set. We are interested in evaluating the membership values of the above members. Saaty [18], [19], and [20] proposed to use a matrix A of rational numbers taken from the set $\{1/9, 1/8, 1/7, ..., 1, 2, 3, ..., 7, 8, 9\}$. Each entry of the above matrix A represents a pairwise judgment. Specifically, the entry a_{ij} denotes the number that estimates the relative membership of element A_i when it is compared with element A_j. Obviously, $a_{ij} = 1/a_{ji}$ and $a_{ii} = 1$. That is, the matrix is a reciprocal one.

Let us first examine the case in which it is possible to have perfect values a_{ij}. In this case it is $a_{ij} = W_i/W_j$ (W_s denotes the actual value of element s) and the previous reciprocal matrix A is consistent. That is:

$$a_{ij} = a_{ik}a_{kj} \quad (i,j,k = 1,2,3,...,n, \text{ where n is the number of elements in the fuzzy set}) \qquad (1)$$

It can be proved that A has rank 1 with $\lambda = n$ to be its nonzero eigenvalue. Then we have:

Determining Membership Values in Fuzzy Sets

$$Ax = nx \quad \text{where x is an eigenvector} \tag{2}$$

From the fact that $a_{ij} = W_i/W_j$ the following are obtained:

$$\sum_{j=1}^{n} a_{ij} W_j = \sum_{j=1}^{n} W_i = nW_i \quad i = 1,2,3,...n \tag{3}$$

or:

$$AW = nW \tag{4}$$

Equation (4) states that n is an eigenvalue of A with W a corresponding eigenvector. The same equation also states that in the perfectly consistent case (i.e., $a_{ij} = a_{ik} a_{kj}$) the vector W, with membership values of the elements 1,2,3,...,n, is the principal right-eigenvector (after normalization) of the matrix A.

2.2. The Eigenvalue Approach

In the non-consistent case (which is the most common) the pairwise comparisons are not perfect, that is, the entry a_{ij} might deviate from the real ratio W_i/W_j (i.e., from the ratio of the real membership values W_i and W_j). In this case, the previous expression (1) does not hold for all the possible combinations. Now the new matrix A can be considered as a perturbation of the previous consistent case. When the entries a_{ij} change slightly, then the eigenvalues change in a similar fashion [20]. Moreover, the maximum eigenvalue is close to n (greater than n) while the remaining eigenvalues are close to zero. Thus, in order to find the membership values in the non-consistent case, one should find an eigenvector that corresponds to the maximum eigenvalue λ_{max}. That is to say, to find the principal right-eigenvector W that satisfies:

$$AW = \lambda_{max} W \quad \text{where:} \quad \lambda_{max} \approx n$$

Saaty estimates the reciprocal right-eigenvector W by multiplying the entries in each row of matrix A together and taking the nth root (n is the number of the elements in the fuzzy set). Since we desire to have values that add up to 1.00 we normalize the previously found vector by the sum of the above values. If we want to have the element with the highest value to have membership value equal to 1.00, we divide the previously found vector by the highest value.

Under the assumption of total consistency, if the judgments are gamma distributed (something that Saaty assumes is the case), the principal right-eigenvector of the resultant reciprocal matrix A is Dirichlet distributed. If the assumption of total consistency is relaxed, then Vargas [31] states that the hypothesis that the principal right-eigenvector follows a Dirichlet distribution is accepted if the consistency ratio is 0.10 or less.

The consistency ratio (CR) is obtained by first estimating λ_{max}. Saaty estimates λ_{max} by adding the columns of matrix A and then multiplying the resulting vector with the vector W. Then he uses what he calls the consistency index (CI) of the matrix A. He defines CI as follows:

$$CI = (\lambda_{max} - n) / (n-1)$$

Then, the consistency ratio CR is obtained by dividing the CI by the Random Consistency index (RC) as given in the following table:

Table 1. Random consistency indices.

n	1	2	3	4	5	6	7	8	9
Random Consistency index (RC)	0	0	0.58	0.90	1.12	1.24	1.32	1.41	1.45

Each RC is an average random consistency index derived from a sample of size 500 of randomly generated reciprocal matrices with entries from the set {1/9, 1/8, 1/7, ..., 1, 2, 3, ..., 7, 8, 9} to see if its CI is 0.10 or less.

2.3. Some Minimization Approaches

If the previous approach yields a CR greater than 0.10 then a re-examination of the pairwise judgments is recommended until a CR less than or equal to 0.10 is achieved.

Chu [2] observed that given the data a_{ij} the values W_i to be estimated are desired to have the property:

$$a_{ij} \approx W_i / W_j \quad (5)$$

This is true since a_{ij} is meant to be the estimation of the ratio W_i/W_j. Then, in order to get the estimates for the W_i given the data a_{ij}, they propose the following constrained optimization problem:

$$\text{minimize } S = \sum_{i=1}^{n}\sum_{j=1}^{n}(a_{ij} W_j - W_i)^2 \quad (6)$$

subject to:

$$\sum_{i=1}^{n} W_i = 1$$

$$W_i > 0 \quad (i = 1,2,3,...,n)$$

They also give an alternative expression, S_1, that is more difficult to solve numerically. That is,

$$S_1 = \sum_{i=1}^{n}\sum_{j=1}^{n}(a_{ij} - W_j/W_i)^2 \quad (7)$$

Federov, et al., in [6] proposed a variation of the above least squares formulation. For the case of only one decision maker they recommend the

Determining Membership Values in Fuzzy Sets

following models:
$$\ln a_{ij} = \ln W_i - \ln W_j + \Psi_1(W_i, W_j) \varepsilon_{ij} \quad (8)$$
$$a_{ij} = (W_i/W_j) + \Psi_2(W_i, W_j) \varepsilon_{ij} \quad (9)$$

Here W_i and W_j are the true (and unknown) membership values; $\Psi_1(X,Z)$, $\Psi_2(X,Z)$ are given positive functions (when $X, Z > 0$). The random errors ε_{ij} are assumed independent with zero mean and variance one. Using these two assumptions they are able to calculate the variance of each individual estimated membership value. However, they do not propose any way of selecting the appropriate positive functions.

2.4. The Human Rationality Factor

According to the Human Rationality Assumption (given by Triantaphyllou, et al., in [28]) the decision maker is a rational person. Rational persons are defined here as individuals who try to minimize their regret [23], to minimize losses, or to maximize profit [33]. In the membership evaluation problem, minimization of regret, losses, or maximization of profit could be interpreted as the effort of the decision maker to minimize the errors involved in the pairwise comparisons.

As it is stated in previous paragraphs, in the inconsistent case the entry a_{ij} of the matrix A is an estimation of the real ratio W_i/W_j. Since it is an estimation, the following is true:
$$a_{ij} = (W_i/W_j)d_{ij} \qquad i,j = 1,2,3,\ldots,n \quad (10)$$

In the above relation d_{ij} denotes the deviation of a_{ij} from being an accurate judgment. Obviously, if $d_{ij} = 1$ then the a_{ij} was perfectly estimated. From the previous formulation we conclude that the errors involved in these pairwise comparisons are given by:
$$\varepsilon_{ij} = d_{ij} - 1.00$$
or (using (10), above):
$$\varepsilon_{ij} = a_{ij}(W_j/W_i) - 1.00 \quad (11)$$

When a fuzzy set contains n elements, then Saaty's method requires the estimation of the following $n(n-1)/2$ pairwise comparisons:

(W_2/W_1)	(W_3/W_1)	(W_4/W_1)	….	(W_n/W_1)
	(W_3/W_2)	(W_4/W_2)	….	(W_n/W_2)
		(W_4/W_3)	….	(W_n/W_3) (12)
			.	
			.	
			.	
				(W_{n-1}/W_n)

The corresponding $n(n-1)/2$ errors are (using relations (11) and (12)):
$$\varepsilon_{ij} = a_{ij}(W_j/W_i) - 1.00 \quad (13)$$
$$\text{where: } i,j = 1,2,\ldots,n$$
$$\text{and } j > i$$

Since the W_i's are degrees of membership that add up to 1.00 the following relation (14) should also be satisfied:

$$\sum_{i=1}^{n} W_i = 1.00 \qquad (14)$$

Apparently, since the W_i's represent degrees of membership we also have: $W_i > 0$ ($i = 1, 2, 3, ..., n$).

Relations (13) and (14), when the data are consistent (i.e., all the errors equal to zero), can be written as follows:

$$B W = b \qquad (15)$$

Above, the vector b has zero entries everywhere except the last one that is equal to 1.00, and the matrix B has the following form (blank entries represent zeros):

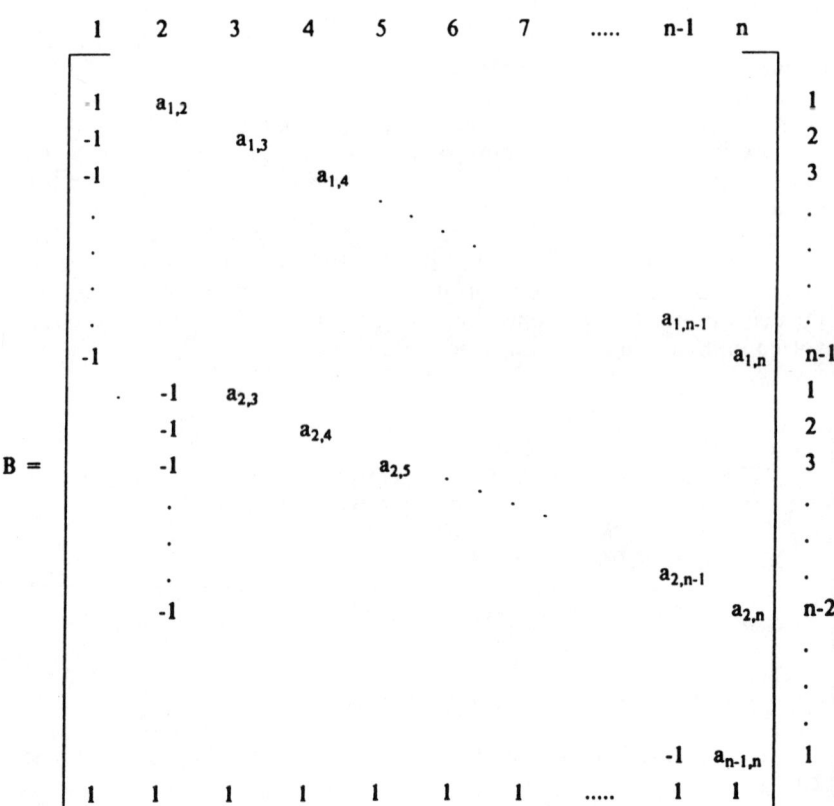

Determining Membership Values in Fuzzy Sets 203

The error minimization issue is interpreted in many cases (Regression Analysis, Linear Least Squares problem) as the minimization of the sum of the squared errors [24].

In terms of the formulation (15) this means that in a real life situation (i.e., errors are not zero any more) the real intention of the decision maker is to minimize the expression:

$$f^2(x) = \|b - BW\|_2^2 \tag{16}$$

which is, a typical Linear Least Squares problem!

If we use the notation described previously then the quantity (6) that is minimized in Chu [2] becomes:

$$S = \sum_{i=1}^{n}\sum_{j=1}^{n}(a_{ij}W_j - W_i)^2 = \sum_{i=1}^{n}\sum_{j=1}^{n}(\varepsilon_{ij}W_i)^2$$

and the alternative expression (7) becomes:

$$S_1 = \sum_{i=1}^{n}\sum_{j=1}^{n}(a_{ij} - W_j/W_i)^2 = \sum_{i=1}^{n}\sum_{j=1}^{n}(\varepsilon_{ij}(W_i/W_j))^2$$

Clearly, both expressions are too complicated to reflect in a reasonable way the intentions of the decision maker.

The models proposed by Federov [6] are closer to the one developed under the Human Rationality Assumption. The only difference is that instead of the relations:

$$\ln a_{ij} = \ln W_i - \ln W_j + \Psi_1(W_i, W_j)\varepsilon_{ij}$$

$$a_{ij} = (W_i/W_j) + \Psi_2(W_i, W_j)\varepsilon_{ij}$$

the following simpler expression is used:

$$a_{ij} = (W_i/W_j) d_{ij}$$

or :

$$a_{ij} = (W_i/W_j)(\varepsilon_{ij} + 1.00) \tag{17}$$

However as it is illustrated in [28], the performance of this method is greatly dependent on the selection of the $\Psi_1(X,Z)$ or $\Psi_2(X,Z)$ functions and now these functions are further modified by (17).

2.5. The Concept of the Closest Discrete Pairwise Matrix

The following forward error analysis is based on the assumption that in the real world the membership values in a fuzzy set take on continuous values. This assumption is a reasonable one since it attempts to capture the majority of the real world cases.

Let $\omega_1, \omega_2, \omega_3, ..., \omega_n$ be the real (and thus unknown) membership values of a fuzzy set with n members. If the decision maker knew the above real values then he would be able to have constructed a matrix with the real pairwise comparisons. In this matrix, say matrix A, the entries are given by: $\alpha_{ij} = \omega_i/\omega_j$. This matrix is called the Real Continuous Pairwise matrix, or the RCP matrix [29]. Since in the real world the ω_i's are unknown so are the entries α_{ij} of the previous matrix. However, we

can assume here that the decision maker instead of the entries α_{ij} is able to determine the closest values taken from the set $\{1/9, 1/8, 1/7, ..., 1, 2, 3, ..., 7, 8, 9\}$. That is, instead of the real (and unknown) value α_{ij} one is able to determine a_{ij} such that:
$$a_{ij} = \min\{|a_{ij} - x/y|\}$$
where: $x \in \{1/9, 1/8, 1/7, ..., 1, 2, 3, ..., 7, 8, 9\}$

In other words, one's judgments about the values of the pairwise comparisons of the ith element when it is compared with the jth one, is so accurate that in real life is the closest (in absolute value terms) to the values one is supposed to choose from. Apparently, this assumption favors the values of the failure rates derived in this paper. This fact indicates that even the failure rates in this paper are smaller than the actual ones.

The matrix with entries the a_{ij}'s that we assume the decision maker is able to construct, has entries from the discrete and finite set $\{1/9, 1/8, 1/7, ..., 1, 2, 3, ..., 7, 8, 9\}$ This matrix is called the Closest Discrete Pairwise matrix or the CDP matrix.

The following example illustrates a case where we deal with a fuzzy set with three members. The real membership values are assumed to be known. Then, the RCP and CDP matrices are derived. When the error minimization method is applied on this problem the resulted membership values are significantly different than the real ones. More importantly, the ranking is altered too, indicating a case where the error minimization method fails.

EXAMPLE

Let us assume that the real (and unknown) membership values, after normalization, of a fuzzy set with three members are: $\omega_1 = 0.73207 \omega_2 = 0.13366$, and $\omega_3 = 0.13427$
Obviously, the corresponding ranking by magnitude of these members is: $\rho_1 = 1, \rho_2 = 3$, and $\rho_3 = 2$.

Then, the RCP (and unknown) matrix with the pairwise comparisons is:
$$RCP = \begin{bmatrix} 1 & 5.47735 & 5.45216 \\ 0.18257 & 1 & 0.99540 \\ 0.18341 & 1.00462 & 1 \end{bmatrix}$$

It can be verified with a simple exhaustive enumeration that the corresponding CDP matrix is:
$$CDP = \begin{bmatrix} 1 & 5/1 & 5/1 \\ 1/5 & 1 & 5/5 \\ 1/5 & 5/5 & 1 \end{bmatrix}$$

This is the matrix we assume the decision maker has determined for this example. Clearly, this is the best comparisons a decision maker can make for the fuzzy set of this example.

The formulation (5) that corresponds to this example is:

Determining Membership Values in Fuzzy Sets

$$\begin{bmatrix} -1 & 5/1 & 0.0 \\ -1 & 0.0 & 5/5 \\ 0.0 & -1 & 5/5 \\ 1 & 1 & 1 \end{bmatrix} \begin{bmatrix} W1 \\ W2 \\ W3 \end{bmatrix} = \begin{bmatrix} 0 \\ 0 \\ 0 \\ 1.0 \end{bmatrix}$$

The vector W that minimizes the above Least Squares is calculated to be :

$$W = (0.71429 \quad 0.14286 \quad 0.14286)$$

and the corresponding ranking is: $R_1 = 1$, $R_2 = 2$, and $R_3 = 2$.

Obviously, these results contradict the real membership values and ranking of the members of the fuzzy set of this example.

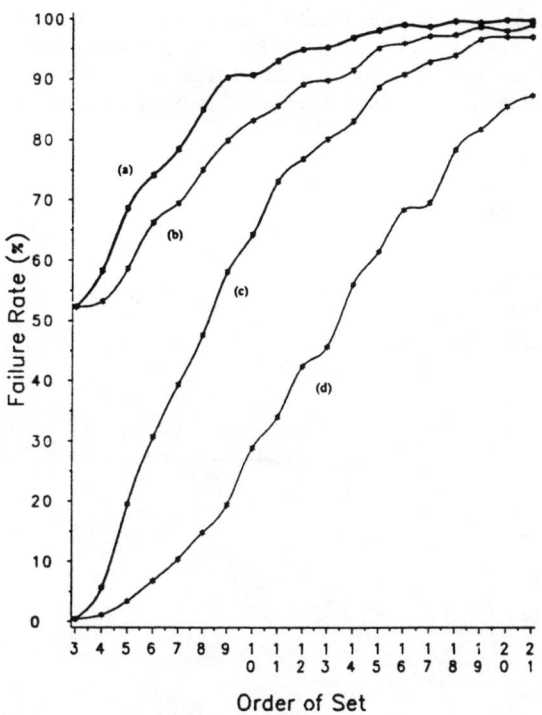

Figure 1.
Failure rates of the Error Minimization and Eigenvalue Method for fuzzy sets of different order
(results are based on one thousand observations).

3. ANALYSIS OF THE FAILURE RATES YIELDED BY THE ERROR MINIMIZATION METHOD

The error minimization method was tested in a similar fashion to the eigenvalue method tested in [28]. More specifically, random problems of different sizes were generated and tested as in the above example. For each such test problem real membership values were generated randomly from the continuous interval (0, 1). However, because the Saaty matrices use values from the set {1/9, 1/8, 1/7, ..., 1, 2, 3, ..., 7, 8, 9} only the random problems that had RCP matrices with entries within the continuous interval [1/9, 9/1] were considered. The RCP matrix, with the real pairwise comparisons was constructed, and after that the CDP matrix was determined.

The error minimization and eigenvalue approaches then were applied. Any contradiction between the real ranking and the ones derived by the error minimization and the eigenvalue methods were recorded as failures. Two kinds of ranking inconsistency were recorded. The first kind is "ranking reversal". For example, if the real ranking of a set of three members is (1, 3, 2) and one method yields (1, 2, 3) then a case of a ranking reversal occurs. The second kind is "ranking indiscrimination". For example, if the real ranking of a set of three members is (1, 3, 2) and one method yields (1, 2, 2) then a case of ranking indiscrimination occurs.

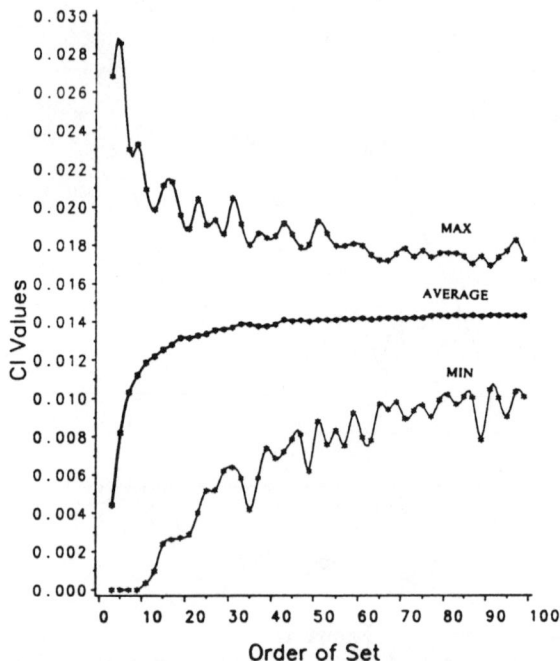

Figure 2.
Maximum, Average, and Minimum CI values of random CDP matrices.

Determining Membership Values in Fuzzy Sets

This simulation approach was implemented in FORTRAN with the appropriate IMSL subroutines. Problems of size 3,4,5,...,21 were considered. For each case the number of random matrices generated was 1,000. This number was large enough to ensure that the means converged to within a small error tolerance. The failure rates for fuzzy sets of order: 3, 4, 5,..., 21 are presented in Table 2 and depicted in Figure 1. In Figure 1 curves (a) and (b) represent the total number of failures (adding ranking reversals and ranking indiscriminations) yielded by the eigenvalue and error minimization methods, respectively. Curves (c) and (d) give the number of ranking reversals yielded by the error minimization and eigenvalue methods, respectively. It is important to note here that the CI values of the previously generated CDP matrices are very small [30]. In particular, the CI's are less than 0.029 with an average value equal to 0.0145 (see also Figure 2). These low CI values guarantee that the Dirichlet criterion, as stated by Vargas [31], is satisfied.

Table 2. Failure rates of the Eigenvalue and Error Minimization Method for fuzzy sets of different order (results are based on one thousand observations)

order of set	Eigenvalue Method		Error Minimization Method	
	inversion rate	total failure rate	inversion rate	total failure rate
3	0.4	52.3	0.4	52.3
4	1.1	58.3	5.7	53.2
5	3.4	68.7	19.5	58.7
6	6.8	74.2	30.7	66.3
7	10.4	78.5	39.4	69.5
8	14.8	85.0	47.6	75.1
9	19.4	90.3	58.1	79.9
10	28.9	90.7	64.3	83.2
11	34.0	93.0	73.1	85.6
12	42.4	94.9	76.8	89.1
13	45.6	95.3	80.1	89.8
14	56.0	96.9	83.1	91.5
15	61.4	98.1	88.6	95.1
16	68.3	99.0	90.7	95.9
17	69.5	98.7	92.8	97.1
18	78.3	99.6	93.9	97.3
19	81.7	99.4	96.6	98.6
20	85.5	99.8	97.0	98.0
21	87.3	99.7	97.0	99.0

Although the performance of the error minimization and eigenvalue methods may change if a different set of values for pairwise comparisons is considered, it is interesting to observe that a decision-maker always is restricted to select from a finite set of choices. In 1846 Weber stated his law regarding a stimulus of measurable magnitude. According to his law a change in sensation is noticed if the stimulus is increased by a constant percentage of the stimulus itself [20]. That is, people are unable to make choices from an infinite set. For example, people cannot distinguish between two very close values of importance, say 3.00 and 3.01. Psychological experiments have also shown that individuals cannot simultaneously compare more than seven objects (plus or minus two) [14]. This is the main reason why Saaty uses 9 as the upper limit of his scale, 1 as the lower limit of his scale and a unit difference between successive scale values.

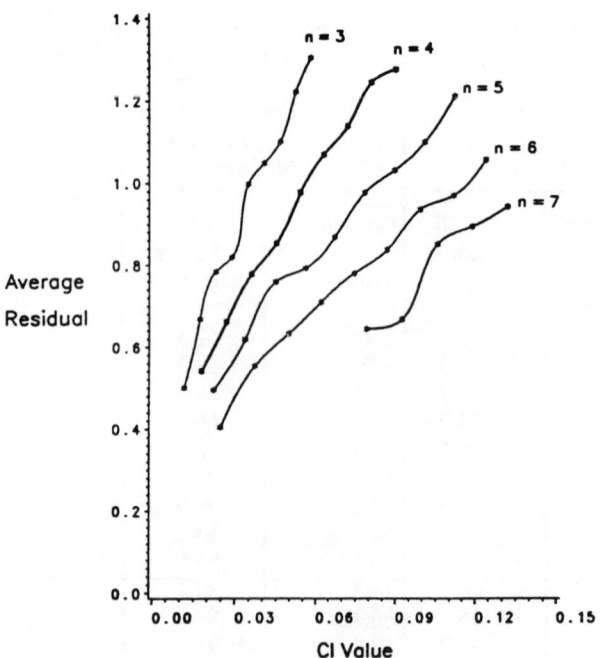

Figure 3.
Average residual and CI versus order of fuzzy set when the Eigenvalue Method is used
(results are based on one hundred observations).

Determining Membership Values in Fuzzy Sets

The findings in [29] and the present tests reveal that the failure rate for the two methods increases with the number of members in the fuzzy set. Table 2 presents the failure rates of both the eigenvalue and error minimization methods when the above simulation approach was used. The results suggest that the eigenvalue method yields, on the average, more failures than the error minimization approach. However, the number of ranking reversals is significantly higher with the error minimization method. In Triantaphyllou, et. al [28], the two methods were evaluated using a least squared residuals criterion. These results are given in Figures 3 and 4, as they appeared in [28]. Although the error minimization method does better as far as the least squared residuals and the total number of failures is concerned, it does poorly in terms of the number of ranking reversals. Therefore, we cannot conclude that there is a single best method.

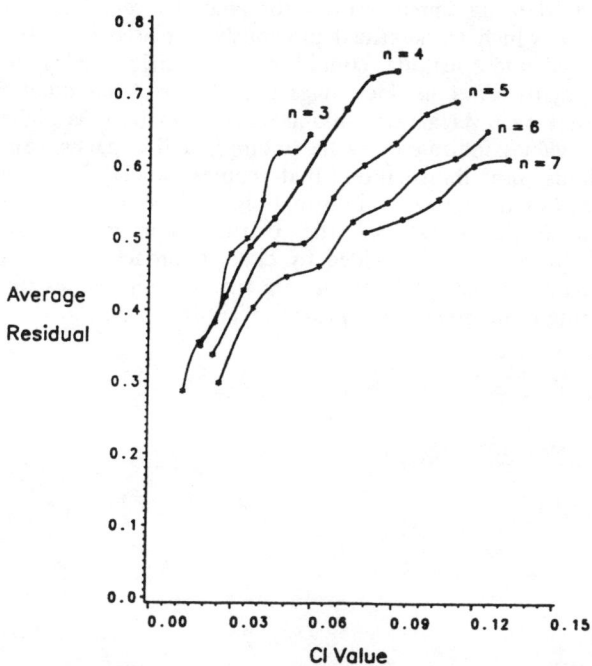

Figure 4.
Average residual and CI versus order of fuzzy set when the
Error Minimization Method is used
(results are based on one hundred observations).

4. CONCLUDING REMARKS

In this paper we considered various techniques for estimating membership values in fuzzy sets for real world problems when only one decision-maker is involved. The forward error analysis presented here yields a mechanism for testing the effectiveness of methods that evaluate membership values in fuzzy sets when pairwise comparison matrices are used as input data. The same analysis also reveals the magnitude of the problem of correctly evaluating membership values. The error minimization approach, with regard to the total number of failures (Figure 2), seems to be more effective than the eigenvalue approach, though still yielding high failure rates.

In the present paper the assumption is made that the decision-maker deals with a CDP matrix. That is, at any moment the decision-maker is able to determine a value from the set $\{1/9, 1/8, 1/7, ..., 1,2,3, ..., 7, 8, 9\}$ that is closest to the actual value (which in reality is unknown). The high failure rates derived using the forward error analysis were based on the above assumption which as described previously is biased in favor of the methods. However, under real life conditions the decision-maker may not deal with CDP matrices. This fact suggests that in real life situations the possibility for ranking reversals or ranking indiscriminations is higher than the already high values indicated by the findings of this paper. Since the examined methods use information that comes solely from pairwise comparisons, additional sources of information, such as order information [32], may enhance the power of the pairwise comparison methods. Currently, the high failure rates yielded by either approach in combination with the importance of accurately evaluating membership values make the need for developing more powerful approaches an urgent one.

REFERENCES

1. BELLMAN, R.E., and L.A. ZADEH, "Decision Making in a Fuzzy Environment", Manag. Sci., Vol. 17, No 4, 1970, pp. 140-164.

2. CHU, A.T.W., KALABA, R.E., and SPINGARN, K., "A Comparison of Two Methods for Determining the Weights of Belonging to Fuzzy Sets", Journal of Optimization Theory and Applications, Vol. 27, No 4, 1979, pp. 531-538.

3. CHANG, S.K., "Fuzzy Programs", Proceedings of the Brooklyn Polytechnical Institute Symposium on Computers and Automata, Vol. XXI, 1971.

4. DUBOIS, D. and H. PRADE, "Fuzzy Sets and Systems: Theory and Applications", Academic Press, New York, 1980.

5. GUPTA, M.M., RAGADE, R.K. and YAGER, R.Y., editors, "Fuzzy Set Theory and Applications", North-Holland, New York, 1979.

6. FEDEROV, V.V., KUZ'MIN, V.B., and VERESKOV, A.I., "Membership Degrees Determination from Saaty Matrix Totalities", Institute for System Studies, Moscow, USSR. Paper appeared in: "Approximate Reasoning in Decision Analysis", M. M. Gupta, and E. Sanchez (editors), North-Holland Publishing Company, 1982, pp. 23-30.

7. KHURGIN, J.I., and V.V. POLYAKOV, "Fuzzy Analysis of the Group Concordance of Expert Preferences, defined by Saaty Matrices", Fuzzy Sets Applications, Methodological Approaches and Results, Akademie-Verlag Berlin, 1986, pp. 111-115.

8. KHURGIN, J.I., and V.V. POLYAKOV, "Fuzzy Approach to the Analysis of Expert Data", Fuzzy Sets Applications, Methodological Approaches and Results, Akademie-Verlag Berlin, 1986, pp. 116-124.

9. KIM, J.B., "Fuzzy rational choice functions", Fuzzy Sets and Systems, 10, 1983, pp. 37-43.

10. KUZ'MIN, V.B., and S.V. OVCHINNIKOV, "Design of group decisions II. In spaces of partial order fuzzy relations", Fuzzy Sets and Systems, 4, 1980, pp. 153-165.

11. LEE, C.T.R., "Fuzzy Logic and the Resolution Principle", Second International Joint Conference on Artificial Intelligence, London, 1971, pp. 560-567.

12. LEE, S.N., Y.L. GRIZE, and K. DEHNAD, "Quantitative Models for

Reasoning under Uncertainty in Knowledge Based Expert Systems", International Journal of Intelligent Systems, Vol. II, 1987, 15- 38.

13. MARINOS, P.N., "Fuzzy Logic and Its Applications to Switching Systems", IEEE Trans. on Computers, c-18, 4, 1969, pp. 343-348.

14. MILLER, G.A. "The Magical Number Seven Plus or Minus Two: Some Limits on Our Capacity for Processing Information", Psychological Review, 13, 1956, pp. 81-97.

15. NURMI, H., "Approaches to collective decision making with fuzzy preference relations", Fuzzy Sets and Systems, 6, 1981, pp. 249- 259.

16. PRADE, H., and C.V. NEGOITA (Eds.), "Fuzzy Logic in Knowledge Engineering", Verlag TUV Rheinland, 1986.

17. RAMSAY, A., "Formal Methods in Artificial Intelligence", Cambridge University Press, Cambridge, 1988.

18. SAATY, T.L., "A Scaling Method for Priorities in Hierarchical Structures", Journal of Mathematical Psychology, Vol. 15, No 3, 1977.

19. SAATY, T.L., "Exploring the interface between hierarchies, multiple objects and fuzzy sets", Journal of Fuzzy Sets and Systems, Vol. 1, 1978, pp. 57-68.

20. SAATY, T.L., "The Analytic Hierarchy Process", McGraw Hill International, 1980.

21. SHAPIRO, D.I., "Decision-making in conditions of fuzzy uncertainties and opposing factors, using human specifics", Artificial Intelligence and Information-Control Systems of Robots, Ed. I. Plander, North-Holland Publ. Comp., Amsterdam 1984, pp. 333-336.

22. SHAPIRO, D.I., "A Model for Decision-Making Under Fuzzy Conditions", Computers and Artificial Intelligence, 4, No 6, 1985, pp. 481-494.

23. SIMON, H.A., "Models of Man", John Wiley & Sons, INC., (second edition), New York, 1961.

24. STEWART, S.M., "Introduction to Matrix Computations", Academic Press, New York, 1973.

25. TANINO, T., "Fuzzy preference orderings in group decision making", Fuzzy Sets and Systems, 12, 1984, pp. 117-131.

26. TANINO, T., "Fuzzy Preference Relations in Group Decision

Making", Lecture Notes in Economics and Math. Systems, 301, Spring Verlag, 1988.

27. TRIANTAPHYLLOU, E. and MANN, S.H., "An Examination of the Effectiveness of Multi-Dimensional Decision-Making Methods: A Decision-Making Paradox", International Journal of Decision Support Systems, Number 5, pp. 303-312, 1989.

28. TRIANTAPHYLLOU, E., PARDALOS, P.M., and MANN, S.H., "A Minimization Approach to Membership Evaluation in Fuzzy Sets and Error Analysis", Journal of Optimization Theory and Applications, to appear 1990.

29. TRIANTAPHYLLOU, E. and MANN, S.H., "An Evaluation of the Eigenvalue Approach for Determining the Membership Values in Fuzzy Sets", Fuzzy Sets and Systems, Number 2, Volume 35, to appear 1990.

30. TRIANTAPHYLLOU, E., PARDALOS, P.M., and MANN, S.H., "The Role of Perfectly Consistent Saaty Pairwise Matrices in Evaluating Membership Values in Fuzzy Sets", working paper, Pennsylvania State University, 1990.

31. VARGAS, L.G., "Reciprocal Matrices with Random Coefficients", Mathematical Modeling, Vol. 3, 1982, pp. 69-81.

32. WHITE, C.C. III, "A Posteriori Representations Based on Linear Inequality Descriptions of A Priori and Conditional Probabilities", IEEE Transactions on Systems, Man, and Cybernetics, Vol. 16, No. 4, 1986, pp. 570-573.

33. WRITE, C., and TATE, M.D., "Economics and Systems Analysis: Introduction for Public Managers", Addison-Wesley, Reading, Massachusetts, 1973.

34. XIE, L.A., and S.D. BERDOSIAN, "The Information in a Fuzzy Set and the Relation Between Shannon and Fuzzy Information", Proceedings of the Seventeenth Annual Conference on Information Sciences and Systems, 1983.

35. ZADEH, L.A., "Fuzzy Sets", Information and Control, Vol. 8, 1965, pp. 338-353.

36. ZADEH, L.A., "Fuzzy Algorithms", Information and Control, 12, 1968, pp. 94-102.

37. ZADEH, L.A., K.S. FU, K. TANAKA, and M. SHIMURA, Eds., "Fuzzy Sets and their Applications to Cognitive and Decision Processes", New York: Academic, 1975.

38. ZADEH, L.A., "Fuzzy sets as a basis for a theory of possibility", Fuzzy Sets and Systems, 1, 1978, pp. 3-28.

39. ZADEH, L.A., "A Theory of Approximate Reasoning", Machine Intelligence, Vol. 9, J. Hayes, D. Michie, and L. I. Mikulich (Eds.), Wiley, New York, 1979, pp. 149-194.

40. ZADEH, L.A., "The concept of a linguistic variable and its applications to approximate reasoning -I, II, III", Information Sciences, 8, 199-249 (1975), 8, 301-357 (1975), 9, 43-80 (1976).

41. ZADEH, L.A., "Can Expert Systems be Designed Without Using Fuzzy Logic?", Proceedings of the Seventeenth Annual Conference on Information Sciences and Systems, 1983.

42. ZADEH, L.A., "Fuzzy logic as basis for the management of uncertainty in Expert Systems", Fuzzy Sets and Systems, 10, pp. 395-460.

43. ZIMMERMANN, H.J., "Fuzzy Set Theory - and Its Applications", Kluwer Academic Publishers, Netherlands, 1985.

IV. DECISION ANALYSIS AND DECISION SUPPORT

This section is comprised of three papers. The first, by Farquhar ("Applications of Utility Theory in Artificial Intelligence Research"), has as its purpose to improve intelligent systems capabilities for handling novel situations, inconsistent information, and incomplete heuristics. Utility theory is a natural way of integrating learning elements in the development of intelligent systems. The paper focuses on applying adaptive utility and multiattribute utility theories to machine learning of heuristic evaluation functions.

Evaluation functions are critical components for learning in many intelligent systems. Use of these utility theories enables the construction of evaluation functions from basic principles in only a few hours; current construction methods often require months of iterative search.

The paper also describes inherent limitations in the popular use of linear, additive functions for evaluation; truly expert systems often require nonadditive functions. Although these more general types of nonadditive utility functions have been slow to be applied in decision analysis, there are substantial gains in expertise from using such functions in AI systems designed for repetitive decision making.

With regard to the second paper in this section ("A Multicriteria Stratification Framework for Uncertainty and Risk Analysis" by Barlow and Glover), it is noted that uncertainty and risk pervade business decision-making, but are very difficult to handle in an intelligent and systematic manner. Attempts to incorporate these issues into optimization models lead to complex nonlinearities even in very simple situations. In the types of situations that may be expected in the real world, present theoretical models are often intractable or inapplicable. The usual alternative is to abandon optimization models and resort to simple ranking and sequential decision schemes, in some cases augmented by decision tree analysis, without a solid means of accommodating the interdependencies created by attendant objective functions and constraints.

This paper provides a framework for treating risk and uncertainty that retains the ability to make use of optimization models, but without introducing the nonlinearities and simplifying assumptions of standard mathematical approaches. The authors establish a link to multicriteria optimization that allows interactive decision making with user feedback to provide progressively improved outcomes. The model used is adaptively structured to respect the user's actual preferences for trade-offs between risk and gain, rather than relying on armchair speculations about what these preferences should be. By making use of this framework, decision-makers are given the ability to see the consequences of their decisions, and therefore to enhance their understanding of the underlying relationships as a basis for developing improved responses to practical problems.

The last paper in this section by Sycara ("Dispute Mediation: A Computer Model") addresses an important topic in group decision support: conflict resolution. Conflict and conflict resolution though negotiation/mediation is ubiquitous in human activities. Unfortunately, conflict resolution is a complex and time consuming process that up to the present time has defied all attempts at solution via analytic techniques. There are many intangibles that enter the process such as the skill and experience of the participants, their beliefs, values and perceptions. These intangibles can best be modelled symbolically. This paper's contribution consists in (a) integrating OR and AI techniques to model all parts of the conflict resolution process, (b) implementing the model into a computer system and thus automating the process, and (c) achieving flexibility and robustness of the system due to the integration of these methodologies. This system is a data point that provides evidence of the power of a hybrid methodology to address complex problems.

The proposed model could be used as a tool of understanding conflict resolution. The negotiation/mediation process as carried out by people is opaque, in the sense that it is impossible to see the underlying mechanisms. The computer program, PERSUADER makes explicit what knowledge is needed in negotiation, how it is represented and organized, and how it is used to make decisions. In addition, the model makes explicit the role and utility of the analytic and AI techniques. The PERSUADER provides a normative reference with which to compare and evaluate actual negotiations and makes possible testable hypotheses that might help in understanding negotiation and conflict resolution.

Applications of Utility Theory in Artificial Intelligence Research*

Peter H. Farquhar

Graduate School of Industrial Administration
Carnegie Mellon University
Pittsburgh, Pennsylvania 15213

ABSTRACT

This paper examines recent applications in the construction of evaluation functions for intelligent computer systems. The purpose is to demonstrate the usefulness of utility theory for these research activities in artificial intelligence and to promote future exchanges between these two fields.

INTRODUCTION

Research in artificial intelligence has grown enormously over the past decade. Intelligent systems, for example, have been developed and applied in electronics, manufacturing, medicine, military science, and many other fields (e.g., see Waterman (1986)). In designing intelligent systems, one usually encounters

---

*This paper updates and expands an earlier paper of the same title published in Toward Interactive and Intelligent Decision Support Systems, Vol. 2 (1987), edited by Y. Sawaragi, K. Inoue, and H. Nakayama (Springer-Verlag: New York), 155-161.

at some stage the problem of evaluating objects with respect to multiple performance criteria. The evaluation methods often vary in form and complexity depending on the nature of the objects themselves, the structure of the problem domain, the importance of the specific task, and other factors. In some recent applications, multiattribute utility theory has been used to generate evaluation functions in intelligent systems. Therefore, we examine the connection between multiattribute utility theory and artificial intelligence systems in more detail.

The next section reviews three basic approaches for constructing evaluation functions in artificial intelligence systems. Recent applications of multiattribute utility theory illustrate each approach. The third section of the paper comments on these approaches and suggests how more advanced topics in utility theory are applicable in machine learning. Exciting opportunities emerge for further research in both fields.

This short paper is neither a review of multiattribute utility theory nor an introduction to artificial intelligence research. Keeney and Raiffa (1976), Farquhar (1977, 1980, 1983), and Bell and Farquhar (1986) do the former; Barr and Feigenbaum (1981, 1982), Cohen and Feigenbaum (1982), Winston (1984), and Charniak and McDermott (1985) do the latter. Although it is certainly desirable in examining the interface with artificial intelligence to enlarge the focus from utility theory to all of decision analysis, this paper adopts a narrower perspective. Topics such as problem structuring, option generation, uncertainty analysis, information valuation, inference, expectations, and others are not discussed (e.g., see Humphreys and Berkeley (1985), Lehner, Probus, and Donnell (1985), Shortliffe, Buchanan, and Feigenbaum (1979), Weiss (1985), Wellman (1985), Wisudha (1985), among others). White (1989) describes more generally the integration of decision analysis and expert systems. Instead, this paper emphasizes recent utility theory applications in artificial intelligence.

APPLICATIONS

The use of evaluation functions in artificial intelligence (AI) dates to Samuel's (1959) checker-playing program which used linear polynomials of board position features to evaluate prospective moves. Much of the later AI work has determined evaluation functions using simple heuristic methods. Indeed, some twenty years later, Berliner (1979, p. 53) remarks

> The AI literature does not contain much information about how to construct evaluation functions; only the work of Samuel (1959) attempts to shed light on how the construction characteristics of the function (rather than the content) bear on the performance of the program using the function.

The AI field, however, has moved quickly in the past few years to develop better means of constructing evaluation functions. This section reviews three approaches based upon multiattribute utility theory (MAUT) that have been employed in recent applications.

Lehner, Probus, and Donnell (1985) note an important difference between single decision problems, where a utility function is constructed to satisfy an appropriate set of axioms for a problem, and repetitive situations, where the same evaluation function is used

Utility Theory and Artificial Intelligence

across applications. Such an evaluation function is at best an approximation and may encounter serious difficulties as the problem structure varies. AI systems are designed for repetitive decision situations and often are able to adjust quickly and easily to problem changes. Thus the challenge to MAUT is to provide methods not only for the construction of evaluation functions, but also for modifying these functions as the problem structure or situation changes.

Three basic MAUT approaches have been reported for modifying evaluation functions in intelligent systems. The first approach assumes a fixed functional form across situations and a procedure for translating information about a given situation into values for the scores and weights in the evaluation function. Lehner et al. (1985) report the development of a rule-based system for parameter assignment in an additive utility function, though few details are given. Similar architectures are found among simple game-playing systems in AI and have limitations which we discuss later.

The second approach involves learning the evaluation function, where the form in presumably known beforehand. Three steps are distinguished in this process: (1) **feature discovery** (i.e., what are all the relevant attributes for evaluation), (2) **feature selection** (i.e., what are the determinant attributes in a given situation), and (3) **weight adjustment** (i.e., what are the trade-offs among the selected attributes). For the latter step, Samuel (1959) describes how textbook training and self-play are used to calibrate the polynomial evaluation function in his checker player. Likewise, Madni, Samet, and Purcell (1985) report using a multicategory pattern classifier for learning the weights in an additive utility function that evaluates messages in military command, control, and communication situations. Lee and Mahajan (1988) also use pattern classification methods for evaluation function learning.

The third approach assumes that the form of the evaluation function is not known initially. Wellman (1985, 1986) describes a promising system for generating multiattribute utility functions from a knowledge base of common independence axioms. The system deduces an appropriate functional form from user responses to various queries about the axioms and other properties.

These basic approaches do not exhaust the applications of MAUT. White and Sykes (1986) describe the use of an additive utility function for evaluating simultaneously activated rules in a knowledge base and resolving conflict among them. Thus a utility function replaces the meta-rule approach to rule selection. By offering an expert user the opportunity to employ his or her preferences in a particular situation, the system gains flexibility in specific tasks. White (1989) also describes a few other MAUT applications.

DIRECTIONS

The above efforts to provide for changes in the evaluation functions do not account for research on adaptive utility theory reported by Cyert and DeGroot (1975), Tesfatsion (1980, 1982), DeGroot (1983), Farquhar (1986, 1987), and others. This research assumes that although the form of the utility function is known, there is uncertainty associated with some of the parameters. Bayesian learning, adaptive control, and other methods are used to converge on a precise utility function by updating the parameters from experienced outcomes. This approach can be contrasted with other techniques which interactively elicit preferences by placing increasingly tighter bounds on the utility function until a decision becomes clear.

On the other hand, the form of the utility function may be misspecified. Cohen and Axelrod (1984) illustrate how adaptation can be used to cope with complex environments where feature discovery and selection are on-going processes. Many applications have wished away these issues; further research is warranted. Lee and Mahajan (1988) explore some of these issues in a Bayesian learning approach to evaluation in their expert system for playing Othello.

Another possibility is that the utility function is precisely known and correctly specified at each stage in a dynamic decision process. The problem is that the utility function can change over time or across situations, so future preferences are uncertain. White (1984) considers stochastically changing weights in an additive utility function, while Farquhar (1981), Farquhar and Fishburn (1981), and others examine situation dependencies in multivalent preference structures. Farquhar (1986) discusses research on adaptation in preferences in more detail.

Although one can criticize the applications mentioned in the first two approaches of the previous section, the opportunity still exists to demonstrate that adaptive utility methods are improvements over simpler methods. Preliminary research suggests that such is the case (cf., Holland and Reitman (1978), Smith (1984), Holland (1986)).

There are some surprising parallels between the development of evaluation functions in expert systems and the development of representation theorems for multiattribute utility functions. For example, Berliner (1979, 1980) describes successive improvements in the evaluation function of his backgammon-playing system, which has defeated the world champion. His first attempt at constructing the evaluation function used a "linear polynomial in which each term represented a particular feature of a backgammon position and the constant coefficient of the term indicated the importance of that feature. The sum of the linear polynomial gave the value of the position" (pp. 66-67). This attempt produced rather limited success, as did Samuel's (1959) checker player and other such game-playing systems.

The next attempt was to group the board positions into classes and develop a separate evaluation function for each class. "This approach led to an improvement in the program's playing strength, but it too had a drawback. The relative values of two positions in different classes were sometimes inaccurate because the evaluation functions yielded significantly different values for positions near the common border of the classes where the values should have been quite close" (p. 67).

The third attempt yielded great success. Berliner abandoned the classes and replaced the constant coefficients in the linear polynomial with slowly changing, smooth variables that provided "a great deal of context but avoided the rough boundaries between

Utility Theory and Artificial Intelligence

different contexts. We called this new approach SNAC, for smoothness, nonlinearity, and application coefficients" (p. 67). These application coefficients were calibrated by textbook training, self-play, and competitive play. The result was a world-class expert system for playing backgammon.

Independent research in multiattribute utility theory parallels these striking developments. Berliner's first evaluation function, which had limited playing ability, is an additive utility model. His next attempt corresponds to the multivalent additive utility model in Farquhar and Fishburn (1981). The problem that Berliner noted with discontinuities at the boundaries of classes is addressed by a theorem in this paper. Finally, the evaluation function in Berliner's SNAC approach is equivalent to the multiadditive utility function in Fishburn and Farquhar (1982). Moreover, the indifference spanning approach described by Farquhar and Fishburn (1983) shows how to construct such evaluation functions in general. The next logical step is to determine if evaluation functions can be further improved using one of the fractional hypercube utility functions (Farquhar (1975, 1977, 1980)).

The evaluation function is a key element not only in game-playing programs but also in other expert systems that rely on efficient search techniques for generating good options. AI researchers are quickly discovering that static evaluation functions based upon linear, additive combinations of features do not perform at the level of human experts. While linearity is easily relaxed, the assumption of additivity poses some difficulties.

Berliner (1979, 1980), Lee and Mahajan (1988), and others who have built truly expert systems note that higher levels of performance often require a deep understanding of feature interdependencies as well as their representation in an evaluation function. The challenge for AI researchers is to identify crucial feature interactions and then to specify them in a nonadditive evaluation function. Farquhar (1989) explores the construction of such functions using representation theorems from MAUT; the methodology is generally applicable and not domain specific.

In summary, MAUT and AI are closely related in several ways. Although more general utility representations have been slow to be applied in decision analysis, there are substantial gains in expertise from using these evaluation functions in AI systems. On the other hand, MAUT might be extended by drawing on new AI paradigms for evaluation (e.g., Berliner and Ebeling (1986), Michalski, Carbonell, and Mitchell (1983, 1986)). We look forward to more exchange between these two fields in the future.

ACKNOWLEDGMENTS

This research was supported by the Office of Naval Research under Contracts #N00014-84-K-0558, #N00014-87-K-0201, and #N00014-89-J-1568.

REFERENCES

Barr, A., and E. A. Feigenbaum, eds. (1981). The Handbook of Artificial Intelligence, Volume I, William Kaufmann, Los Altos, California.

Barr, A., and E. A. Feigenbaum, eds. (1981). The Handbook of Artificial Intelligence, Volume II, William Kaufmann, Los Altos, California.

Bell, D. E., and P. H. Farquhar (1986). "Perspectives on Utility Theory," Operations Research, 34.

Berliner, H. (1979). "On The Construction of Evaluation Functions for Large Domains," Proceedings of The Sixth International Joint Conference on Artificial Intelligence, Tokyo, Japan, 53-55.

Berliner, H. (1980). "Computer Backgammon," Scientific American, June, 64-72.

Berliner, H., and C. Ebeling (1986). "The SUPREM Architecture: A New Intelligent Paradigm," Artificial Intelligence, 28, 3-8.

Charniak, E., and D. McDermott (1985). Introduction to Artificial Intelligence, Addison-Wesley, Reading, Massachusetts.

Cohen, M. D., and R. Axelrod (1984). "Coping with Complexity: The Adaptive Value of Changing Utility," American Economic Review, 74, 30-42.

Cohen, P. R., and E. A. Feigenbaum, eds. (1982). The Handbook of Artificial Intelligence, Volume III, William Kaufmann, Los Altos, California.

Cyert, R. M., and M. H. DeGroot (1975). "Adaptive Utility," in R. H. Day and T. Groves, eds., Adaptive Economic Models, Academic Press, New York, 223-246.

DeGroot, M. H. (1983). "Decision Making with an Uncertain Utility Function," in B. P. Stigum and F. Wenstop, eds., Foundations of Utility and Risk Theory with Applications, Reidel, Dordrecht, Holland, 371-384.

Farquhar, P. H. (1975). "A Fractional Hypercube Decomposition Theorem for Multiattribute Utility Function," Operations Research, 23, 941-967.

Farquhar, P. H. (1977). "A Survey of Multiattribute Utility Theory and Applications," TIMS Studies in The Management Sciences, 6, 59-89.

Farquhar, P. H. (1980). "Advances in Multiattribute Utility Theory," Theory and Decision, 12, 381-394.

Farquhar, P. H. (1981). "Multivalent Preference Structures," Mathematical Social Sciences, 1, 397-408.

Farquhar, P. H. (1983). "Research Directions in Multiattribute Utility Analysis," in P. Hansen, ed., Essays and Surveys on Multiple Criteria Decision Making, Springer-Verlag, New York, 63-85.

Farquhar, P. H. (1984). "Utility Assessment Methods," Management Science, 30, 1283-1300.

Farquhar, P. H. (1986). "Adaptive Preferences in Decision Making," working paper, Carnegie-Mellon University, Pittsburgh, Pennsylvania.

Farquhar, P. H. (1989). "Machine Learning of Heuristic Evaluation Functions," working paper, Carnegie-Mellon University, Pittsburgh, Pennsylvania.

Farquhar, P. H., and P. C. Fishburn (1981). "Equivalences and Continuity in Multivalent Preference Structures," Operations Research, 29, 282-293.

Farquhar, P. H., and P. C. Fishburn (1983). "Indifference Spanning Analysis," in B. P. Stigum and F. Wenstop, eds., Foundations of Utility and Risk Theory with Applications, Reidel, Dordrecht, Holland, 443-459.

Fishburn, P. C., and P. H. Farquhar (1982). "Finite-Degree Utility Independence," Mathematics of Operations Research, 7, 348-353.

Holland, J. H. (1986). "Escaping Brittleness: The Possibilities of General-Purpose Learning Algorithms Applied to Parallel Rule-Based Systems," in R. S. Michalski, J. G. Carbonell, and T. M. Mitchell, eds., Machine Learning: An Artificial Intelligence Approach, volume II, Morgan Kaufmann, Los Altos, California.

Holland, J. H., and J. S. Reitman (1978). "Cognitive Systems Based on Adaptive Algorithms," in D. A. Waterman and F. Hayes-Roth, eds., Pattern-Directed Inference Systems, Academic Press, New York, 313-329.

Humphreys, P., and D. Berkeley (1985). "Handling Uncertainty: Levels of Analysis of Decision Problems," in G. Wright, ed., Behavioral Decision Making, Plenum, New York, 257-282.

Keeney, R. L., and H. Raiffa (1976). Decisions with Multiple Objectives: Preferences and Value Tradeoffs, John Wiley & Sons, New York.

Lehner, P. E., M. A. Probus, and M. L. Donnell (1985). "Building Decision Aids: Exploiting The Synergy Between Decision Analysis and Artificial Intelligence," IEEE Transactions on Systems, Man, and Cybernetics, SMC-15, 469-474.

Lee, K.-F., and S. Mahajan (1988). "A Pattern Classification Approach to Evaluation Function Learning," Artificial Intelligence, 36, 1-25.

Madni, A., M. Samet, and D. Purcell (1985). "Adaptive Models in Information Management," in S. J. Andriole, ed., Applications in Artificial Intelligence, Petrocelli, Princeton, New Jersey, 279-294.

Michalski, R. S., J. G. Carbonell, and T. M. Mitchell, eds. (1983). Machine Learning: An Artificial Intelligence Approach, Tioga, Palo Alto, California.

Michalski, R. S., J. G. Carbonell, and T. M. Mitchell, eds. (1986). Machine Learning: An Artificial Approach, Volume II, Morgan Kaufmann, Los Altos, California.

Samuel, A. L. (1959). "Some Studies in Machine Learning Using The Game of Checkers," IBM Journal of Research and Development, 3, 211-230.

Shortliffe, E. H., B. G. Buchanan, and E. A. Feigenbaum (1979). "Knowledge Engineering for Medical Decision Aids," Proceedings of The IEEE, 67, 1207-1224. Reprinted in W. J. Clancey and E. H. Shortliffe, eds., Readings in Medical Artificial Intelligence: The First Decade, Addison-Wesley, Reading, Massachusetts, 1984.

Smith, S. F. (1984). "Adaptive Learning Systems," in R. Forsyth, ed., Expert Systems: Principles and Case Studies, Chapman and Hall, London.

Tesfatsion, L. (1980). "A Conditional Expected Utility Model for Myopic Decision Makers," Theory and Decision, 12, 185-206.

Tesfatsion, L. (1982). "A Dual Approach to Bayesian Inference and Adaptive Control," Theory and Decision, 14, 177-194.

Waterman, D. A. (1986). A Guide to Expert Systems, Addison-Wesley, Reading, Massachusetts.

Wellman, M. P. (1985). "Reasoning about Preference Models," Technical Report 340, Laboratory for Computer Science, Massachusetts Institute for Technology, Cambridge, Massachusetts.

Wellman, M. P. (1986). "Reasoning about Assumptions Underlying Mathematical Models," in J. S. Kowalik, ed., Coupling Symbolic and Numerical Computing in Expert Systems, North-Holland, Amsterdam.

Weiss, J. J. (1985). "GenTree: A Content-Oriented Aid to Decision Problem Structuring," in S. J. Andriole, ed., Applications in Artificial Intelligence, Petrocelli, Princeton, New Jersey, 453-477.

White, C. C. (1984). "Sequential Decisionmaking under Uncertain Future Preferences," Operations Research, 32, 148-168.

White, C. C. (1989). "A Survey on the Integration of Decision Analysis and Expert Systems," working paper, University of Virginia, Charlottesville, Virginia.

White, C. C., and E. A. Sykes (1986). "A User Preference Guided Approach to Conflict Resolution in Rule-Based Expert Systems," IEEE Transactions on Systems, Man, and Cybernetics, SMC-16, 276-278.

Winston, P. H. (1984). Artificial Intelligence, second edition, Addison-Wesley, Reading, Massachusetts.

Wisudha, A. D. (1985). "Design of Decision-Aiding Systems," in G. Wright, ed., Behavioral Decision Making, Plenum, New York, 235-256.

A MULTICRITERIA STRATIFICATION FRAMEWORK FOR UNCERTAINTY AND RISK ANALYSIS

Judith Barlow
Department of Computer Science
University of Liverpool
Liverpool, England L69 3BX
email: jab@mva.cs.liv.ac.uk

Fred Glover
Graduate School of Business
University of Colorado
Boulder, Colorado 80309-0419

We present a new decision framework for handling uncertainty and risk based on stratifying uncertainty in a model whose structure parallels that of multicriteria optimization. The approach incorporates multiple risk perspectives without homogenizing them into one representative function. Rather than requiring a priori omniscience about how factors will interact, it generates progressively amended decision templates based on model outcomes.

INTRODUCTION

Multicriteria optimization methods have recently been used to design more efficient expert systems. They have been used to to effectively reduce the size of the solution space in expert systems with large knowledge bases.[DuBois1989] Multicriteria optimization methods have also been used in domains where the combined expertise of a panel of experts is required.[Brans1986] The nature of expert systems is to provide advice in areas where the effects of risk and uncertainty are important considerations. Certainty calculi and heuristics are often used to compensate for uncertainty.[Cohen1983] Multicriteria optimization provides an adaptive and evolutionary approach for handling uncertainty and risk.

Uncertainty and risk are among the most challenging problems in decision making. Because major decisions in real world settings must be made in the face of uncertain information, the proper handling of uncertainty is crucial for effective operation. Often the approach chosen for handling uncertainty involves dangerous oversimplification. Competing investment alternatives are frequently compared, for example, in terms of their relative expected values, disregarding other features of the populations they represent, including the attendant element of risk. In order to make intelligent choices among different investment alternatives, the determining factors underlying uncertainty and risk must somehow be captured and accounted for.

The approaches to uncertainty and risk analysis currently in favor assume that it is appropriate to use a single (point) estimate of price, cost, or profitability coupled with an associated risk label. Often this estimate represents a weighted average of values derived from some simulated or guessed at collection of possible future conditions. The limitations inherent in collapsing the range of potential outcomes into a single estimate (whose behavior is presumably governed by a probability distribution which in many cases is unknown), becomes compounded in customary models by another serious flaw. The effects of risk and uncertainty are defined without reference to model interactions generated by objectives and constraints -- interactions that can render conclusions based on simple probability distributions meaningless.

Rationality and risk orientation are not isolated or pure attributes, but depend on the manner in which inputs get transformed into outputs. This in turn depends on a collection of technical and economic interrelationships. Our approach attempts to take account of such interrelationships by means of a mathematical programming model that treats uncertainty by a "divide and conquer" strategy of stratifying its influence. Rational behavior requires experience with how the first tentative suppositions turn into prescriptions, and the levels of performance that result. To the extent that the real world itself is not subject to manipulation, this experience is gained in our approach by direct interaction with the model.

To implement the process, the decision maker selects inputs that intuition suggests will lead to the outputs desired. The model then provides outputs that more closely reflect what reality may be expected to provide, and these will typically be somewhat different than those originally anticipated -- insofar as the decision maker cannot be expected to fathom the complex influence of the constraints (any more than one would be expected to guess the solution to a complex system of equations). The model accordingly provides an essential, if somewhat chastening, form of feedback. On the basis of this feedback, the decision maker is progressively able to revise her mental map of the setting in which she operates, to probe the connections between tentative choices and outcomes, and eventually to refine intuition until good choices are found. The process transforms blind groping into rational behavior.

A MCDM Framework for Uncertainty and Risk Analysis

The core of our approach is to adapt the framework of multicriteria optimization to provide a way to treat uncertainty without relying on special assumptions about an underlying probability distribution. The approach may be applied either to a single decision maker or to a decision making panel. A fundamental tenet is that the perspectives of different experts with different risk postures need not be a source of irreconcilable conflict, nor reduce decision making to the act of flipping a coin, but may in fact offer a source of creative improvement in the decision process.

TRANSLATING UNCERTAINTY AND RISK INTO "UNCERTAINTY STRATA"

We describe our approach in the context of an investment decision problem. In this setting, a human or combined human/machine expert (e.g., making use of computer data bases or forecasting models) is invited to create stratifications of uncertainty by making different estimates of the return from each investment, according to varying levels of optimism and expectation. By means of these stratifications, the common notion of a "most likely" estimate is replaced by a collection of estimates, each gauged most likely for the range -- or perceived optimism level -- under consideration. Because optimism and expectation are simultaneously involved, these estimates unite the two elements of risk and uncertainty.

The resulting strata naturally lend themselves to analysis in the framework of multicriteria optimization. In particular, the goal of determining a solution that is optimal under the estimated conditions corresponding to each of these strata may be expressed as follows, corresponding to a common representation of a multiple objective problem:

$$\text{OPTIMIZE } Z_h = C_{h1}X_1 + C_{h2}X_2 + \cdots + C_{hn}X_n, \quad h = 1, \ldots, p$$

In the investment planning context, n identifies the number of different investments and p identifies the number of different strata, or levels, of estimates. Thus, in particular, the coefficient C_{hj} represents the return for the j-th investment at level h. (In a setting that employs a decision making panel, the returns associated with different values of h may embody the proposals of different experts whose estimates are applied to a shared set of levels.) We follow the convention whereby the investment decision variables, represented by the X_j 's, take on values between zero and one, identifying the degree to which an investment is chosen. Fractional values, while admissible in the present context, would be ruled out in a discrete optimization framework.

Restrictions governing feasible collections of investments in this example setting are expressed by linear inequality constraints of the form:

$$a_{i1}X_1 + a_{i2}X_2 + \cdots + a_{in}X_n \leq b_i, \; i = 1, ..., m.$$

The coefficient a_{ij} represents the amount of resource i used by investment j and b_i represents the maximum availability of resource i. Such constraints may represent budget limits, usage restrictions for component materials, joint restrictions on subsets of investments, user-defined portfolio diversification requirements and risk exposure limits.

To elaborate the manner in which the objectives may embody different risk perspectives, suppose the decision maker chooses to make three different types of estimates, by identifying most likely return at optimistic, pessimistic and neutral levels, thus giving rise to three objective functions. The conservative objective, if optimized without consideration for the others, would result in a solution which is optimal relative to a pessimistic or risk-averse perspective. The optimistic level objective would be expected to result in a different "optimal" solution compatible with a risk-seeking perspective. The remaining objective, which represents a middle-of-the-road or risk-neutral orientation, will typically result in an optimal solution that differs from those obtained for the other objectives.

METHODOLOGY

To adapt multiple criteria optimization to the treatment of risk and uncertainty by means of our formulation, there are a wide variety of approaches on which to draw.[Dyer1973, Evans1973, Mulvey1984, Zionts1976, Zeleny1982] We propose a hybrid approach deriving from our previous applications of multicriteria optimization to practical decision problems.[Glover1987, Barlow1985] Our approach, which is an extension of the STEM method, [Benayoun1971, Steuer1985] appears especially suited for adaptation to the risk and uncertainty context. The first phase of the procedure may be expressed by means of the following simple outline:

Phase I

Step 1. Evaluate each of the objectives (risk perspectives) individually, solving the linear program which optimizes the objective without reference to the others.

Step 2. Identify optimal values of the investment decision variables for the different risk perspectives, and evaluate each such solution across all other objectives by plugging it into the alternative objective functions.

Step 3. If a solution is found that is optimal for all objectives (evaluation criteria), then this is presented to the decision maker for scrutiny and the procedure enters the interactive mode (described subsequently).

Step 4. If none of the individual solutions is optimal for all objectives, as may generally be expected, the procedure identifies a compromise solution. This solution is generated to be equitable in terms of the trade-offs between the different objectives and to be efficient in terms of assuring that no single objective can attain an improved value without entailing that another must receive a value that is worse.

The key to applying the procedure of Phase I rests on identifying the form of the compromise solution of Step 4. A compromise solution which meets the requirements expressed in this step is obtained by introducing a new decision variable D which is weighted and incorporated into each of the original objectives to produce the following transformed problem.

MINIMIZE D

subject to

$$C_{h1}X_1 + C_{h2}X_2 + \cdots + C_{hn}X_n + U_hD = Z_h, \ h = 1, ..., p$$

$$a_{i1}X_1 + a_{i2}X_2 + \cdots + a_{in}X_n \leq b_i, \ i = 1, ..., m$$

$$1 \geq X \geq 0, \ j = 1, ..., n$$

The formulation above is a goal program in which the Z_h values identify the optimal level of each of the original problem objectives. Under the assumption that the original objectives involve maximizing return, D can only attain a minimum value of 0 if all such objectives simultaneously reach their maximum. The coefficient U_h of D associated with the objective at level h is assigned a positive value, which can simply be 1 for all h if the model represents minimizing the greatest deviation from any objective, and otherwise can be a factor that inversely reflects the relative importance the decision maker attaches to the level under consideration -- that is, the degree to which one is concerned about obtaining a final solution that takes account of this level of optimism. By this means, the model can be adjusted to handle different types of risk postures, offering a flexibility not customarily found among approaches to risk and uncertainty. The U_h values are among the parameters that may usefully be adjusted during later stages of refinement.

From the preceding observations, it follows that whenever the joint attainment of all of the goals (Z_h values) is impossible, the solution of the goal program balances trade offs among the competing objectives of the original problem to achieve a best compromise. By our approach of representing different risk perspectives within the framework of these objectives, the compromise solution yields a suggested investment mix which is equitable with respect to each of the different evaluation criteria, and hence with respect to each of the original stratifications of uncertainty.

The concept of equitability employed here is somewhat broader than that of a simple "minmax" notion because of the ability to impart varying degrees of influence to the different objectives by means of the U_h coefficients. The initial compromise solution then becomes the starting point for the second phase, which introduces a process of interactive refinement.

Phase II -- Interactive Refinement

The compromise solution inherited from Phase I gives the decision maker a frame of reference that discloses the consequences of interactions among most likely measures of return at different risk levels (derived from the decision maker's choice of "levels of optimism" in defining the original uncertainty strata). This information provides a basis for further exploration in several ways.

The decision maker will customarily find that the compromise solution generated in Phase I will cause certain constraints to come into play that were not previously anticipated to be relevant. These may include constraints whose specific form could not have been known before generating and solving the model. For example, the investment mix obtained by solving the model in Phase I may yield too small a return on investment from the conservative perspective, once this return is compared to the optimum return that is attainable according to the objective function defined by this same perspective. In response, the decision maker may introduce a constraint which sets a lower acceptable bound on the return from the conservative (risk-averse) estimate.

The objectives deriving from the other perspectives may be bounded in much the same way. Similarly, other constraining requirements may be introduced to fine-tune the model. Phase II thus incorporates these changes in the model solved in Step 4 of Phase I, and resolves the model. Once this occurs, the process is open to be repeated if the outcome discloses additional constraints the decision maker would seek to impose.

New constraints may also be introduced provisionally, as a means of carrying out a "what if" analysis. Likewise, the decision maker is free to adjust the U_h values -- or to incorporate a further level of the objective function -- in the process of discovering those levels that achieve satisfactory or unsatisfactory ranges, or whose influence appears too great or too small.

This interactive approach allows for model revision based on model outcome and does not require a priori guesses about the way factors will interact. It allows new estimates and bounds to be generated in response to information provided by the model solutions. The result of this interaction is a clearer understanding of the interrelationships in the decision environment.

Applied to a panel of decision makers, this approach may be employed as a tool to induce those with different perspectives to progress toward group understanding and consensus on levels of uncertainty and risk orientation. By experimenting and observing interactions in the model, the decision makers can assess the realism of the different estimates and the sensibleness of how they weight and bound their investment options.

Uncertainty Stratification Applied to Project Selection

The interactive decision framework is illustrated using a project selection example involving fifteen projects. Stratification yielding estimates of profitability for the projects are given from three different risk perspectives: conservative or risk-adverse, liberal or risk-seeking and moderate or risk-neutral. This information is depicted in Table 1.

	Project Number														
	1	2	3	4	5	6	7	8	9	10	11	12	13	14	15
Optimistic	100	112	130	100	162	101	112	98	84	114	85	105	65	87	98
Neutral	55	70	84	68	108	81	90	84	78	109	70	93	63	78	92
Conservative	30	48	59	47	81	56	65	61	55	77	63	90	60	72	90

Table 1: Profitability estimates for optimistic, neutral, and conservative risk perspectives.

Potential projects also have budget requirements which are summarized in Table 2.

PROJECT NUMBER															
	1	2	3	4	5	6	7	8	9	10	11	12	13	14	15	Budget
Cost	10	12	14	11	18	12	14	13	12	17	13	18	11	14	17	75

Table 2: Project budget requirements.

When the fifteen projects are evaluated with respect to the liberal (risk-seeking) profit estimates and the constraining requirements, the project mix which maximizes expected profit includes projects 1, 2, 3, 4, and 5 at an investment level of 1 and project 6 at an investment level of .833. Optimization using the moderate estimate yields projects 6, 7, 8, 9, and 10 at a level of 1 and project 4 at a level of .6364. The conservative perspective yields an optimal project mix consisting of projects 11, 12, 13, 14, and 15 at a level of 1 and project 8 at a level of .1538. This information is summarized in Table 3.

Because the three different risk perspectives each suggest different courses of action, Phase I of our approach undertakes to determine a compromise solution which is equitable with respect to the three different risk perspectives and feasible with respect to budget limitations and project selection rules. We have selected the U_h each to have the value of their objectives at optimality, thus weighting all strata with respect to the relative magnitude of its optimal attainment level. The compromise solution that results by solving this formulation includes projects 5, 6, 7, and 13 at level 1, project 3 at level .6757 and project 15 at level .62. This initial compromise solution similarly is shown in Table 3.

The decision framework presented here has been implemented on a VAX-11/780. The decision makers are prompted for their evaluations of the potential investment alternatives under different risk perspectives. They are also allowed to define customized constraining relationships and/or choose from a menu of common constraint types. Among the constraint types available on the menu are budget requirements (both minimum and/or maximum amounts), resource supply and demand relationships, specifying the minimum (or maximum) number of projects to select, specifying projects which can never be chosen together (mutually exclusive) or those which must be chosen together (dependent projects), and certain additional broader project compatibility requirements.

A MCDM Framework for Uncertainty and Risk Analysis

	OPTIMISTIC OBJECTIVE	NEUTRAL OBJECTIVE	CONSERVATIVE OBJECTIVE	COMPROMISE SOLUTION
MAXIMUM PROFIT	688.17	485.27	384.39	----
WORST CASE PROFIT	455.07	408.92	321.83	----
optimal objective:	conservative	conservative	optimistic	----
Basic values at Optimality:				
Project 1	1	0	0	0
Project 2	1	0	0	0
Project 3	1	0	0	.6757
Project 4	1	0	0	0
Project 5	1	.6364	0	1
Project 6	.8333	1	0	1
Project 7	0	1	0	1
Project 8	0	1	.1538	0
Project 9	0	1	0	0
Project 10	0	1	0	0
Project 11	0	0	1	0
Project 12	0	0	1	0
Project 13	0	0	1	1
Project 14	0	0	1	0
Project 15	0	0	1	.6200

Table 3: Optimal Project Mix Solutions.

Output includes a graphical display for each objective that presents a scale which ranges from the optimum objective value at the upper end to a "relative minimum" at the lower end, where the latter is the least value for that objective over the feasible region determined by the current constraining relationships. The current compromise solution is graphed as a point on the scale which lies between these extreme values. As the decision makers adjust the solution by adding new constraining requirements, the effect on each of the objectives is easily detected by the change in the current value of each objective as depicted on the new graphical output.

One consequence of a change may be to induce an expert who has a special interest in one set of objectives -- corresponding to her perception of how uncertainty ought to be stratified -- to "bargain" with other experts (interested in other objectives and uncertainty strata) by testing the effects of potential compromises. The outcome of this process is a simulation of the dynamics of group decision making leading to a group consensus through enlightened bargaining.

The interactive process of defining new constraints and/or removing or relaxing existing constraints continues until a solution is chosen. The mix of projects chosen reflects a compromise which the decision maker (or competing experts) deems to be "best" with respect to the multiple criteria that embody the alternative stratifications of uncertainty.

CONCLUSIONS

The approach we propose gives the decision maker an expanded context for responding to uncertainty and risk. Expressing these factors by means of multiple, simultaneously interacting objectives based on stratifying uncertainty offers a conceptual framework whose flexibility transcends adherence to a presupposed risk posture (such as minimize regret, Laplace-Savage, etc.). Additional power to explore and shape an informed decision strategy is provided in this approach by allowing the decision maker to progressively redefine the objectives that express the links between risk and uncertainty and to control their interrelationship by the imposition of bounds on performance relative to specific strata (as embodied in optimistic, moderate, or pessimistic projections).

In short, our approach is adaptive and evolutionary. It does not require an implicitly supposed omniscience about the interaction of uncertainty with problem goals and constraints, but lets new estimates and bounds be generated in response to information supplied to the decision maker about the result of that interaction. The decision maker is not compelled by the presence of uncertainty to settle for a shot-in-the-dark decision, but is given tool to "divide and conquer" the uncertain environment by means of stratifying its performance evaluations into a multicriteria framework, and then t probe the interrelationships disclosed by generating a monitored sequence c best compromise solutions.

References

Barlow1985.
Judith Barlow and Ruth Maurer, "Application of a Multicriteria Solution Method to Problems in the Public Sector," *Colorado School of Mines Working Paper Series, ME-WP 1013*, November 1985.

Benayoun1971.
R.J. de Montgolfier Benayoun, J. Tergny, and O. Laritchev, "Linear Programming with Multi-Objective Functions: Step Method (STEM)," *Mathematical Programming, 1, 3*, pp. 366-375, 1971.

Brans1986.
J. P. Brans, Ph. Vincke, and B. Mareschal, "How to Select and How to Rank Projects: The PROMETHEE Method," *European Journal of Operational Research, 24, 6*, pp. 228-238, 1986. q.

Cohen1983.
P Cohen and M Grinberg, "Theory of Heuristic Reasoning About Uncertainty," *AI Magazine, 4, 2*, pp. 17-24, 1983.

DuBois1989.
Ph. DuBois, J. P. Brans, F. Cantraine, and B. Mareschal, "MEDICIS: An Expert System for Computer-Aided Diagnosis Using the Promethee Multicriteria Method," *European Journal of Operational Research, 39, 3*, pp. 284-292, April 1989.

Dyer1973.
James S. Dyer, "An Empirical Investigation of a Man-Machine Interactive Approach to the Solution of Multiple Criteria Problem," in *Multiple Criteria Decision Making*, ed. M. Zeleny, pp. 202-216, University of South Carolina Press, Columbia, 1973.

Evans1973.
J. P. Evans and R. P. Steuer, "Generating Efficient Extreme Points in Linear Multiple Objective Programming: Two Algorithms and Computing Experience ," in *Multiple Criteria Decision Making*, ed. M. Zeleny, pp. 54-72, University of South Carolina Press, Columbia, 1973.

Glover1987.
Fred Glover and Fred Martinson, "Multiple Use Land Planning and Conflict Resolution by Multiple Objective Linear Programming," *European Journal of Operations Research, 28, 3*, pp. 343-350, 1987.

Mulvey1984.
John M. Mulvey, Marsha D. Anderson, and Roderic G. March, "Solving Multiobjective Problems Via Interactive Dialog," *Princeton University Working Paper EES-81-8*, July 1984.

Steuer1985.
> R. E. Steuer, *Multiple Criteria Optimization: Theory, Computation, and Application*, John Wiley & Sons, New York, 1985.

Zeleny1982.
> Milan Zeleny, *Multiple Criteria Decision Making*, McGraw-Hill, New York, 1982.

Zionts1976.
> S. Zionts and J. Wallenius, "An Interactive Programming Method for Solving the Multiple Criteria Problem," *Management Science, 22, 6,* pp. 652-663, 1976.

DISPUTE MEDIATION: A COMPUTER MODEL

Katia Sycara

School of Computer Science
Carnegie Mellon University
Pittsburgh, PA 15213

ABSTRACT

This paper integrates Artificial Intelligence techniques with decision theoretic methods to address the problem of finding compromise solutions to multi-agent conflicts through mediation/negotiation. This is a difficult problem since conflict resolution is an ill-defined and complex process, the compromise choices that a problem solver has for continuum-valued issues are infinite, and the agents need to be persuaded to shift their positions during problem solving. Previous approaches have been based on quantitative models that are inflexible and inaccessible to the practitioners. The proposed conflict resolution model integrates reasoning from past cases similar to the current conflict, the use of multi-attribute utility theory, and reasoning to accommodate idiosyncratic behavior of the agents. The model has been implemented in a computer program, the PERSUADER, that functions as a mediator in hypothetical labor negotiations. It suggests appropriate settlements to the disputants. If a suggested compromise is rejected, the PERSUADER either improves the compromise or generates persuasive arguments to change the opposing party's "view" of the settlement.

INTRODUCTION

The main processes through which disputes are resolved are forms of negotiation, mediation, arbitration, or adjudication. In negotiations, the two or more sides interact to try to come to an agreement. The agreement may be considered mutually desirable or may be forced by one party on the other through threats of negative sanctions. In mediation, a non-disputant tries to convince the disputants to reach an agreement. If a non-disputant has the power to impose a solution on the parties, and he/she has been chosen by the disputants, then he/she is referred to as an arbitrator. If the non-disputant has been imposed on the parties, then

---

[1]This research was funded in part by the Army Research Office under contract No. DAAG 29-85-K-00230.

he/she is referred to as a judge. Mediators generally seek solutions that will be considered mutually satisfying by the parties. Arbitrators and judges seek solutions that are considered correct in accordance with a body of past practice, common sense or rules, and the solutions need not be satisfying to the parties.

The negotiation/mediation process is ill-structured primarily because there is no recipe for successful negotiations. The outcome of the process depends on the negotiator's skills, the parties' behaviors, and on the exhaustive and systematic analysis of the problem.

A famous labor mediator, William Simkin, in a semifacetious mood, once listed the following qualities as desirable in the "ideal" mediator [Simkin 71]:

1. The patience of Job
2. The sincerity and bulldog characteristics of the English
3. The wit of the Irish
4. The physical endurance of the marathon runner
5. The broken-field dodging abilities of a halfback
6. The guile of Machiavelli
7. The personality-probing skills of a good psychiatrist
8. The confidence-retaining characteristics of a mute
9. The hide of a rhinoceros
10. The wisdom of Solomon

In a more serious tone, he extended the list to include:

11. Demonstrated integrity and impartiality
12. Basic knowledge and belief in the collective bargaining process
13. Firm faith in voluntarism in contrast to dictation
14. Fundamental belief in human values and potentials, tempered by ability to assess personal weaknesses as well as strengths
15. Hard-nosed ability to analyze what is available in contrast to what might be desirable
16. Sufficient personal drive and ego, qualified by willingness to be self-effacing

An automated mediator could clearly possess at least qualities 1, 2, 4, 8, 9, 11, 15, the first half of 12 and the second half of 16. Automation of the mediation function has many advantages for the more efficacious resolution of conflicts. These advantages include objectivity, fairness, confidentiality, the ability to help the parties maintain rationality in decision making avoiding emotional clouding of issues, the ability to explore a much larger set of potential compromises, and the rapid generation of proposals.

Decision science has given rise to various quantitative models of the negotiation process. Depending upon the type of assumptions considered, it is possible to apply multiobjective decision-making [Jarke 87], goal programming [Kersten 85], game theory and meta-game analysis [Frazer

84], aspiration theory [Kersten 86], and stochastic control [Fogelman 83]. These approaches result in the development of well-defined modes and procedures which are formally elegant but could be inflexible and inaccessible to the practitioners [Raiffa 82]. Combining decision theoretic approaches with AI techniques allows models to take advantage of the strength of both approaches.

We believe that an integration of qualitative and quantitative models will result in systems that are both robust and flexible. We have used Artificial Intelligence techniques for problem structuring, representation and processing, and relaxed assumptions of human rationality integrated with the use of multi-attribute utility theory to create a model of resolution of multi-agent conflicts through negotiation/mediation. In contrast to knowledge-based work on negotiation/mediation that has concentrated on providing support for human negotiators [Matwin 87, Shakun 88], our work concentrates on modeling the complexities and dynamic aspects of negotiation, and automating the gradual modification of negotiating positions through tradeoffs and persuasive argumentation to achieve a compromise satisfactory to all parties. Our model has been implemented in a computer program, the PERSUADER which resolves labor management disputes [Sycara 87a]. The PERSUADER, emulating the behavior of human mediators, negotiates with each of the disputants to arrive at a mutually satisfying settlement.

The PERSUADER performs its task by integrating the following reasoning methods: (a) Case-Based Reasoning (CBR), which consists in retrieving from memory and adapting previous compromises of similar disputants, (b) Preference Analysis, which use the utilities that the disputants associate with the issues under negotiation to rank possible compromises, (c) Situation Assessment, which recognizes exceptional situations in terms of their abstract causal structure and accesses knowledge structures, called SAPs, that embody the causal knowledge and provide domain independent compromise strategies, and (d) Use of Rules/heuristics to come up with appropriate compromise modifications, when previous cases are not available.

The PERSUADER facilitates dispute resolution by performing the following functions:

- **removing communication barriers** by providing a confidential channel and by communicating the mediator's assessment of the rigidity of positions of a party to its opposite side. The above two functions are especially important when the atmosphere in the negotiations is hostile, the parties' positions have become polarized and emotions are running high.

- **assessing the party's priorities** with respect to the dispute issues/goals of each party. This assessment is the result of discussions the mediator has with the disputants as well as knowledge of the domain the mediator possesses. In our model, the assessment of the parties' priorities is done using a combination of Preference Analysis, previous mediation cases, and feedback from the disputants.

- **deflating extreme positions**. The most difficult cases are the ones where high priorities of both sides point in opposite directions. The parties' positions are extreme but both sides have to give something if an agreement is to be reached. The mediator, in separate meetings with the parties, tries to deflate their positions, so that a compromise can be effected. In our work, this function is performed through persuasive argumentation.

- **recommending a "package" settlement**, i.e. a settlement that expresses potentially acceptable compromises and tradeoffs of the parties. In our work, the generation of a settlement is the result of using a combination of Case-Based Reasoning and Preference Analysis.

- **justifying recommendations**. Often the disputants, blinded by their values and criteria, cannot recognize why a proposed compromise may be the best under the circumstances. If a mediator is able to generate justifications for the desirability of the compromise from a disputant's point of view, then the compromise has more chance of being accepted. In our model justification is provided through presentation of appropriate previous settlements.

- **exploring feasible alternatives** so as to optimize the proposed compromise. A memory for past mediation cases provides a rich repository of such alternatives.

- **offering solutions in exceptional circumstances** by relaxing assumptions about the agents' rationality and taking into consideration idiosyncratic behavior of the agents. In our work, this is done using Situational Assessment.

- **modifying a rejected compromise** to make it more acceptable to the rejecting party without making it unacceptable to the party that had previously accepted it. This is done using previous cases and modification rules. A modified compromise is evaluated using an appropriate criterion of improvement.

The values that enter the search for a suitable compromise are so many that mediators themselves have not been able to elucidate a well-defined procedure that a mediator can follow. There is no *typical* or *model* mediator behavior that can be codified and emulated. In our work, we have tried to capture the non-typicality of mediator behavior by using a combination of reasoning methods (case-based reasoning, preference analysis, situational assessment) and knowledge structuring mechanisms (graphs, generalized episodes, SAPs, utility curves).

MEDIATION KNOWLEDGE

A mediator uses two kinds of knowledge to perform his functions: (a) domain knowledge and (b) reasoning knowledge. Domain knowledge includes knowledge about disputes, disputants, disputants' goals, dispute context and dispute settlements. Reasoning knowledge includes the mediator's goals, knowledge he needs to assess the "fairness" of a solution, and knowledge he needs to improve a rejected solution.

Some of the reasoning knowledge that a mediator uses in our model has its conceptual origins in standards that are commonly used to justify arbitration awards [Elkouri 72]. The standard that is most frequently used is the *prevailing practice* standard, namely the bargaining behavior of similar disputants. In our work, this standard gets abstracted to *reasoning from precedent cases*. This is a suitable method for the domain since previous cases represent good solutions to the difficulties that are endemic to finding acceptable compromises in multi-agent conflicts.

Knowledge acquisition in the PERSUADER is based on four main sources: (1) the labor relations literature including books, journals, newspapers and magazines, (2) published arbitration awards, where the arbitrator cites the facts and criteria used in the decision, and (3) two human expert mediators[2], and (4) the PERSUADER's Case-Based Reasoning process. Nine months were spent tapping these sources and making decisions on the knowledge to be initially incorporated in the Case Knowledge Base. The system is started with a set of cases that are placed in its Knowledge Base. At the end of each problem solving session the PERSUADER's memory is updated with information from the newly resolved case as well as any new information that has been acquired during problem solving (e.g., from the parties' feedback). Thus, knowledge acquisition and learning is ongoing since it is a by-product of problem solving.

The PERSUADER's memory contains successes, i.e. cases that resulted in mutually accepted settlements in past similar circumstances. They are used as basis in resolving similar disputes. The PERSUADER also stores failures, i.e. cases where no settlement has been agreed upon in the allotted time, and their failure reason (if one can be found). Failures can be recalled in situations that have similar features to the one where the failure occurred. Previous similar failures warn a problem solver about potential difficulties that might arise in the current case. Successes are indexed under salient domain features. Failures have three additional indices, an index indicating "failure", the failure cause, and the negotiation issue(s) that was involved. For more details on memory update, see [Sycara 87a]. As the PERSUADER sees more disputes, its

---

[2]One is a professor of Economics and the other came up through the ranks of the International Machinists' Union.

experiential base is enriched, making available a great variety of previous cases it can reason from. Thus, learning from experience is incorporated in the problem solving improving the quality of solutions. This mirrors real life problem solving situations.

Cases are organized hierarchically in memory around important concepts in the problem domain. In order to perform Case-Based Reasoning, cases need to be retrieved in terms of conceptual similarity. The basic idea behind conceptual similarity between two concepts is that they have important common attributes. Similar concepts are organized into larger groupings based on their similarities, and differentiated from each other in terms of their differences. The high level knowledge structure that we use to organize similar concepts in memory is called a *generalized episode* [Kolodner 84]. Generalized episodes organize cases into a hierarchical network whose nodes are either another generalized episode or an individual case. Each concept instance is retrieved based on its individual characteristics.

Another source of reasoning knowledge in mediation is the parties' utilities with respect to a proposed compromise. Mediators estimate the importance that various issues under discussion have for the disputants as well as the utilities that the disputants associate with various alternatives for each issue, and use this knowledge to estimate the overall satisfaction that a compromise will give each disputant. In the PERSUADER, this is done by a linear combination of a party's utilities with respect to the issues involved in the negotiation. Using an additive function is appropriate since it is *compensatory*, i.e. an increase in the utility of one issue can compensate for the decrease in the utility of any other issue. Thus, such a function models well the tradeoffs that a party is willing to make. Additive functions are the ones most frequently used in practice [Keeney 76].

Another source of knowledge that the PERSUADER uses is experiences from domains other than labor relations. Such knowledge is useful in problems that are exceptional, in the sense that they violate a reasoner's expectations. This knowledge is structured in representational units called Situation Assessment Packets (SAPs) that capture the causal structure of the present situation and contain domain independent conflict resolution strategies. For example, SAP IDEOLOGY which describes situations where an agent foregoes his chance of obtaining tangible benefits because of ideological motivations, violates a reasoner's assumptions about the economic rationality of agents.

A mediator uses domain knowledge and reasoning knowledge to come up with a potentially acceptable settlement that she proposes to the disputants. Very often, one of the disputants will reject the compromise. The situation resulting from this rejection is a *mediation impasse*. To resolve the impasse, a mediator either (a) tries to convince the rejecting party to accept the settlement, or (b) re-analyzes the situation and proposes a (hopefully) improved settlement. The interleaving of persuasive argumentation and re-analysis simulates the process of incrementally narrowing the differences of the disputants until an

Dispute Mediation: A Computer Model 255

agreement has been reached. Figure 1 presents the Mediation/Negotiation Process.

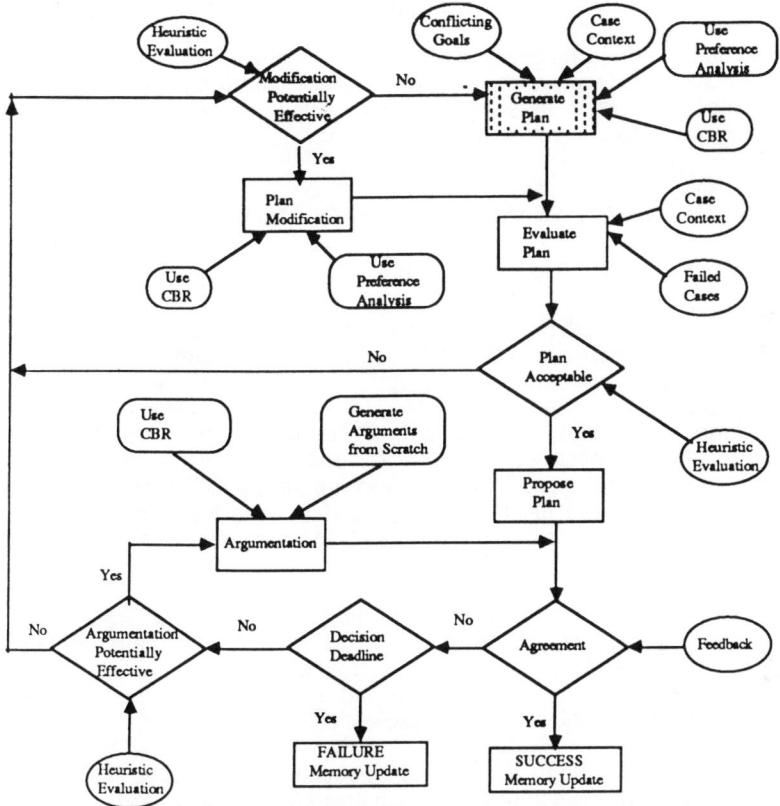

Figure 1: The Mediation/Negotiation Process

Frequently, when a party rejects a proposed contract it states the issues and reason for disagreement. A reason for rejecting proposed economic concessions, for example, may be that the company cannot afford them. Knowing the reason for contract rejection is useful for a mediator since it can be used to generate persuasive arguments or to modify the contract to make it more acceptable to the rejecting party.

The PERSUADER's architecture is particularly suited to negotiation, a task characterized by lack of a strong domain model, many and complex planning steps, and lack of certain or complete knowledge. Previous disputes successfully resolved can be re-used in a current conflict, thus improving performance. In addition, re-using a solution that has worked in similar previous circumstances (even if the reasoner does not know exactly why), provides a quality of solutions unobtainable by problem solvers that depend on strong domain models. Previous failures warn the problem solver of potential difficulties and offer recovery advice. Use of multi-attribute utilities enables the PERSUADER to evaluate case-based compromises and provides a computational framework for generation of persuasive arguments. Situation Assessment provides a method of accommodating irrational and idiosyncratic behavior of the agents. The PERSUADER's architecture makes it robust and flexible allowing it to access knowledge in various forms: generalized (inferential rules), specific (cases), quantitative (utilities).

DETERMINING AN INITIAL COMPROMISE

When Previous Cases Are Available

Mirroring the behavior of practicing mediators, the PERSUADER uses Case-Based Reasoning (CBR) [Kolodner et al. 85, Hammond 86, Sycara 87a] in trying to come up with acceptable compromises. It first retrieves appropriate previous cases from memory and selects the most appropriate from those retrieved. Then, based on similarities and differences between the current case and the selected precedent, it constructs a first approximation to a compromise, called the "ballpark" compromise. The ballpark compromise is evaluated and incrementally adapted so that it could be a potentially good compromise for the current dispute.

To retrieve a set of cases similar to the current one, the PERSUADER uses a set of salient features of the domain as indices (memory probes). The idea is that the tradeoffs and compromises that similar parties under similar circumstances were willing to make can constitute a first approximation to the sought after present compromise.

Of particular interest is the indexing structure of contracts, arguments, proposals and counterproposals. Contracts are predominantly used in construction of an initial proposal, previous arguments are used in the argumentation phase and previous proposals and counterproposals are used during improvement of a rejected settlement. Because the features that are used for retrieval are different in each task that the PERSUADER performs, the previous experiences that it reasons from are potentially different for each task. However, the case-based reasoning process behaves in the same way.

In the PERSUADER's memory, contracts are indexed by a multiplicity of features including the industry to which the company belongs, the geographical location, several features descriptive of the economic and political context within which the negotiations take place, features descriptive of the financial situation of the industry and the company, features describing the composition of the bargaining unit, and the international union to which the local belongs. Economic context includes attributes such as boom or recession, degrees of foreign and domestic competition, supply and demand of the job classifications in the

Dispute Mediation: A Computer Model 257

bargaining unit in the geographical area of interest. Negotiation context includes the attitudes of the parties towards each other (degrees of trust or hostility), the kind of relationship between the local and the international union, the historical context of the parties' relationship, the historical and current attitude of the union members (degree of militancy).

Arguments are indexed by (a) the goal/issue to which they pertain, (b) the persuadee type (union or company), (c) the persuader's argumentation goal and strategy. Argumentation goals (e.g. "change the importance that the persuadee attaches to an issue") are associated with the ways that a persuadee's beliefs can be affected by an argument. Argumentation strategies (e.g. "indicate possible unpleasant consequences of a persuadee's demand") are used to achieve the argumentation goals. Additional information associated with an argument is a list of goals to which the issue contributes. Moreover, information about the effectiveness of the argument depending on attributes of the parties and the negotiation environment is included.

The PERSUADER also keeps information about the process by which agreement was reached. The negotiation history is represented by a list of *impasses* that record the PERSUADER's proposals, the parties' feedback (acceptance or rejection), the rejection reason (if one can be found), the concessions that the parties are making at each step, the arguments that were used with the notation of whether they were successful or unsuccessful, and the PERSUADER's repairs to improve a rejected proposal. When memory is updated with the new case at the end of problem solving, the negotiation process is also part of the case and can be subsequently accessed. Impasses are indexed by the contract issue to which they pertain and by the cause (if known) of rejecting the proposal concerning the issue. Impasses are used to access possible repairs to a rejected proposal and to avoid potential failures during proposal evaluation.

Since a concept is characterized by many salient features, some of which are the same and some different for two concepts, *partial matches* between concepts result. CBR relies on similarity-based retrieval of cases. It is therefore necessary to be able to evaluate the degree of similarity (partial match) between two cases.

The criteria that the PERSUADER uses to evaluate whether a case is similar enough to the current dispute are salient features/dimensions of the domain. Each of the important domain features (e.g., industry, geographical location, job classification of union members, competitive position of the company) constitutes a similarity criterion. For each similarity criterion, a similarity hierarchy is constructed. The nodes at each level of this hierarchy represent concepts that are generalizations of their children nodes. A *similarity class* is defined at each level of the hierarchy and consists of the collection of siblings at that level. The members of a similarity class at a particular level exhibit a greater degree of similarity than members of two distinct similarity classes of that level. The members of a similarity class at the leaf level are the most similar since (a) they are similar (by belonging to the same similarity class), and (b) they contain the most specific information available in the domain model. The dynamic nature of attribute values adds another level of complexity. The PERSUADER stores the date of the current negotiation and the dates when contracts were signed. The closer in time a contract is to the current negotiation, the most similar it is considered to be.

In complex domains, each concept belongs to many semantic hierarchies, each of which is formed by a salient feature and its specializations. For example, a company belongs to the hierarchy defined by the "industry" dimension. Since the company has some geographical location, it also belongs to the semantic hierarchy defined by the "geographic location" dimension. To evaluate degrees of similarity, the PERSUADER uses an algorithm that takes into consideration both the importance of a dimension and the location of the concept along a similarity hierarchy. The algorithm is as follows:

1. Consider the most important problem dimension.

2. Use as precedents the cases that are siblings of the current case along the most important problem dimension. Out of those prefer the ones that are most recent. Go to step 4.

3. If step 2 returns the empty set, then use as precedents
   in a preorder traversal the cases that are ancestors of
   the current case along the most important problem dimension
   according to the following scheme:
   (a) prefer the closest ancestors to the current case
   (b) use more remote ancestors (up to six levels)
      up the hierarchy only if there are no closer ones

4. LOOP for problem dimensions taking values from 2 to n,
   out of the cases returned
   select the ones that would be siblings of the current case
   along the ith most important problem dimension

5. If step 4 returns more than one cases, select the first one
   from the list as most appropriate

This algorithm directs search first to competitors' contracts selecting the most recent ones for further consideration. The cutoff temporal value is three years[3]. If competitors' contracts are unavailable, similarity of industry is considered. There is a cutoff value of six levels in the similarity hierarchy, since it was felt that beyond that level similarity would be meaningless.

---

[3]This cutoff value was suggested by our experts as reasonable, since within three years it is likely that the situation will have changed so as to invalidate similarity comparisons.

Dispute Mediation: A Computer Model   259

The criteria for prioritization of the problem dimensions are domain-specific. The PERSUADER uses as the most important problem dimension the industry to which the company belongs. The industry dimension subsumes many other features, such as shared goals (e.g., to produce product x) and economic pressures. Secondary problem dimensions (in decreasing priority) are the job description of the workers, the economic climate (for the industry, the nation, and international, if applicable), conditions of competition, political considerations of the dispute, the geographical location of the company, features of the local and the company, and features of the international union of which the local is a member.

After the selection of the most similar precedent case, knowledge is extracted from its solution part (the contract) and adjusted through standard adjustments to form the "ballpark" solution. The values of the various issues in the precedent contract are checked to determine (a) whether the considered value violates any known applicability conditions, and (b) whether the value had resulted in failure. When the value under consideration has been deemed unacceptable for extraction, the PERSUADER decides whether the flaw lies in its reasoning or in unusual circumstances that make the case *atypical*.

By the time a mediator gets involved in a typical dispute the initial, potentially inflated demands, have been paired down to a reasonable level. The criterion used to determine case typicality is that if the demands under discussion are still within four to five times the value they have in a precedent (or industry average), then the case is typical, otherwise atypical[4]. For non-numerical demands, such as seniority and subcontracting, we have used an ordinal scheme with range from 1 to 10 to rank values of these demands. The scheme is based on the language used to qualify non-numerical demands in a survey of about fifty contracts (real and textbook examples) that we studied. For example, unlimited subcontracting has the value 10; subcontracting for limited time periods 9, subcontracting only if work is not lost to union members 8 etc. If the situation is atypical, case-based and analytic methods are inappropriate and the method of Situation Assessment is used. If the case is not atypical, the ballpark compromise is further adapted to the specifics of the current situation using both CBR and heuristics.

Before proposing the adapted compromise to the agents, the PERSUADER tries to *anticipate* potential difficulties with the contemplated compromise, so that it can avoid them. This is done through *intentional reminding* [Schank 82] of failed compromises where the conjunction of the solution's features are used as indices to retrieve failures that have the same features as the contemplated compromise. The knowledge that a solution has failed in the past can suggest to a reasoner the potential for failure if the solution is adopted in the current situation. If an associated repair is stored along with the retrieved

---

[4]This heuristic was suggested by our domain experts.

failure, the reasoner can apply the repair to the compromise to get a compromise that avoids the difficulty.[5]

An Example. Consider, for example, the PERSUADER trying to find a compromise for Muriel's Apparel Inc., a company that makes women's dresses and its union. The union wants 13% increase in piece rates, 7% increase in pensions, and no subcontracting. The company wants no increase in piece rates, no pension increase, and unlimited subcontracting. The PERSUADER searches memory to ascertain prevailing practice. The best contracts to reason from are contracts of competitors. Out of the competitors' contracts that it retrieves, the PERSUADER selects the recently negotiated contract of the Elegant Girl Inc. company using as additional selection criteria the company's locational similarity to the location of Muriel's Apparel Inc. (Elegant Girl is located in Florida and Muriel's Apparel in Georgia (both Southeastern States)), job classifications of the employees and political history of the dispute. In both cases, the relations of the union and management have been amicable, strikes have been a rare occurrence, the company managements have been in position for a long period and are respected by the rank and file. The relations of the locals in the two cases with the international union are good. The fact that the Elegant Girl Inc. contract is current, ensures that the economic climate of the two disputes is similar (the program checks for abrupt economic changes). The Elegant Girl Inc. contract provided 11% increase in piece rates, 5% pension increase and unlimited subcontracting.

The PERSUADER checks to see whether the situation of Muriel's Apparel Inc. is typical or atypical. It finds out it is typical, so it proceeds to adjust the case-based solution to fit the current situation. The PERSUADER considers appropriate adjustments to the Elegant Girl Inc. contract. Precedent adjustment is done using known heuristic modifications in labor mediation, namely adjustments with respect to the competitors' position in industry, and area wage differentials between Florida and Georgia. These adjustments result in a ballpark compromise with 9% increase in piece rates, 4% increase in pensions and limited subcontracting only when extra work is available.

The PERSUADER now adapts the ballpark compromise to the current situation. Checking the financial situation of the company, it finds out that Muriel's Apparel Inc. has suffered 4% losses in the past three years. It searches memory for similar cases, selects the most similar and applies the heuristic used in that case.

```
Searching memory with index GARMENT-INDUSTRY,
CONTINUOUS-LOSS...
3 cases found
```

---

[5]Since the PERSUADER does not assume the existence of a strong domain model, or the full cooperativeness of the agents, it may be known that a compromise has failed, but no explanation or repair was found.

Dispute Mediation: A Computer Model 261

```
Select case2
based on similarity of features
Apply heuristic used in this case
Decrease increases in piece rates by half percentage
of losses
Increase in piece rates becomes 7%
```

Before proposing the updated compromise, the PERSUADER searches memory to discover potential problems with the contemplated subcontracting language (with indices "failure", "subcontracting language, "limited to extra work"). It retrieves a case where the union had filed a grievance protesting that the company, having extra work, resorted to subcontracting for long periods of time instead of hiring more workers. The arbitrator in that case did not vindicate the union because no time limitation was written in the contract but proposed that the union get a time limitation for its next contract.

```
Searching memory with index FAILURE, SUBCONT-LANG,
LIMITED-EXTRA-WORK...
1 case found
Apply repair used in this case
Put time limit in subcontract language
```

The PERSUADER modifies the subcontracting language to impose a time limit to the company's right to subcontract, and proposes the resulting initial compromise to the parties.

When Previous Cases Are Not Available

If previous similar cases are not available, the PERSUADER uses Preference Analysis [Sycara 87a, Sycara 88] to find suitable compromises. Preference Analysis is based on Multi-Attribute Utility Theory [Keeney 76] and is used in our model as the underlying formalism for portraying the parties' preferences. Utility theory models the process through which a decision maker evaluates a set of alternatives, so that he can choose the best one. It has also been used in aiding a decision maker to structure his problem in such a way that evaluation of the alternatives is easily accomplished [Whitmore 74, Keeney 75]. We concentrate on the ways that utility theory can be exploited by a third party problem solver to: (1) generate potentially acceptable solutions to be proposed to the parties, (2) measure the quality of a modification to a rejected settlement, and (3) determine the effectiveness of persuasive argumentation.

The finite set of alternative settlements is obtained by range subdivision of each of the attribute values and combination of the resulting values. Utilities are computed at the endpoints of the subdivisions. For each of these alternative settlements, the overall utility, the *payoff*, of each agent is calculated and used to rank compromises from each agent's point of view.

One way to obtain utility curves is to question the decision maker directly. A variety of assessment procedures to that end has been reported in the literature [Johnson 77]. These procedures are time consuming and often impractical (e.g., they presuppose trust on the part of the decision maker towards the questioner). The PERSUADER uses two ways to obtain utility curves: (a) retrieval from memory of the utility curves of similar decision makers, and (b) hypothesizing the utility curves with the aid of domain-specific heuristics. The PERSUADER constructs the overall utility function for each party by asking the parties directly for the weights they attach to the issues. Retrieving the utility curves of similar parties is tried first. If no such past experiences are available, the PERSUADER hypothesizes the curves using a set of domain-specific heuristics to select the appropriate curves from a set of curves that it knows about.

In our implementation, the utilities of past persuadees are stored in the profile frames of the interacting agents. We assume that the utility curves of similar agents for a particular attribute will have the same functional form. This assumption is supported by various experimental studies (e.g., [Swalm 66, Spetzler 68]). The domain of each utility curve is the interval whose lower end-point is the company value and higher endpoint the union value for the issue under consideration. The range varies over the interval [0, 100]. For example, if the union's wage demand is 20% increase and the company's proposal 2% increase, then the domain interval for wage utilities is [2, 20]. These intervals are normalized.

Hypothesizing the utility curves of the parties relies heavily on domain-specific heuristics that pertain to the disputants. For example, the factors that are used in the heuristics to select utilities for a union include the economic state of the industry, the unemployment rate for the bargain unit's job classification in the area, and the structure of the bargaining unit (e.g., proportion of skilled vs. unskilled workers, young vs. old). Figure 2 shows how the factor of economic boom or recession impacts the health-benefits curve of a union.

Notice that in both cases the union will not be satisfied at all if it is not given any health benefits increases, and it will be 100% satisfied if it is given the maximum increase. Under boom, the union will be less than 50% satisfied if it is given an increase of magnitude $\frac{1}{2}Max-increase$, whereas under recession, it will be more than 50% satisfied if the conceded increase is $\frac{1}{2}Max-increase$. Thus, the two curves in figure 2 reflect qualitatively the realities of a union's satisfaction under two different economic conditions. Elementary calculus gives analytic expressions for these two curves.

We have considered two heuristic criteria that seem reasonable and can guide the problem solver in selecting the "best" for both parties compromise solution: (1) maximizing the joint payoff, and (2) minimizing the payoff difference of the parties. To apply the first criterion, one can proceed as follows: By range subdivision of each of the attribute values and combination of the resulting values, a finite set of alternatives is constructed. For simplicity, we present the case where

Dispute Mediation: A Computer Model

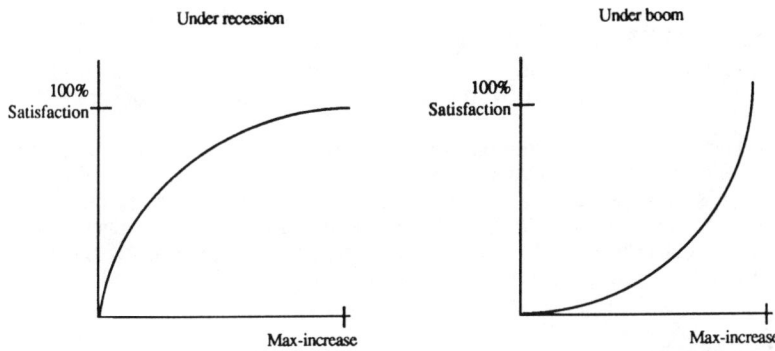

Figure 2: Possible union utility curves for health benefits

there are two issues, x_1 and x_2 under consideration. The company's (and union's) utility curves can be expressed by the general formula

$$v(x_1, x_2) = \alpha v_1(x_1) + (1-\alpha) v_2(x_2),$$

where α and $(1-\alpha)$ are the weights and $v_1(x_1)$ and $v_2(x_2)$ utility curves for issues x_1 and x_2 for each respective party. The joint payoff of the parties is given by the general formula

$$U(x_1, x_2) = u_1(x_1, x_2) + u_2(x_1, x_2)$$

where $u_1(x_1, x_2)$ is the company's utility curve and $u_2(x_1, x_2)$ the union's for settlement (x_1, x_2). The alternative that gives the maximum of these values could be selected and proposed.

Another possible criterion could be to select the most fair solution, namely the one with the smallest difference in the parties' payoffs. This is done as follows: Once the parties payoffs for each alternative have been calculated, the difference

$$U_d(x_1,x_2) = |u_1(x_1,x_2) - u_2(x_1,x_2)|$$

is calculated. The alternative that minimizes this difference is selected.

In order to decide which criterion the PERSUADER would use, we ran examples using each one. Maximizing the joint payoff very often gave contracts with quite unequal utilities for the parties. In those experiments, the payoff of one party would be so low as to practically guarantee rejection of the settlement by that party. On the other hand, minimizing the difference can lead to absurd results. For instance, this criterion would not be able to differentiate between alternatives one of which gives both parties a payoff of 40, and another that gives both parties payoff 70 (since in both cases the payoff difference is 0). Hence, we chose to combine the two criteria. The compromise that the PERSUADER selects to propose is the one that *maximizes the joint payoff of the agents and minimizes the payoff difference*.[6] This criterion combines maximal gains with equity. For more details, see [Sycara 88].

The utilities portray the possible tradeoffs that can be made among the conflicting goals to arrive at an acceptable compromise. Knowing an agent's utilities helps a planner predict which compromise(s) an agent will be most willing to accept. Since the utilities of the agents are heuristically derived, they represent a rough approximation of the true utilities. Hence, the proposed compromise might not be accepted.

An Example. In this example, we ran the PERSUADER on the Muriel's Apparel Inc. case having removed from memory similar cases in order to force the system to use Preference Analysis. The current impasse in negotiations is presented. The union wants a 13% increase in piece rates, 7% increase in pensions, and no subcontracting while the company proposes a 0% increase in piece rates, no pension increase and unlimited subcontracting.

```
Searching memory for competitors' contracts...
Failed to find competitors' contracts...
Searching memory for contracts of similar industries...
Failed to find contracts of similar industries...
Searching memory for contracts with same
job classifications...
Failed to find contracts with same
job classifications...
```

After its failure to find precedent cases, the PERSUADER considers Preference Analysis. It searches memory for similar parties and transfers their utility curves for wages, pensions and subcontracting.

```
Subdividing the interval [13, 0] of the values for
piece rates
into 9 pieces
Subdividing the range [7, 0] of pension values
into 9 pieces
Subdividing the interval [0, 10] of language values
for subcontracting
into 9 pieces
Generating the contracts that result from the combinations
of the values for the three issues
There are 1000 contracts under consideration.
```

---

[6]The PERSUADER resolves conflicts between two agents. The selection criterion can be generalized by taking pairwise differences of the agents' payoffs and selecting the compromise that maximizes the joint payoff and minimizes the greatest number of payoff differences.

Dispute Mediation: A Computer Model

The PERSUADER then asks the parties for the importance that they attach to the issues. By linear combination of the individual utilities of each party, it calculates the overall utility of each party for each contract. Then the criterion of maximizing the joint payoff and minimizing the payoff difference is used to select the best overall compromise.

```
Calculating joint payoff for each of 1000 contracts...
Calculating difference payoff for each of 1000 contracts...

Calculating difference between the
joint payoff and the difference payoff
for each of the 1000 contracts.
Selecting contract with 10% increase in piece rates,
4% increase in pensions
and 8 language ranking for subcontracting
("subcontracting only if no work is lost to union members")
since this contract minimizes the difference
and maximizes the joint payoff
```

When A Situation Is Exceptional

Example1: During contract negotiations, Southern Airlines presents its employees with the ultimatum that, if they don't take wage cuts of 8%, the company which has become non-competitive will go bankrupt. The employees protest and a mediator is called in. The mediator finds out that Southern Airlines has been loosing money because of mismanagement in an industry where other airlines are making money. She proposes that the employees accept 5% wage cuts and that the company give stock to the employees as as well as have employees sit on the board of directors.

The above example illustrates a situation where the solution to the problem was constructed neither through adaptation of a case-based solution nor by analyzing utilities, but by introducing a novel alternative. Judging the Southern Airlines negotiations as atypical, the mediator came up with a solution neither in the realm of prevailing practice or predicted by payoff adjustments. To distinguish a typical from an atypical situation, the PERSUADER evaluates the present case vis a vis prevailing practice. As has been mentioned, this evaluation happens during applicability checking for knowledge extraction from a selected precedent to construct a ballpark solution. The result of evaluation is an indication of whether the case can be considered typical or not. If so, adaptation of the ballpark solution takes place. If not, Situation Assessment, the process of generating a creative compromise, is activated.

The most important characteristic of atypical cases is that they violate a reasoner's expectations. These expectations usually stem from the following sources: (a) prevailing behavior of similar agents (prevailing practice), (b) beliefs about the rationality of the agents, (c) beliefs about the temporal continuation of a state, and (d) roles and relationships among the disputants. In labor mediation the violation of expectations arising from prevailing practice is the most widely used criterion for evaluating dispute typicality.

To evaluate case typicality in the above example, the PERSUADER checks that the air transport industry average for wages is 5% increase. Asking for a wage *cut* of 8%, the company is far outside the range of four to five times lower than the average (which would be an *increase* of 1%). It concludes that the case is atypical and activates the process of Situation Assessment. Situation Assessment classifies atypical situations in categories characterized in terms of the expectation violation they embody. The Southern Airlines negotiations violate the mediator's expectations that a company in a prosperous industry should not be loosing money. The cause of this expectation violation is mismanagement on the part of the company. Causes of violation expectations are represented in knowledge structures called Situational Assessment Packets (SAPs). SAP recognition rules are associated with each expectation violation category. Using the recognition rule "If an agent's situation violates prevailing practice and the agent is to blame, then activate SAP MISMANAGEMENT", the PERSUADER accesses SAP MISMANAGEMENT which provides general strategies for generating an appropriate compromise. The solution suggested by the SAP is to have the guilty party (the company in this case) bear the brunt of the cost of its mismanagement action by being denied a prerogative or a reward. In the example of Southern Airlines, the company is partially denied its prerogative of control by having employees sit on the board of directors. SAPs also store justifications along with the resolution strategies. In the example, the theme of "just desserts" is used.

SAPs are mechanisms for introducing departures from a model based on rational agents. SAP MISMANAGEMENT embodies equity theory and the principle of distributive justice which hold that human agents will seek equity and/or proportionality in payoffs even at some sacrifice to themselves.[7] Besides providing a problem solver with access to novel solutions and justifications, SAPs act as a source of preventive and recovery advice. In addition, SAPs describe and organize situations in terms of causally related abstract features that facilitate reminding across different domains. For more details on SAPs' role in problem solving, see [Sycara 87b, Sycara 87a].

IN THE FACE OF COMPROMISE REJECTION

No matter how the initial compromise has been calculated, by using cases or utilities, it is very rare that it will be accepted by both parties. The utility functions of the parties are not completely accurate [Johnson 77, Shepard 64]. Moreover, the preference structures of the agents can change during problem solving [Bartos 74, Swingle 70]. Cases and

---

[7]This has been borne out in real situations, as for example the 1985 settlement of Eastern Airlines with the International Pilots' Association.

SAPs are heuristic devices that may not capture all the requisite information for agreement. A mediator needs to be able to incrementally narrow the differences of the disputants so that a compromise will be mutually acceptable. The PERSUADER's reaction to rejection of a compromise is twofold: changing the rejecting agent's evaluation of the compromise through persuasive argumentation [Sycara 85a, Sycara 85b, Sycara 87a], and modifying/repairing the compromise so that it will be more acceptable. Persuasive argumentation is tried first, since, if the objecting agents can be convinced to accept the compromise, then a successful resolution has been found. If, on the other hand, a rejected compromise is modified/repaired, the repair may make it objectionable to agents that had agreed before. Thus, only after persuasive argumentation is no longer judged effective (i.e. all applicable arguments that the system could generate have been tried and rejected), is repair tried.

The Gentle Art of Persuasion

The PERSUADER's aim during argument generation is to change the belief structure of another agent, the *persuadee*, with respect to a proposed compromise. Since an agent rejects a compromise if it gives him low payoff, convincing him to change his evaluation of the rejected compromise is modeled as producing an argument to increase the payoff that the compromise gives him. Hence, the task of a persuader can be viewed as finding the most effective argument that will increase the agent's payoff. Since an agent's payoff can be approximated by a linear combination of his utilities, his payoff can be increased by (a) changing the importance (coefficient) the agent attaches to an issue, and (b) by changing the utility value of an issue. These constitute a persuader's argumentation goals. In labor mediation, the mediator is the persuader and the union or company the persuadee. Our argumentation model makes provisions both for selecting and adapting previously used arguments through CBR [Sycara 85b], and for constructing arguments from scratch [Sycara 85a, Sycara 89].

After a persuadee has been presented with an argument he can either accept it or reject it. If he accepts the argument, the model assumes that the change of the quantity (weight or utility) that the argument was aimed at has been accomplished and it updates the quantity. The heuristic value of a 3% change is used by the PERSUADER. If the argument is not accepted by the persuadee, no updating takes place. Either a new argument is tried or the rejected compromise is improved. Accepting an argument does not necessarily mean that a persuadee will accept the proposed compromise since the payoff increase may not be sufficient. In this case, the PERSUADER calculates the resulting payoff and tries to find additional arguments to increase the payoff even further. If all potential arguments have been exhausted and the persuadee still refuses to accept the proposed compromise, modification of the rejected compromise is tried next.

To construct arguments from scratch, the PERSUADER models the goals of a persuadee in a directed acyclic graph, called the persuadee's *belief structure*. It is searched and updated during argument generation.

The nodes represent goals with the associated importance, utility value, and desired direction of change (increase or decrease). The arcs represent the percent contribution of a goal to each of its ancestor goals. For example, an increase in wages contributes to increases in total company labor costs. In contrast, the subgoal of decreasing employment contributes to a decrease in labor cost. The argument, addressed to a union that has refused a proposed wage increase, "If the company is forced to grant higher wage increases, then it will decrease employment" is meant to decrease the importance the union attaches to wage increases by pointing out unpleasant consequences for the union of forcing an unwanted by the company wage increase.

To generate the above argument, the PERSUADER matches the wage goal in the company's belief graph. It propagates the wage increase that the union wants to force to the parent of the wage goal (total labor cost). Children of this node might indicate subgoals that the company can fulfill to counteract the wage increase. Such a counteracting action that violates a union goal that is more important than the union wage increase constitutes an argument that is aimed at reducing the importance that the union attaches to wage increase. The PERSUADER has two ways of obtaining the importance that a party attaches to an issue: (1) via feedback from the agent, and (2) by transferring the importance value of a most similar agent from its case memory, if the current agent does not want to disclose this information. For more details on the algorithm for argument generation see [Sycara 85a].

To continue with the Muriel Apparel Inc. negotiations, the PERSUADER suggests the compromise derived through Case-Based Reasoning to the parties. The company agrees saying that the increase in piece rates is the highest it can afford but the union wants a higher increase. The PERSUADER's argumentation goal becomes to convince the union to accept the proposed increase. By examining the union's payoff structure, the PERSUADER ascertains that to increase the union's payoff, it needs to decrease the importance the union attaches to piece rate increases.[8] The union's belief structure, starting from the wage goal, is traversed and the argument "If the company is forced to grant higher piece rate increases, then it will decrease employment" is produced.

```
Importance for wage-goal1 is 6 for union1
Searching company1 goal-graph...
A increase in wage-goal1 by company1
will result in a increase in labor-cost1
To compensate, company1
will decrease employment1
which violates employment-goal1 of union1
Importance of employment-goal1 is 8 for union1

Since importance of employment-goal1 >
importance of wage-goal1
One possible argument found
```

---

[8]For details of this calculation, see [Sycara 87a].

Dispute Mediation: A Computer Model   269

When the argument generating process produces more than one potential argument, the best order of argument presentation must be chosen. The PERSUADER uses the strategy of presenting the "weakest" (less convincing) argument first, presenting "strong" arguments only if the weak ones have been rejected. We have developed a hierarchy of argument types according to their convincing power [Sycara 87a].

Improving Rejected Compromises

If persuasive argumentation proves unsuccessful, the PERSUADER tries to improve the rejected compromise by appropriate modifications. Each contemplated modification is evaluated using the criterion that it has to *increase the rejecting agent's payoff more than it might decrease the payoff of the agents who have agreed to the compromise*[9].

Having a criterion for solution improvement is very necessary because otherwise, a problem solver could fall into a loop of proposing modifications that do not narrow the parties' differences. To see how the parties' payoff can be used as a means of calculating whether a modified settlement has improved its chance of acceptability, consider the following example: Suppose that a proposed contract with 40 cents increase in wages and 10 cents increase in pensions and with payoffs 52% for the company and 62% for the union, is rejected by the company. The mediator proposes a 3 cents pension reduction resulting in the contract (40, 7). The mediator calculates the payoffs of the parties for the contract (40,7). Suppose these payoffs are 61% for the company (an increase of 8%) and 58% for the union (a decrease of 4%). The contract (40, 7) fulfills the evaluation criterion and is therefore suggested to the parties.

Using the evaluation criterion, a problem solver does not waste time in proposing solutions that are inferior to rejected ones. The incremental solution improvement process is akin to *hill climbing* and the evaluation criterion affords the test for proceeding. In our implementation, we assume that a solution that affords both parties a payoff greater than or equal to 70% will be accepted by the parties[10]. The parties might of course choose to accept a settlement that gives less than 70% payoff.

To generate appropriate modifications, the PERSUADER ascertains from the rejecting agent's feedback objectionable issues, the reason for the rejection and the importance the agent attaches to the issues. The issues and reason for rejection are used as indices into the case memory

---

[9]In most conflict situations if a compromise that was acceptable to one party is modified in favor of the opposing party, the resulting compromise will give the party that had accepted it a smaller payoff than the previous resolution.

[10]This number has been checked for approximate accuracy by practicing mediators. The reason that 100% satisfaction with a resolution is not necessary is that the parties are assumed to be reasonable, in the sense that they know that they need to compromise.

to select impasses with the same stated impasse issue and impasse cause as in the present impasse. The selected impasse supplies modifications that will hopefully improve the rejected solution. If no appropriate impasses can be found, the PERSUADER uses standard heuristics that it knows about.

In the Muriel's Apparel Inc. case, confronted with the union's refusal to accept an increase in piece rates less than 7%, the PERSUADER tries to boost pensions in the hope that the union will accept the new "package". It searches memory for impasses where pensions needed to be increased. The most suitable impasse from the ones retrieved is selected and the associated modification is tried. The modified compromise is evaluated according to the criterion for improvement stated above and, if the criterion holds, the new compromise is proposed. If the contemplated modification does not constitute an improvement, modifications from other retrieved cases are tried.

```
Searching memory with index FAILURE, PENSION, TOO-LOW
5 impasses found
Select impasse1
since it is same industry, same job classification,
same area...
Looking at modification1 "increase pension by
additional 2%"
from impasse1
Since the majority of workers are older
modification1 seems applicable

Apply improvement criterion
Success

Contract3  which resulted
from applying modification1
will be proposed
```

SUMMARY AND CONCLUSIONS

We have presented the PERSUADER as a model of resolving multi-agent conflicts through mediation/negotiation. The PERSUADER plans iteratively by interacting with the agents, using their feedback in refining and repairing compromises, and in generating persuasive arguments. The PERSUADER plans for labor mediation, a domain full of uncertain knowledge and changing circumstances. Such characteristics typify most real world domains, such as international relations, law, labor relations, management, and manufacturing.

Unlike rule-based expert systems that solve each problem from scratch (thus expending the same effort to solve the same problem a second time), the PERSUADER updates its memory with each new experience. Memory update provides automatic knowledge acquisition/learning. Having an experiential memory provides the PERSUADER with the ability to improve its efficiency and quality of solutions. Moreover, using

Dispute Mediation: A Computer Model 271

previous cases for explanation and justification of a proposed solution is more acceptable to a user than invocation of a rule. The PERSUADER's ability to produce persuasive arguments is a capability unique to our model.

The integration of case-based and analytic methods makes the PERSUADER robust and flexible. It does not break down when cases and rules are not applicable. Moreover, it has the flexibility to use whichever method is more natural to the particular problem solving stage it is engaged in.

Our model could be used as a tool in understanding conflict resolution. The negotiation/mediation process carried out by people is opaque, in the sense that it is impossible to see the mechanisms working. The PERSUADER makes explicit what knowledge is needed in negotiation, how it is represented and organized, and how it is used to make decisions. The PERSUADER provides a normative reference with which to compare and evaluate actual negotiations. By making the knowledge and mechanisms explicit, the PERSUADER makes possible testable hypotheses that might help in understanding the process of negotiation and conflict resolution.

References

[Bartos 74]   Bartos, O.
              Process and Outcome of Negotiations.
              Columbia University Press, New York, N.Y., 1974.

[Elkouri 72] Elkouri, F. and Elkouri, E.
              How Arbitration Works.
              The Bureau of National Affairs, Washington, DC, 1972.

[Fogelman 83] Fogelman-Soulie, F., Munier, D., and Shakun, M.F.
              Bivariate Negotiations as a problem of Stochastic Terminal Control.
              Management Science 29:840-855, 1983.

[Frazer 84]   Frazer N.M, and Hipel, K.W.
              Conflict Analysis, Models and Resolutions.
              North Holland, New York, 1984.

[Hammond 86] Hammond, K.J.
              CHEF: A model of case-based planning.
              In *Proceedings of AAAI-86*, pages 267-271. Philadelphia, PA, 1986.

[Jarke 87]   Jarke, M., Jelassi, M.T., and Shakun, M.F.
              MEDIATOR: Towards a Negotiation Support System.
              European Journal of Operational Research , 1987.

[Johnson 77] Johnson, E.M., and Huber, G.P.
              The Technology of Utility Assessment.
              IEEE Transactions on Systems, Man and Cybernetics SMC-7:311-325, 1977.

[Keeney 75]   Keeney, R.L., and Nair, K.
              Decision Analysis for the siting of nuclear power plants-The relevance of multiattribute utility theory.
              Proceedings of the IEEE 63:494-500, 1975.

[Keeney 76]   Keeney, R.L. and Raiffa, H.
              Decisions with Multiple Objectives.
              John Wiley and Sons, New York, 1976.

[Kersten 85] Kersten, G. E.
              NEGO - Group Decision Support System.
              Information and Management 8:237-246, 1985.

[Kersten 86] Kersten, G.E., and Szapiro, T.
Generalized Approach to Modelling Negotiations.
European Journal of Operational Research
26:142-149, 1986.

[Kolodner 84] Kolodner, J.L.
Retrieval and Organizational Strategies in Conceptual Memory: A Computer Model.
Lawrence Erlbaum Associates, Hillsdale, NJ, 1984.

[Kolodner et al. 85]
Kolodner, J.L., Simpson, R.L., and Sycara-Cyranski, K.
A Process Model of Case-Based Reasoning in Problem Solving.
In *Proceedings of IJCAI-85*, pages 284-290. Los Angeles, CA, 1985.

[Matwin 87] Matwin, S., Szpakowicz, S., Kersten, G., Michalowski, W., Koperczak, Z.
Logic-based Tools for Negotiation Support.
Technical Report 87-10, University of Ottawa, 1987.

[Raiffa 82] Raiffa, H.
The Art and Science of Negotiation.
Harvard University Press, Cambridge, Mass., 1982.

[Schank 82] Schank, R.C.
Dynamic Memory.
Cambridge University Press, Cambridge, 1982.

[Shakun 88] Shakun, M., F.
Evolutionary Systems Design: Policy Making Under Complexity and Group Decision Support Systems.
Holden-Day, Oakland, CA., 1988.

[Shepard 64] Shepard, R.N.
On subjectively optimal selection among multiattribute alternatives.
In Shelley, M.W. and Bryan, G.L. (editor), *Human Judgement and Optimality*. Wileay and Sons, New York, N.Y., 1964.

[Simkin 71] Simkin, W.E.
Mediation and the Dynamics of Collective Bargaining.
The Bureau of National Affairs, Washington, D.C., 1971.

[Spetzler 68] Spetzler, C.S.
The development of a corporate risk policy for capital investment decisions.
IEEE Transactions on Systems, Science and Cybernetics SSC-4:279-300, 1968.

[Swalm 66] Swalm, R.D.
Utility theory insights into risk taking.
Harvard Business Review 44:123-136, 1966.

[Swingle 70] Swingle, P., (Ed.).
the Structure of Conflict.
Academic Press, New York, N.Y., 1970.

[Sycara 85a] Sycara-Cyranski, K.
Arguments of persuasion in labor mediation.
In *Proceedings of IJCAI-85*, pages 294-296. Los Angeles, CA, 1985.

[Sycara 85b] Sycara-Cyranski, K.
Persuasive argumentation in resolution of collective bargaining impasses.
In *Proceedings of the Seventh Annual Conference of the Cognitive Science Society*, pages 356-360. Irvine, CA, 1985.

[Sycara 87a] Sycara, K.
Resolving Adversarial Conflicts: An Approach Integrating Case-Based and Analytic Methods.
PhD thesis, School of Information and Computer Science Georgia Institute of Technology, 1987.

[Sycara 87b] Sycara, K.
Finding creative solutions in adversarial impasses.
In *Proceedings of the Ninth Annual Conference of the Cognitive Science Society.* Seattle, WA, 1987.

[Sycara 88] Sycara, K.
Utility Theory in Conflict Resolution.
Annals of Operations Research 12:65-84, 1988.

[Sycara 89] Sycara, K.
Argumentation: Planning Other Agents' Plans.
In *Proceedings of IJCAI-89.* Detroit, Mich, 1989.

[Whitmore 74] Whitmore, G.A. and Cavadias, G.S.
Experimental Determination of community preferences for water quality-cost alternatives.
Decision Sciences 5:614-631, 1974.

V. MATHEMATICAL PROGRAMMING AND AI

There are two papers that comprise this section. The first paper, by Sklar et al. ("Eliciting Knowledge Representation Schema for Linear Programming Formulation"), remarks that linear programming (LP), a modeling technique of wide applicability and great importance, is not applied as often as it could be. Since solution algorithms are widespread, it is possible that this underuse comes from decision-makers being uncomfortable with the LP problem formulation. The use of LP should increase when automated formulation aids become available. At this time, however, such aids are not available. Furthermore, the appropriate techniques for constructing such aids are not known.

This paper explores the classification and knowledge representation schemes used by domain experts in the field of LP formulation. These are issues that can have significant influence on the design, performance, and feasibility of a knowledge-based LP formulation system. The authors attempt to study classification and knowledge representation in a more rigorous way than practicing knowledge engineers have the time or resources to do. From this study evidence exists to suggest that LP formulation experts use a classification scheme when formulating LP problems, that experts classify according to problem type, and that a frame representation scheme adequately expresses the classification method. These conclusions can serve as the basis for the design and implementation of an automated formulation aid.

The methodology used in this study is also of interest. Most published reports of expert systems development use concurrent verbalization and/or interview techniques for knowledge acquisition. This study demonstrates the feasibility of context focusing, an attractive alternative in some situations.

The second paper ("A Knowledge Base for Integer Programming - A Meta-OR Approach" by Zahedi) introduces the concept of "meta-OR" and applies it to integer programming applications. Meta-OR is based on the metaknowledge concept in artificial intelligence. The idea is that the qualitative knowledge of the applications of OR/MS methods should be formalized and stored in a permanent and organized inventory (or knowledge base). Otherwise, one can hardly generalize from the reported applications, and hence the lessons learned and reported in these applications will be lost to the future practitioners in the field.

Furthermore, as structured modeling takes hold and the steps of problem solving become increasingly computerized, the theoretical characteristics of a problem will not be sufficient for model selection. The knowledge obtained from the wealth of experience should have an equal role in the selection of a model. Therefore, one must be able to formalize and capture the lessons learned from experience for the future use in model selection.

This paper shows how the AI approach can be used in formalizing the qualitative results of applying the MS/OR methods to various problems. The paper applies this approach to the integer programming applications published since 1980 and reports 103 rules that are grouped into seven categories. These are 103 lessons learned from experience, which can be stored as metaknowledge to guide the process of model selection in the future. They also constitute data on the quality and nature of applications that can be generalized, categorized, or, in short, scientifically analyzed.

In any scientific field, data from application feeds the development of theory. The concept of "meta-OR" may perform this role for the OR field, and bring some of the neglected qualitative aspects of the OR practice to the attention of OR theorists.

ELICITING KNOWLEDGE REPRESENTATION SCHEMA FOR LINEAR PROGRAMMING FORMULATION

Margaret M. Sklar
Department of Management, Marketing and CIS
School of Business Administration
Northern Michigan University
Marquette, MI 49855

and

Roger Alan Pick
G. Brian Vesprani
James R. Evans
Department of Quantitative Analysis
and Information Systems
University of Cincinnati
Cincinnati, OH 45221-0130

Abstract: The Linear Programming (LP) model is widely-used, having applications in many decision-making situations. To date, building LP formulations still requires the expertise of an Operations Research (OR) analyst. We feel that this expertise can be incorporated into an automated formulation system that will make the LP model available to managers who are not OR experts. Our study involves eliciting methods used by OR experts in building LP formulations for textbook problem statements. We pay special attention to eliciting the knowledge representation (KR) schema used by OR experts. Previous work in this area elicits methods only, leaving the KR schema up to the Knowledge Engineer's discretion. We argue that a more "natural" system can be developed if we know more about expert KR schema, and describe a study to elicit these schema.

INTRODUCTION

Linear Programming (LP) has been accepted as an important modeling technique since its development in the 1940's. However, while the method is widely used, translation of real-world LP problem statements into the matrices or algebraic representations required by LP solvers is still considered a difficult task for decision makers who are not experts in LP formulation. It is interesting to note that some decision makers experience this difficulty even though they may be expert enough to recognize that LP is appropriate for the task at hand (Orlikowski & Dhar, 1987; Sklar, Pick & Evans, 1987). As a result, LP modeling is not used as much as it could be. The end result of our research will be an expert system which will emulate expert knowledge representation (KR) schema for LP and the methods experts use to manipulate these KR schema. Such a system would free the user from the tedium of translating problem features into the required LP formulation format and allow him to focus, instead, on the salient features of the problem at hand. This should encourage wider use of the LP model, particularly by users with some, but not expert, knowledge of the formulation process (Sklar, Pick and Evans, 1987; Stohr, 1987). While we emphasize LP formulation, our major interest is in the investigation of methods to elicit knowledge required to build expert systems in Operations Research (OR) domains in general.

We have focused our efforts on discovering the methods that experts use in building LP formulations from problem statements. As a first step in building the system, we have devised a study of expert representation schema. We believe that the organization of knowledge is as important as the methods that experts use in problem solving situations. The purpose of this paper, therefore, is two-fold. First, we wish present an argument in favor of the necessity of including KR knowledge as well as modeling knowledge in such an expert system. In order to do so fully, we review some knowledge acquisition (KA) techniques which appear particularly useful for eliciting knowledge related to OR problem environments in general. Second, we present the results of the KR stage of our study. The rest of the paper is organized as follows: first, we argue the importance of KR; next, we discuss the KA techniques which we used for eliciting LP problem solving knowledge; finally, we present the results of the KR schema elicitation study.

KNOWLEDGE REPRESENTATION

In developing expert systems, considerable effort goes into eliciting methods or "tricks of the trade" that experts use for problem-solving within their domains. There is considerable evidence in the cognitive science literature that experts have not only different problem solving methods, but also different knowledge organization schema from those of non-experts. (McKeithen et al, 1981; Newell & Simon, 1982) However, there is very little attention paid to eliciting knowledge organization schema in the Knowledge Engineering literature. In fact, while expert methods are carefully elicited, the Knowledge Representation scheme is usually left to the discretion of the Knowledge Engineer (KE).

While the KE often has some domain knowledge, he is rarely considered a domain expert. It seems reasonable that methods used to manipulate knowledge would operate more efficiently and more naturally if the KR scheme emulates the expert's organization just as the rule base emulates the expert's methods.

This attitude is expressed by others. Fink & Lusth (1987, p. 341) argue that the way knowledge is represented in a system is a "key to shortcomings in current expert system technology." Chandrasekaran (1986, p. 24) states that "knowledge acquisition is often directed toward strategies for conflict resolution, whereas they ought to be directed to issues of knowledge organization." Likewise, Dolk & Konsynski (1984) and Fedorowicz and Williams (1986) discuss the importance of KR schema for intelligent assistance in Decision Support and Model Management Systems. Guenthner et al (1986, p. 39) conjecture that once knowledge representation for a system "has been decided, the remaining tasks are 'easy' to do." Yet it is precisely the remaining tasks that many expert systems developers attempt, ignoring knowledge representation as an issue requiring expert input.

Types of Knowledge Used in LP Formulations

In an earlier study (Sklar, Pick & Evans, 1987), we reported that experts use several different kinds of knowledge in building LP formulations, namely: syntactic LP knowledge, semantic LP formulation knowledge, semantic LP domain knowledge and semantic world knowledge. Syntactic LP knowledge consists of the rules for building grammatically valid LP constraints and objective functions. Semantic LP formulation knowledge pertains to interrelationships between different parts of an LP formulation. This includes information necessary to translate inputs into problem parts as well as information necessary to connect problem parts, such as inventory balance constraints or flow capacity constraints. Semantic LP domain knowledge includes information about specific LP domains, such as production management. (Binbasioglu, 1986) Semantic world knowledge is knowledge about the world in general, not directly about the LP formulation process. Examples of world knowledge are facts such as that the component parts of a product must meet demand requirements for that product or that milling and assembly belong to a general class called processes. World knowledge is not explicitly stated in problems and therefore cannot be deduced from the problem statement. Omitting any of these types of knowledge imposes limits on the usefulness of an LP formulator.

Likely Knowledge Representation Schemes for LP

Frames.

Finding a knowledge representation scheme for LP formulation is a matter of deciding how best to represent the above types of knowledge. Dolk & Konsynski (1984) suggest a frame representation for model management systems in general. A frame provides a structured way of representing information about an object or a class of objects. (Dolk & Konsynski, 1984; Fikes & Kehler, 1985) Frames store information in "slots" which may contain either simple data or procedures. In

LP formulation, frames can be used to identify a problem type by matching given data with the inputs needed for that particular problem type. Each frame stores semantic information about the decision variables, the objective function and the constraint types necessary for that frame. If experts classify by problem type, a frame representation scheme is likely to be appropriate for representing their classificatory knowledge. Identification of the correct frame would then dictate a set of expectations about the types of information still needed and guide the search for further input.

Inheritance Networks and Production Rules.

An inheritance network is a hierarchical representation of objects in which objects at lower levels may inherit properties of their predecessor. Evans & Camm (1987) apply inheritance network concepts to representing interrelationships between LP problem parts. Inheritance networks can also be used to store world knowledge and to serve as an overall structure to store information about interrelationships between individual frames.

Production rules (IF/THEN rules) are used to draw inferences about the data, and to manipulate data. For LP formulation, production rules would be used:

> To classify problem type by matching inputs against the frame structure.

To traverse the inheritance network to identify synonyms, subclasses, etc.

To identify the need for and handle unit conversions (for example, minutes to hours)

Within frames, to direct the search for needed input and to translate that input into LP constraints and the objective function.

In short, production rules, together with inheritance network and frame representations have the potential ability to store the syntactic, semantic and procedural knowledge required for LP formulation. At this point, it seems that a hybrid scheme including these types of information is consistent with expert knowledge organization. Our research studies the appropriateness of such a scheme.

KNOWLEDGE ACQUISITION

Knowledge Acquisition is considered the most difficult and time-consuming aspect of expert system construction. Expertise is learned over a long period of time and is sometimes so completely internalized that an expert is unaware of exactly what rules or processes are being used. (Wright & Ayton, 1987) There are a number of common KA techniques that have been more or less successful for extracting cognitive models of expert knowledge in various expert system applications. In OR applications, expertise is evidenced by actually building models rather than talking about the domain. We find that combining verbal elicitation techniques with working through LP formulations to be a particularly useful approach. Think-aloud sessions allow us to elicit those techniques that experts employ which can be verbalized while the expert performs the task. The KA techniques we employ include concurrent and retrospective verbalization, semi-structured interviews and context focusing. We find that this combination has the potential to elicit a more complete model than any individual technique.

Concurrent Verbalization

Concurrent verbalization is considered to be the strongest evidence that an expert uses a particular rule or process. (Ericsson & Simon, 1984; Wright & Ayton, 1987). In this technique, the expert is asked to think aloud, reporting his thoughts, plans and intentions while working through a problem example. Actually working through the problem helps the expert focus his thoughts, and is considered easier for the expert than trying to discuss methods in abstract terms. (Hart, 1985) These verbal protocols are recorded and later transcribed and analyzed. Analysis of the protocols gives a rather fine-grained view of the rules, methods and KR schema to which the expert attends while working the individual problems. This seems an excellent way to elicit a good deal of working knowledge, but is limited by the expert's reporting abilities.

Concurrent verbalization provides evidence that the expert is using a particular rule or process. It can, therefore, be translated into a cognitive map of the expert.

(Ericsson & Simon, 1984; Wright & Ayton, 1987) For our purposes, however, we needed to establish a balance between psychological validity and a usable system that could be produced in finite time. Using concurrent verbalization alone, we found that our experts reported what they were aware of thinking. There were obvious gaps, however, which we needed to fill.

Retrospective Verbalization and Interviews

While interviews are generally considered time consuming and inadequate for eliciting detailed domain knowledge, (Slatter, 1987) adding an interview at the end of a think-aloud session gives the KE the opportunity to delve for information to fill some of the gaps. The KE can ask for further clarification of a point in the current problem, or ask for comparisons between this and previous problems, thus encouraging retrospective verbalization. Combining concurrent and retrospective verbalizations seems to be a more natural setting for our experts. During think-aloud sessions, they repeatedly referred to problems worked during previous sessions or from other experiences. Retrospection can lead to rules and metarules missed using concurrent verbalization alone.

Context Focusing

In situations where the KE suspects a classification scheme, context focusing can be used to get a quick idea of the knowledge structure and the rules the expert uses to search the structure. (Wright & Ayton, 1987)

Context focusing can be considered an inverse interview technique, with the expert asking the questions. His goal is to try to discover a problem state imagined by the KE. For example, if a student has problems with a computer program, the teacher is likely to ask first if there were compile-time errors, and then if there were run-time errors, etc. Each question narrowing in on the problem while eliminating branches of the search tree. It can be noted in this example, that the order in which nodes of the tree are tested also indicates the priority the expert places on them.

We found context focusing to be an excellent method to slow down experts as they processed problems that were, to them, trivial. This allowed us to capture knowledge that could not be retrieved through concurrent verbalization.

TOWARD BUILDING A GENERAL-PURPOSE LP FORMULATOR

We are currently in the process of building a prototype expert system which will accept data about textbook LP problems from non-expert users and build the LP formulations for these problems. There has recently been a great deal of interest in automating the LP formulation process, most notably research done by Binbasioglu (1986) and Ma, Murphy and Stohr (1988a, 1988b). The major difference between these systems and ours is that our research is more exploratory in nature. Our main interest is in the investigation of KA methods which seem promising for developing expert systems in OR problem domains.

LP is an excellent starting point for these investigations. LP is an interesting and economically important modeling tool; it is widely used, yet not readily available to non-experts; there is a common body of LP knowledge available in numerous textbooks; LP methods are semi-structured in that there are commonalities among LP problems or problem parts, but specific applications or problem statements must be formulated individually. These features combine to form a problem domain that is interesting to study and one that promises a high potential for success in expert system development. (Harmon & King, 1985)

Our purposes in this research are both to actually build a formulation system and to explore techniques which will capture more complete knowledge than have previous systems. We feel that developing a general purpose LP formulation system that can be used by non-experts is feasible given current technology. Paying attention to expert KR schema should enable system builders to design an interface which will prove to be highly user-tolerant. Furthermore, we feel that capturing expert KR schema is an important consideration in developing such a system. In LP formulation, this expert knowledge includes syntactic knowledge about the format of LP objective functions and constraints, as well as semantic knowledge about particular LP domains and interrelationships between problem parts. In addition, expertise includes a large amount of general, or world, knowledge that is used to interpret problem statements. A primary goal in this research is to study expert KR schema, particularly that used for the LP formulation process. We know of no other formulation system which has studied knowledge representation in its own right.

ELICITING A KNOWLEDGE REPRESENTATION SCHEME

Most current expert systems are developed by eliciting domain knowledge and rules for operating upon this knowledge from experts. The KR scheme is usually left up to the KE to devise. Many researchers now recognize the importance of eliciting an expert's knowledge representation scheme as well as domain knowledge. Discovering the KR scheme can help to make the formulation system seem more natural to the user. It can also make further knowledge elicitation easier, giving the KE a framework on which to base interview questions. In the case of LP formulations, it seems likely that experts follow predefined frames of reference for much of the model formulation. For example, if a problem can be first identified as having features of a product mix problem, the decision variables, objective function and constraints are expected to have a certain format. By contrast, if the problem is identified as an assignment problem, the decision variables, objective function, and constraints will be defined very differently.

Expert Classification of LP Problems

Earlier work has indicated that experts perform some type of classification before formulating LP expressions. This classification step has been demonstrated in studies comparing physics experts with novices. (Chi et al, 1981; Konst et al, 1983) Orlikowski and Dhar (1987) indicate that some of the same differences exist between LP experts and novices. Since we are interested in KR from the point of view of building a system, we felt that a more formal approach to the LP classification question was important to our research. Concurrent and retrospective protocols collected during an earlier study (Sklar, Pick & Evans, 1987) indicated that experts were processing a considerable amount of information before they began working the LP formulations aloud. Furthermore, this processing took place almost instantaneously, and probably unconsciously. When asked whether they classified by problem type, experts were divided in their answers, although their approaches to the problem formulation were nearly identical. It was not obvious that classification was occurring, but we suspected that this might be the case.

A Study to Discover Knowledge Representation Schemes

We see the classification question as a major key to expert KR in LP formulation. Since this question was not one that the experts themselves seemed able to answer for us, we devised a study to find what classification scheme, if any, was used by the experts. The questions in this study involve: (1) discovering whether classification occurs; and (2) if classification occurs, discovering the taxonomy used.

Research Concerns.

A major research concern in this study was possible bias of the KE, who had strong opinions on appropriate representations for LP knowledge. In order to counteract this bias, a second KE was selected for this part of the study. This KE was then trained in LP formulation and in KA techniques. To guard against the second KE developing similar biases which potentially could influence his interaction with the subjects, he was kept ignorant of the full purpose of the study. This "blind" KE was told that the purpose of this study was to elicit from the expert the preliminary thoughts that are not reported during the think-aloud sessions.

A second concern involved the number of experts to use in the study. In building expert systems, opinion is divided as to the advisability of using a single expert or multiple experts for KA purposes. Using a single expert (the more common approach) has the advantage of giving a non-conflicting rule base on which to build the system. Using multiple experts has the advantage of possibly yielding a richer rule base, but can lead to conflicts or inconsistencies in either rules or methods. We were interested in gaining as full a rule set and as representative a KR scheme as possible. At the same time, practicality demanded that we avoid the possibility of extreme conflict situations. We compromised, deciding to use multiple experts in the study, but not so many experts that conflicts would be irresolvable.

A third concern involved eliciting experts' "compiled" knowledge, or knowledge that is so internalized as to be unnoticed by the expert as it is being used. It has been suggested that a journeyman, who is knowledgeable but not quite expert, may be closer to "first principles" and therefore easier to study in this type of experiment. (Kuipers & Kassirer, 1987) We decided to include journeymen in this study. If they are closer to "first principles," then their KR schema might be not only easier to elicit than those of the experts but also more representative of experts' compiled knowledge.

Description of the Study.

Eight subjects were chosen among the faculty and Ph.D. students from the Department of Quantitative Analysis and Information Systems at the University of Cincinnati. Faculty members were classified as experts; Ph.D. students as journeymen. Context focusing was chosen as the KA method to be used during this study. A second KE was chosen and trained as stated above. His instructions were to allow the subject to ask as many questions as necessary to reach a point where the subject felt that he had acquired enough information to write the LP formulation for each problem. The subjects were to feel comfortable with the problem, not necessarily to do the formulation. Ideally, KA could end as soon as classification was acknowledged or obvious. But, since the KE did not know the purposes of the study, he could not be asked to look for such a terminating point. Consequently, it was suggested that the KE impose a time limit of fifteen minutes on individual problems. Sessions could be terminated before that time if the subject was satisfied with the information he had obtained.

Ten problems were selected from introductory Management Science textbooks. There were two problems of each of the following types: product mix, multiperiod production scheduling, blending, transportation and assignment. Some of the problems had "twists" so that if a subject missed some information, he would be on the way to an incorrect formulation. An eleventh problem that was not readily classifiable as one of the above types was added to the set. Six of the eleven problems were randomly assigned to each subject. These six problems were then presented to the subject in random order.

These sessions were conducted with individual subjects. Unlike the concurrent verbalization sessions, the subjects did not have a copy of the problem statement before them. The problem statement was held by the KE, and the subjects were to "discover" the problem features by questioning the KE. The subjects were told that they were free to ask specific questions about the problem, but that they should not ask for broad information, such as the nature of the decision variables or the constraints. These were to be induced by answers to more specific questions. Subjects were further instructed that they would direct the questioning, and that they should continue until they reached a point where they felt that they had enough information to write the LP formulation--that this judgment, rather than the formulation, was their goal. Sessions were recorded and transcribed.

A problem transcription which contained a statement such as, "Oh, this is a transportation problem," was considered a clear indication that classification by problem type was occurring. If the subject did not volunteer such information, the transcripts were scanned by a panel of judges to see if a predetermined "frame" was being followed. That is, if a point could be determined after which the subject's questions were obviously searching for information that followed the expected format for a particular problem type. If such a point could be identified, that also was accepted as evidence of classification by problem type.

Sample Sessions.

This study was primarily interested in evidence for or against classification by problem type. Analysis of the collected protocols was, therefore, intended to discover frame representations if they existed. We found, as a side benefit, that the context focusing sessions yielded some very useful search paths which occur at a point before that at which concurrent verbalization begins. To further illustrate the context focusing method and its usefulness, we discuss excerpts from some of the sessions here. The excerpts are given in the Appendix.

In session 1, the subject called a halt to the session when he felt (correctly) that he had gathered all of the information needed to build the LP formulation. The subject quickly identified the problem as a product mix type and spent the rest of the session filling in slots appropriate to a product mix frame. When he felt that he had enough information, he stopped the session. For this same problem, other subjects went on to define decision variables and to build the objective function and constraints after gathering the information. This subject built formulations for some, but not all of the problems.

The second session involves a different subject. This subject, also, seems to be working from an internal framework. The excerpt in the Appendix goes up to the point where the subject identified a multiple-period production scheduling problem. After he reached this point, he went on to ask about monthly demands, beginning and ending inventory information, and fluctuation costs. He then proceeded to build the formulation. After completing the formulation, he retrospectively discussed model design issues and finally went through an internal checklist before calling a halt to the session.

The two sessions presented here are typical in that the subjects seemed to gather information by filling slots in a frame. They are atypical in the rapidity of their focusing, however. More often, subjects seemed to be filling slots in one frame and then switching frames when the earlier frame was seen to be inappropriate. In this fashion, they eventually found the needed information.

Logical Design From the Context Focusing Sessions.

Our main purpose in the context focusing sessions was to discover evidence for classification by problem type. Translating these sessions into a decision network format was not an overly difficult task. Most of the questions asked narrowed down the search for information until a particular problem type frame could be identified. A partial decision network derived from the context focusing sessions is presented in Figure 1. In the decision network, paths end with a trial frame. This is to emphasize that our classification is tentative. At individual frame levels, the expected information for that frame is sought. If a problem is more complex, then an individual frame will collect the appropriate information from the user, but other frames may be required to collect the remaining information.

DISCUSSION OF THE STUDY

We found differences in the subjects' individual information-gathering styles rather interesting. All but one of the subjects identified tentative decision variables and frames, some early on in the process, some more slowly and cautiously. These differences did not seem to be attributable to the expert/journeyman difference. We originally expected a clear delineation between questions of a general nature and questions whose purpose was to fill in slots for a particular problem type frame. In twenty-two of the forty-eight trials, the judges were able to identify such a clear delineation. In another twenty-five trials, the change was gradual. But even though such a clear distinction was not made, the judges could identify points where a particular frame was discarded and another frame attempted. Only in two of the forty-eight trials, did the subject fail to classify correctly. These two trials involved the same subject. Even here, it appeared as if the subject were attempting to classify, but missed some important information.

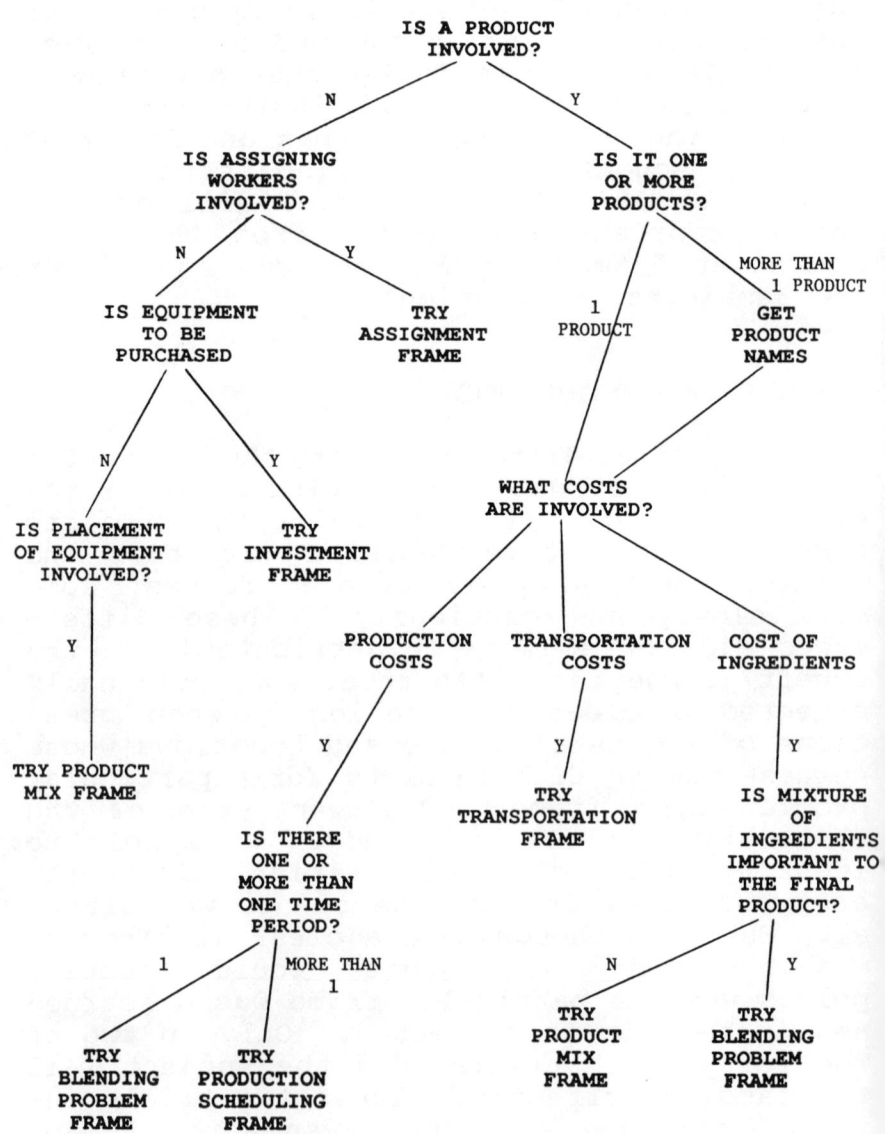

Figure 1. Partial Decision Network Elicited from Context Focusing Sessions

Individual Differences in Subjects' Styles

For identification purposes, we assigned arbitrary numbers to the subjects. In the following discussion, subjects 1 through 4 are faculty members, or "experts;" subjects 5 through 8 students in the Quantitative Analysis doctoral program, or "journeymen."

Subject 1 would quickly identify tentative decision variables and then attempt to fill slots in frames. If necessary (though rarely) this subject would discard a particular frame if it was inappropriate and then begin working on another. After building the basic constraints and objective function consistent with a particular frame, this subject would check for other data consistent with that frame, and add to the constraint set or objective function if necessary.

Subject 2 seemed to have an internal check list available. This subject worked more cautiously, collecting information in a more general way, and narrowing in on a frame, rather than classifying early. He discussed his thinking, talking about the possibilities that each new piece of information opened or closed. When he came around to actually classifying a problem type, he first checked off what information he had acquired to that point, and then listed the information that might still be applicable. Even after building the formulation, he always asked whether there was further information that might be relevant to the problem.

Subject 3, like subject 1, classified early in almost all cases. His method was to fill in some slots of a frame and then check to see if that frame is still applicable. He discussed what problem type he was considering and why. During problem formulation, he checked for extra information, but did not ask as thoroughly as subject 1 for other possible information.

Subject 4 spent less time on information gathering than most of the other subjects. He quickly narrowed in on important problem features and classified accordingly. Once this subject classified, he instructed the KE on what types of information should be available in the problem statement. He also discussed at some length the differences between "real world" and textbook problems. This subject seemed quite interested in not only fulfilling the purposes of the study, but also in giving as much other practical knowledge as possible in the short time allotted.

Subject 5 had a tendency to choose a frame type early on in the process, but not necessarily the correct type. He was the only subject who asked for specific information about the decision variables, for example, whether they were continuous or discrete. This subject seemed to try to classify according to this information, rather than get the other relevant information. This may account for the fact that although this subject classified correctly (eventually), he missed some constraints and even incorrectly formulated an objective function in one instance.

Subject 6 was probably the most cautious. This subject filled in general slots, only very gradually narrowing down to a particular problem type frame. In the opinion of the judges, this subject left nothing unchecked. Only in the case of the subject's last problem, while working on a problem of similar type to the one immediately preceding it, did this subject classify early by problem type. He seemed to be looking for more complex problems, and looked for links between possible problem types after filling in a frame.

Subject 7 was as thorough in information gathering as subject 2. Like subject 2, this subject also seemed to fill in general information first, but unlike subject 2, he quickly narrowed in on problem types, sometimes from just a few keywords. This subject always attempted to classify by problem type and kept to a specific problem-type frame once it was chosen. This subject went over a partial checklist before halting the sessions. This was the only subject who collected all the relevant information without building the formulations.

Subject 8 was the only subject to fail to correctly classify some of the problem types. Even in the four problems he classified correctly, this subject had a tendency to ramble. He seemed to start filling in slots, but switched frames frequently. In the opinion of the judges, this occurred because he misinterpreted some of the information given him. We don't know whether to attribute this to a communication problem between the subject and the KE, an unexpected lack of expertise on the part of this particular subject, or even the possibility that the successes of all of the other subjects were merely accidental. We do discount the likelihood of KE inexpertise in this case, since this subject was chronologically the last of the eight. By the time he worked with this subject, our KE had had sufficient time to "practice."

Table 1. Results from the Classification Study

	BLENDING PROBLEMS		PRODUCTION SCHEDULING PROBLEMS		PRODUCT MIX PROBLEMS		TRANSPORTATION PROBLEMS		ASSIGNMENT PROBLEMS		UNCLASSIFIED PROBLEM
	1	2	1	2	1	2	1	2	1	2	1
SUBJECTS 1	AHA	AHA		AHA		YES			AHA	YES	
2	AHA	AHA	YES		YES		AHA				YES
3	YES	AHA			AHA	AHA	AHA			YES	
4			AHA		YES	AHA	AHA			AHA	YES
5		AHA	AHA			YES	YES		YES	AHA	YES
6			YES	YES	YES	YES	YES			YES	YES
7		AHA	AHA	AHA	YES			YES	YES	AHA	
8	YES	FAIL			YES	YES	YES			FAIL	YES

Our Findings

We defined an "AHA!" experience as happening if the subject clearly and correctly identified the problem type. This happened if either: (1) the subject stated the problem type; or (2) the judges could see a clear point delineating general questions and slot filling for a particular frame. This happened in nearly all the blending problem trials and in forty percent of the other problem trials. In a full 50% of the trials, the subject proceeded more slowly and seemed to be asking more general questions, but along the way picked up the correct information for a frame and realized it belatedly. In these cases, typically, the delineation point came when the subject started to work on decision variables and constraints consistent with a particular problem type, and inconsistent with others. In this case, the subject was also said to be classifying by problem type. Even if the subject attempted to classify by problem type, but failed to do so, a "FAIL" situation was defined. This happened in only two of the forty-eight trials (4%), and with only one of the eight subjects. The above is summarized in Table 1. An entry of "AHA" indicates that the subject correctly classified by problem type, and that the AHA experience as described above was noted. An entry of "YES" indicates that the subject classified correctly by problem type. An entry of "FAIL" indicates that the subject failed to classify correctly, whether or not an initial attempt was made at classification.

It is interesting that in all but two of the trials, the subjects eventually identified the problem types and in all cases where classification by problem type occurred, the subject gathered all the relevant information in the problem. This is true even though some of the problems had little "twists" that required more than the basic information that a particular problem type would ordinarily require.

SUMMARY

This study has LP formulation as its focus domain. But we feel that the issues addressed, especially the elicitation of knowledge representation schemes, is applicable to OR domains in general. These domains are essentially quantitative; hence, expert knowledge for these domains may not be readily verbalized. The problems inherent in using common knowledge acquisition techniques, in particular the inability to give full verbal reports, are often compounded in such domains. We recommend that elicitation techniques for acquiring organizational knowledge as well as those for acquiring procedural knowledge be included in the KA stages of formulating expert systems for these domains. It is likely that the combination of verbal techniques that we have presented is appropriate for eliciting both types of knowledge for OR application areas.

The results of an empirical approach to study expert LP classification have been outlined. This type of research does not lend itself to strict statistical analysis. Usually a single expert is used for KA purposes, since the consensus is that multiple experts lead to unnecessary and unavoidable conflicts and inconsistencies in rule sets and organizational schema. However, we felt that a fuller system would result from consultations with multiple experts. Although there is as yet no statistical tool available to judge results of this type, we feel that the fact that seven out of eight subjects always classified, and classified correctly, is not an accident. It seemed as if many of the subjects were working from internal checklists, filling slots in general frames and narrowing down to a classification by problem type. All subjects at least attempted to classify, and in forty-six of the forty-eight (95.8%) trials classified correctly.

This study is not all-inclusive, nor was it meant to be. It should be repeated, using subjects other than academics, and other knowledge engineers. Our purpose was to add a small amount to what we feel is an overlooked facet in knowledge elicitation. The results of this study lend support to the classification theory. It seems that the problem of automated LP formulation can be supported, at least in part, by a model-driven approach in which frames store expectations about input to be gathered.

Acknowledgement. This research was supported in part by grants from the University Research Council and the College of Business Administration of the University of Cincinnati. We also wish to express our gratitude to the subjects who took part in this study.

APPENDIX: EXCERPTS FROM CONTEXT FOCUSING SESSIONS

The following are excerpts from context focusing sessions. In the first session, the subject rapidly identifies a product mix problem and gathers the appropriate information. The subject calls a halt to the session when he feels that he has all the information necessary to proceed with the formulation. In these excerpts, S represents the subject's questions and/or comments; KE those of the knowledge engineer.

SESSION 1

S: Who's making the decision?

KE: Just a company.

S: Do they have a product?

KE: Yes.

S: Are they trying to decide how much to produce?

KE: Yes.

S: Do they make one or more products?

KE: More than 1.

S: How many products?

KE: 3.

S: Do different raw materials go into the

products?

KE: It doesn't specify.

S: Do they receive money for selling these products?

KE: Yes.

S: Can I have the selling price of these things and their names?

KE: $7 for regular; 15 for super; and $25 for deluxe.

S: Are there costs associated with making them?

KE. Yes. $3 raw materials for regular; $6 for super; and $10 for deluxe.

S: Is there labor involved?

KE: Yes.

S: And this is per unit? Per regular, the labor would be?

KE: 0.1.

S: Is this ten cents? Or what?

KE: One tenth of an hour.

S: OK. Are super and deluxe also given in hours?

KE: Yes. 0.2 hours and 0.5 hours.

S: 0.2 hours; 0.5 hours. Is the labor all

in the same category? Any labor can work with any of the products?

KE: Yes.

S: Are there any demand or guaranteed delivery units of each type?

KE: I have upper limits on weekly sales.

S: So, weekly sales of regulars cannot exceed?

KE: 1000 units.

S: Is there a minimum amount required?

KE: No.

S: Super?

KE: 800.

S: Deluxe?

KE: 300.

S: Is there a plant capacity? In terms of labor available per week?

KE: I have 5 individuals who work up to 40 hours per week.

S: Do they get paid whether they work or not?

KE: Yes. They're salaried.

S: So, if they work 20 hours or 40 hours, they get the same wage?

KE: Yes.

S: Are there any limitations on the amount of raw material available?

KE: No.

S: In terms of their decision, it's how much to produce. Is it how much to produce on a weekly basis? They review this weekly?

KE: Yes.

S: Are there any more costs involved, shipping, overtime, inventory, anything like that?

KE: No.

S: OK. I've got enough to do the problem.

Session 2

In this session, the subject identifies a multiple-period production scheduling problem. After the excerpt presented here, the subject went on to ask about monthly demands, beginning and ending inventory information, and fluctuation costs, and then built the formulation.

S: Am I given per-unit costs, and if so, for what?

KE: You're given per-unit costs, but what do you mean by "for what?"

S: I mean is it doughnuts, or something? Do you know?

KE: Just a product.

S: OK. So I have a cost of a product. Is there more than one product?

KE: No.

S: Am I given production costs? Holding costs?

KE: I have production costs.

S: Production costs per unit?

KE: Yes.

S: Is what?

KE: It varies.

S: So, it must be a time problem. Does it vary by time?

KE: Yes.

S: How many periods?

KE: Six.

S: Six periods. OK. And production costs varies for a single product. So, is there a holding cost involved from period to period?

Eliciting KR Schema for LP Formulation

KE: Yes.

S: Is it the same for all six periods?

KE: Yes. $2 per month per unit.

S: OK. $2. These periods are months?

KE: Yes.

S: OK. So it's a 6-month time horizon. And the production cost varies. Are we given any other kind of cost besides holding and production?

KE: No.

S: We're not given any kind of selling price or anything like that?

KE: No.

S: Is there a limit or capacity as to how much you can produce in a single period?

KE: Yes.

S: Does it vary from period to period?

KE: No.

S: Is there overtime available?

KE: Yes.

S: What is the capacity per period? Regular time?

KE: One hundred units per month on regular

time.

S: How much overtime is available? In terms of units?

KE: Fifteen units per month.

S: Is there an additional cost associated with producing a unit in overtime versus in regular time?

KE: Yes.

S: Is it attributed directly to the unit or to the hours?

KE: To the unit.

S: How much is that?

KE: It's $5 more per unit than overtime.

S: Why don't you go ahead and give me the cost for regular time production in each of the six periods?

In the rest of session 2, the subject went on to ask for more information, built the formulation and finally ended the session with a discussion of model design and a checklist: (X_i is regular time production in month i, Y_i overtime production in month i)

S: And I also note that X_i is always going to be all the way up to 100. In other words, X_i is always going to be saturated up to its limit before a Y_i will become positive, because the cost for overtime is always higher than the cost of regular time.

S: So, if that wasn't true, you might have to force some relationship that said Y_i doesn't become positive until X_i is filled up, but that will automatically happen in the optimization (for this problem).

S: So, capacity's taken care of; inventory's taken care of; demand is taken care of. I think that's it.

REFERENCES

Binbasioglu, Meral. (1986) KNOWLEDGE BASED MODELLING SUPPORT FOR LINEAR PROGRAMMING. Ph.D. Dissertation, Graduate School of Business Administration, New York University.

Chandrasekaran, B. (1986) "Generic Tasks in Knowledge-Based Reasoning: High-Level Building Blocks for Expert System Design." IEEE EXPERT, 1(3), 23-30.

Chi, M. T. H., P. J. Feltovich and R. Glaser. (1981) "Categorization and Representation of Physics Problems by Experts and Novices. COGNITIVE SCIENCE, 5, 121-152.

Dolk, Daniel R. and Benn R. Konsynski. (1984) "Knowledge Representation for Model Management Systems." IEEE TRANSACTIONS ON SOFTWARE ENGINEERING, 1984, 10(6), 619-628.

Ericsson, K. A. and H. A. Simon. (1984) PROTOCOL ANALYSIS: VERBAL REPORTS AS DATA. MIT Press, Cambridge, MA.

Evans, James R. and Jeffrey D. Camm. (1987) "An Expert Modeling Assistant for Production Planning Optimization Problems." Proceedings, DSI National Conference, Boston.

Fedorowicz, Jane and Gerald B. Williams. (1986) "Representing Modeling Knowledge in an Intelligent Decision Support System." DECISION SUPPORT SYSTEMS, 2, 3-14.

Fikes, Richard and Tom Kehler. (1985) "The Role of Frame-Based Representation in Reasoning." COMMUNICATIONS OF THE ACM, 28(9), 904-920.

Fink, Pamela K. and John C. Lusth. (1987) "Expert Systems and Diagnostic Expertise in the Mechanical and Electrical Domains." IEEE TRANSACTIONS ON SYSTEMS, MAN AND CYBERNETICS, SMC-17(3), 340-349.

Guenthner, Franz, Hubert Lehmann and Wolfgang Schonfeld. (1986) "A Theory for the Representation of Knowledge." IBM JOURNAL OF RESEARCH AND DEVELOPMENT, 30(1), 39-56.

Harmon, Paul and David King. (1985) EXPERT SYSTEMS. ARTIFICIAL INTELLIGENCE IN BUSINESS. John Wiley & Sons, New York.

Hart, Anna. (1985) "Knowledge Elicitation: Issues and Methods." COMPUTER AIDED DESIGN, 17(9), 455-462.

Konst, L., B. J. Wielinga, J. J.Elshout and W. N. H. Jansweijer. (1983) "Semi-Automated Analysis of Protocols from Novices and Experts Solving Physics Problems." PROCEEDINGS OF THE EIGHTH INTERNATIONAL JOINT CONFERENCE ON ARTIFICIAL INTELLIGENCE. 8-12 August, Karlsruhe, West Germany, Volume 1, 97-99.

Kuipers, Benjamin and Jerome P. Kassirer. (1987) "Knowledge Acquisition by Analysis of Verbatim Protocols." In Alison L. Kidd, ed. KNOWLEDGE ACQUISITION FOR EXPERT SYSTEMS. Plenum Press, New York, 45-71.

Ma, Pai-chun, Frederic. Murphy and Edward A. Stohr. (1988a) "A Graphics Interface for Linear Programming." Working paper, College of Business & Economics, University of Delaware. March 1, 1988.

Ma, Pai-chun, Frederic. Murphy and Edward A. Stohr. (1988b) "Representing Knowledge about Linear Programming Formulation." Working paper, College of Business & Economics, University of Delaware. April 5, 1988.

McKeithen, Katherine B., Judith S. Reitman, Henry H. Rueter and Stephen C. Hirtle. (1981) "Knowledge Organization and Skill Differences in Computer Programmers." COGNITIVE PSYCHOLOGY, 13, 307-325.

Murphy, Frederic H. and Edward A. Stohr. (1986) "An Intelligent System For Formulating Linear Programs." DECISION SUPPORT SYSTEMS, 2, 39-47.

Newell, Allen. (1982) "The Knowledge Level." ARTIFICIAL INTELLIGENCE, 18, 87-127.

Newell, Allen and Herbert A. Simon. (1972) HUMAN PROBLEM SOLVING. Prentice-Hall, Englewood Cliffs.

Orlikowski, Wanda and Vasant Dhar. (1987) "Imposing Structure on Linear Programming Problems: An Empirical Analysis of Expert and Novice Models." IJCAI-87, Proceedings of the Tenth International Joint conference on Artificial Intelligence, Milan, Italy, August, 308-312.

Sklar, Margaret M., Roger Alan Pick and James R. Evans. (1987) "On the Automatic Formulation of LP Problems." Presented at ORSA/TIMS, St. Louis, October 25-28.

Slatter, Philip E. (1987) BUILDING EXPERT SYSTEMS: COGNITIVE EMULATION. Ellis Horwood, Ltd., Chichester, UK.

Stohr, Edward A. (1987) "Automated Support for Formulating Linear Programs. Working Paper, Information Systems Area, Graduate School of Business Administration, New York University.

Wright, George and Peter Ayton. (1987) "Eliciting and Modelling Expert Knowledge." DECISION SUPPORT SYSTEMS, 3, 13-26.

A KNOWLEDGE BASE FOR INTEGER PROGRAMMING – A META-OR APPROACH

Fatemeh Zahedi

INTRODUCTION

In this paper, we introduce the concept of "meta-OR" for formalizing the qualitative knowledge imbedded in the applications of OR/MS methods. "Meta-OR" is based on the meta-knowledge concept in artificial intelligence, and formalizes the qualitative knowledge of OR applications as "meta-rules".

The formal encoding of qualitative knowledge is necessitated by two main factors in the field. The first factor is the lack of any formal procedure by which one can generalize from the reported applications, thus accumulating the knowledge obtained from such experiences. Secondly, in structured modeling where attempts are made to formalize, and eventually, computerize the process of problem solving, theoretical characteristics of a problem may not be sufficient for the decision on which method should be used to formulate the problem. The knowledge obtained from the wealth of experience has equal, if not more, value in the model selection.

To test the usefulness of the concept of "meta-OR", we have applied it to integer programming applications. The number and diversity of meta-rules obtained show that codifying the qualitative knowledge of the application process could prove helpful in transferring expertise on OR application. Furthermore, such formalization identifies the shortcomings that exist in reporting qualitative information in the application field.

META-OR

The need for a formal method of organizing the qualitative knowledge about OR applications arises from a number of issues: 1) encoding formulation, 2) accumulating application knowledge, 3) formalizing implementation, and 4) structured modeling.

Encoding Formulation. In most problems, there are more than one approach to formulating and modeling a problem. For example, an allocation problem with some integer variables could be formulated either as a mixed integer programming problem, or as a linear programming problem with some approximation. The characteristics of a given problem should dictate which formulation is most appropriate. One of the major sources of knowledge about the match between problem features and formulation methods is the application case. Formalizing such knowledge can help applied workers select the appropriate formulation.

Accumulating Application. The OR/MS community has traditionally reported application cases of various techniques, and considered such reports as major contributions to the field and to the understanding of the method. After four decades, the field has accumulated a voluminous literature on applications of the OR/MS methods. The data produced from such applications are now large enough to warrant the call for going "from micro observations to macro laws" (Little 1986, p. 4). The question is what approach one should use to formalize the qualitative knowledge generated from applications to facilitate moving from "micro observations" to "macro laws."

When we compare the process of knowledge accumulation through theoretical developments to that of application, some major differences become evident:

a. In theory, the knowledge is coded formally. The formalism is mostly mathematical. No such formalism exists in application.

b. The formalism of theory makes it possible to accumulate knowledge in that the next step or extension is built on the previous result. Such accumulation is nonexistent in application.

c. In theory, an implicit standard of presentation has emerged in that the method starts from a set of assumptions or axioms, and uses the existing theoretical knowledge to arrive at a useful conclusion. No such standard exists for application reports.

d. The survey articles, textbooks, and books on special topics emphasize, synthesize and expand on methodology. The application cases, at best, receive a cursory mention. This is due to the lack of formal methods for presenting the qualitative knowledge on application cases.

Due to the lack of any vehicle summarizing and accumulating application experiences, gaining knowledge on the application has been delegated to case studies, field experience, and other individual experiments. Each person has to either start from scratch, reinvent the application wheel himself, or read every application paper in the field; papers that have little similarity in structure and presentation of qualitative information. Neither of the two prospects is a satisfactory choice. This has prompted the recommendation that OR/MS methods should be used to organize and manage the practice of OR/MS (Jones and Smithin 1984; Kennedy 1982).

Formalizing Implementation. The literature on the use of OR methods testifies to the fact that finding the best solution does not guarantee its successful implementation. Although this has been long recognized as one of the most important impediments in the success of OR in organizations, the application papers devote little space to implementation issues. This omission is partly due to the lack of the authors' involvement in the implementation process, and partly to the lack of a formal mechanism to report implementation problems and issues in an orderly and efficient manner.

Structured Modeling. Structured modeling (Geoffrion 1987) strives to shape a modeling structure by breaking the problem into its fundamental components, then formulating and solving it automatically. Such a process requires a knowledge-base of the qualitative knowledge of previous experiences in formulating and solving similar problems.

Formalizing and encoding the qualitative knowledge about the applications of OR methods address the above issues, and make it possible to bring the application from the realm of an art closer to a science. We call this formalism "meta-OR."

META-RULES

"The expression 'meta-X' means 'X about X,' so meta-knowledge means knowledge about knowledge" (Hyes-Roth, Waterman, and Lenat, 1983, p. 220). The concept of meta-knowledge has been applied in the artificial intelligence (Davis 1980). One of the methods of encoding knowledge in AI is production rules, where the knowledge is expressed in the conditional form of IF-THEN. These rules are task specific. For example,

- If   formulating the problem in LP
  then   determine whether it is a max or min problem
       ^ determine the linear objective function
       ^ determine the linear constraints,
  where ^ stands for "and".

This rule is specific to the task of formulating a problem in LP form. The production rules mostly consist of interrelated rules in that the antecedent (IF part) of one rule could be the consequent (THEN part) of another rule. For example,

- If   determine the linear objective function
  then identify the number of variables
       ^ identify the coefficient of each variable in objective function
       ^ form the linear objective function.

In the above example, one of the consequents of the first rule appears as the antecedent of the second rule, whose consequents could be the antecedents of other rules, hence providing a chain of reasoning which ties the rules together, and provides a cohesive knowledge-base, related to the task of formulating a problem in LP.

There are cases where the knowledge does not pertain to the task at hand, but it is related to the selection and use of other rules. For example,

- If   a problem could be formulated in LP or in integer programming
       ^ the size of the problem is large
  then formulate the problem in LP.

This is called a "meta-rule;" it guides the system to the production rules of LP. The collection of the meta-rules is called the "metaknowledge-base" (Zahedi 1987). The meta-rules, while determining the priority of production rules and determining their selection, may not necessarily be interrelated themselves.

We suggest "meta-OR" as the meta-rules for formalizing the qualitative knowledge regarding the application and implementation of OR methods. Meta-OR by its nature is i) experience-based, ii) dynamic, iii) uncertain, and iv) cumulative.

i) Meta-OR is experience-based by definition, in that the method's application environment produces the qualitative and inexact knowledge. Theoretically, in the above example, both LP and IP algorithms lead to the optimal solution. The meta-rule for preferring LP to IP when both could be applied equally, is based on the experience that no IP code runs faster than its LP counterpart for large size problems.

ii) The dynamism of meta-OR results from the fact that new theoretical developments may change the nature and concerns of the application process. For example, if one finds an algorithm for integer programming that works faster than its LP counterpart, then there is no need for the meta-rule recommending LP over IP. Hence, the metaknowledge-base should be continuously updated.

iii) Since the meta-OR is based on experience, the validity of its rules is uncertain. As in any applied knowledge, there is no guarantee that the results have been correctly and completely reported, and that a better approach is not possible.

iv) The above characteristics of meta-OR lead to its cumulative nature, since as the number of application reports containing the same meta-rule increases, so does the validity of that rule. This highlights the importance of accurately encoding and reporting the qualitative aspects of applications.

Due to the diversity of applications, one can categorize the meta-rules of meta-OR in the following categories:

1. The first category of meta-rules pertains to the general system of problem solving. An example of such meta-rules are the rules for computerizing the implementation of an OR model.

2. The second category consists of the meta-rules for the formulation of a particular problem that could be generalized to the formulation of a similar class of problems. An example of such meta-rules is the use of the spider graph approach (tree graph in which only one node may have degree greater than 2) for formulating a transshipment problem.

3. The third category is for the novel formulation of a particular attribute of the problem that may be generalized to any problem where the attribute is present. An example of such a meta-rule is the method of modeling time as one of the attributes of the problem.

4. The fourth category includes the meta-rules for applying a solution technique to a formulated problem that can be generalized to the class of such formulation. An example of such a meta-rule is the following meta-rule:

- If   solving a 0-1 integer programming formulation

        ^ no integer programming package is available
        ^ an LP package is available
        ^ approximate answers may be sufficient
then   add a small value (e_i) to the coefficients of binary variables in the objective function
        ^ remove the binary constraints
        ^ solve the problem using an LP package
(Ellis and Corn 1984).

5. The fifth category consists of the meta-rules about the application of an algorithm or a solution method that can be generalized to cases when the solution method is applied. An example of the meta-rules in this category is

- If   solving an integer programming problem with numerous constraints
       ^ want to gain computational efficiency
  then introduce some constraints only when they are violated
(Sherali et. al. 1987).

6. The sixth category consists of meta-rules about an existing problem for which no solution exists and one needs to make additional assumptions, approximations, or other simplifications in order to solve it. This category also involves the unresolved issues and weaknesses that may exist in various methods. Two examples of such meta-rules are as follows:

- If   solving the setup problem in its general form for a flexible manufacturing system then the problem is intractable (Stecke 1983).

- If solving a large scale mixed integer problem
then hard constraints are a shortcoming (Bell et. al. 1983).

7. The last category includes the meta-rules for computations that can be generalized to similar computational problems. An example of such meta-rules is

- If reduction in CPU memory and time requirement is desired
  ^ the computation involves a sparse matrix of density about 15%
  ^ the elements of matrix are 0 or 1
then replace the matrix with a matrix of row and column indices for elements with value 1
(Haessler and Talbot 1983).

Although the delineation of these categories is not crisp, the classification of the meta-rules into these categories makes it feasible to analyze them. In the next section, we will use these categories to discuss the meta-rules for integer programming applications.

META-OR IN INTEGER PROGRAMMING

To check the applicability and usefulness of the meta-OR concept, we applied it to the case of integer programming applications. Since the novel and important applications are usually published in the OR/MS journals, we reviewed the papers on IP application published in the major OR/MS journals since

1980. These papers were reviewed to derive meta-rules from their reports. The cutting point of 1980 was selected to accommodate the most recent applications and to limit the extent of the review.

The meta-rules derived from these papers should be considered as collected data from the field, data which is dynamic, uncertain, and cumulative. Since the papers were written without any concern for the organization and presentation of the qualitative knowledge, only a few papers contain general conclusions on the qualitative aspects of the applications. Therefore, the generalization from the content of the paper to the meta-rule is the author's interpretation of the papers' contents.

Of the papers on the integer programming applications published since 1980, about 40 papers contained some qualitative information on application, and 103 meta-rules were derived from them, leading to an average of about 2.5 meta-rules per paper.

Interesting is the distribution of the meta-rules in the seven categories discussed in the previous section. As Table 1 shows, the first four categories contain 87 meta-rules, about 84 per cent of all the meta-rules. The second category, the meta-rules for the formulation of a similar class of problem, contains 28, the highest number of meta-rules. This shows the focus of the papers located in formulation.

The category with the second highest number of meta-rules, 24, is the fourth class: meta-rules for generalizing a solution technique for solving the class of similar formulation.

The third category, the novel formulation of an attribute of the problem, is also related to the formulation, and has 16 meta-rules. Hence, the categories for generalizing formulation and solution (categories 2 through 4) contain 66 per cent of all the meta- rules. This documents the common criticism of the MS/OR field – it puts too great an emphasis on formulation and solution aspects of the problems. Even in application reporting, where other phases of problem solving should receive attention, the reported qualitative information involves formulation and solution 66 percent of the time.

Of 40 papers, only nine discussed the general issues of modeling, resulting in 19 meta-rules of the first category. Ideally, this category should have a great deal of implementation meta-rules. However, except for a couple of papers, little information on implementation was provided in the reviewed papers.

The next section reports details of the IP meta-rules.

IP META-RULES

Meta-rules are categorized in 7 categories, and are reported in Appendices A through G. A word in "[]" indicates that it has the potential for generalization, that is, one may replace it with a variable. For example, a paper concerning emergency services has produced a number of meta-rules that may be applicable to other cases as well. In that case, [emergency services] indicates that it has the potential for generalization to other types. Moreover, at the end of every meta-rule, the paper from which the meta-rule has been derived is cited.

Category 1. Meta-rules for general system problem solving

As Appendix A shows, this category contains meta-rules pertaining to the system in general. It is divided into four subcategories:

1.1. Meta-rules for automated and manual systems (6 rules)
1.2. Meta-rules for large-scale systems (3 rules)
1.3. Meta-rules for gaining understanding of the system (3 rules)
1.4. Meta-rules for increasing the usefulness of the system for users and managers (7 rules).

The major concern of the first subcategory is the manual system and the process of automation, and the possible ways of resolving conflicts between automated and manual systems. These rules are helpful in specifying the scope of the problem to be modeled and automated. Given the serious impact of such decisions on the success of the implementation, the small number of such meta- rules should be of major concern.

The second subcategory contains the meta-rules for large-scale automation. Considering the fact that IP formulation and solution in recent years have been greatly concerned with large- scale problems, there is inadequate information on the large systems in general in IP application papers. For more specific aspects of formulation and solution, the concern about the size of the problem is addressed more frequently.

The third subcategory contains meta-rules for exploring the nature of the problem. Only three rules fall into this subcategory. The three meta-rules are technical and solution- oriented. Again, given the orientation of the solution methods towards large-size problems, the number and scope of meta-rules for exploring the nature of the problem are grossly inadequate.

The fourth subcategory addresses the meta-rules for increasing the efficiency of the system. The major theme of the seven meta- rules in this subcategory is the variety of ways for involving the management and users with the system.

Category 2. Meta-rules for the formulation of a similar class of problems

As Appendix B shows, in this category the meta-rules contain the features of the problem which lead to the selection of a particular method of formulation. Obviously, the features listed in the meta-rules are in addition to the technical requirements of the method, and should not be considered an exhaustive list.

This category has four subcategories:

2.1. Meta-rules for network design, siting, and location (15 rules)

2.2. Meta-rules for scheduling (4 rules)

2.3. Meta-rules for screening and assignment (2 rules)

2.4. Meta-rules for other types of problems (7 rules).

The first subcategory, the network, siting, and location problems, has the highest number of meta-rules and the greatest variety of conditions. In most cases, the consequent of the meta- rule recommends using one of the IP formulation types and applying a particular solution method.

One of the common themes of this subcategory is how to reduce the size of the problem; the recommendations include: aggregating suppliers into supply zones (rule 24), deleting impossible combinations (rule 27), application of variable reduction and constraint aggregation methods (rule 32), and using average costs in the objective function (rule 34). Similar concern is noticeable in the second and third subcategories, although the number of the meta-rules in these are too few for any conclusion. The fourth subcategory shows the variety of possible applications that may be generalized to similar classes of problems. In some cases, integer goal programming is suggested to accommodate the problem's multiple objectives.

Category 3. Meta-rules for novel formulation of a particular attribute that one can use in any problem where that attribute is present

This category focuses on the particular attribute of a problem that one may have difficulty formulating (Appendix C). Problems of different types may have a similar attribute, and these meta- rules provide advice on how to model such attributes.

The subcategories of this category are:

3.1. Meta-rules for size attribute (3 rules)
3.2. Meta-rules for time attribute (4 rules)
3.3. Meta-rules for continuous variable attribute (4 rules)
3.4. Meta-rules for nonexact solution attribute (2 rules)
3.5. Meta-rules for stochastic variable attribute (1 rule)
3.6. Meta-rules for estimated variable attribute (2 rules).

Size is the running concern in the formulation and solution of IP applications. The first subcategory addresses the size as an attribute of the formulated IP problem. Although the number of meta-rules is small in this subcategory, other categories do contain meta-rules with size as part of their antecedent. For example, meta-rule 48 in this subcategory has similar consequent to that of meta-rule 27, both recommending the elimination of the dominated options. However, in 27, the recommendation is part of the formulation process, while in 48, the problem is the size of the formulated problem.

The second subcategory (3.2) contains the ways to model time in a problem. Each of the four rules in this subcategory recommend a different treatment of time, discretisizing (rule 51), performing sensitivity analysis (rule 52), using time as decision criterion (rule 53), and having a subscript for time (rule 54). This subcategory is a special case of the next one, the continuous variable case.

The third subcategory (3.3) contains the recommendations for the treatment of continuous variables, ranging from discretisizing (rule 55), combined discretisizing and sensitivity analysis (rule 56), to ranking and using goal programming (rules 57, 58).

The subcategories of stochastic and estimated variables (3.4 and 3.5) have too few rules, but a great potential for future applications.

Category 4. Meta-rules for applying a solution technique to a formulated problem that can be generalized to the class of similar formulation

This category focuses on solution techniques. As Appendix D shows, it has the second largest number of rules, and its meta- rules are divided into the following subcategories:

4.1. Meta-rules for solving large IP problems (7 rules)
4.2. Meta-rules for solving nonlinear IP problems (7 rules)
4.3. Meta-rules for solving IP problems when the most appropriate resource is not available (3 rules)
4.4. Meta-rules for solving multi-objective IP problems (1 rule)
4.5. Meta-rules for IP problems with the integer objective function (1 rule)
4.6. Meta-rules for heuristic design (5 rules)

The first subcategory (4.1) contains the recommendation for solving large size IP problems. The use of variable dominance (in this case, in groups) is again one of the recommended approaches (rule 68). The recommended solutions are the varieties of Lagrangian relaxation (rules 65, 66, 69), X-system (rule 64), and heuristics (rules 67 and 70).

The second subcategory (4.2) addresses nonlinearity in IP. The recommended approaches are simple scaling (rule 71) and linearization (rules 72-77).

The third subcategory (4.3) focuses on cases where the best resources are not available to solve the IP problem. It varies from the case where a routing problem has to be solved manually (rule 78), when IP package is not available but LP package is (rule 79), to when only a small computer is available (rule 80). This subcategory has crucial importance for an applied worker because one rarely has access to the best resources at the start of a project. The small size of the meta-rules in this subcategory is another indication of how lack of formalism can impede the reporting process of qualitative information.

The fourth subcategory (4.4) addresses the multi-objective IP solution methods and has only one meta-rule (rule 81). However, the multi-objective nature of the problem has been addressed as a part of rules in other categories, e.g. rules 33, 45, and 57. This indicates that one meta-rule could be listed under more than one category. Similarly, the case where the IP problem has to have an integer objective function has only one meta-rule, recommending a simple heuristics.

The last subcategory (4.6) is not strictly IP oriented, and all rules of this subcategory are based on Hillier (1983). Since a number of meta-rules contain the design and use of heuristics, this subcategory was included here.

Category 5. Meta-rules about the application of an algorithm or a solution method that can be

generalized to cases where the solution method is applied

As Appendix E shows, the number of meta-rules in this category is limited to 4, and methods are for the X-system (rule 88), Hop, Skip, and Jump (rules 89, 90), and heuristics (rules 91 and 92). The small number of meta-rules for the use of the algorithms and solution methods might have been caused by the cutting date of 1980, before which the meta-rules for the use of the conventional IP algorithms are covered. However, heuristics has been used more frequently after 1980, and one would expect to find more qualitative information from the applications of such methods.

Category 6. Meta-rules about shortcomings and unresolved issues

This category is one of the most illuminating, in that it identifies the unresolved issues from the application point of view, and should constitute the feedback from the field to those working in pure theory. It is, therefore, disappointing to see only 5 meta-rules in this category (Appendix F).

Of the five rules in this category, the first four (rules 93-96) are more in the nature of negative recommendation and shortcoming than of unresolved issue. It is the last rule (rule 97) that identifies a problem as intractable. This lack of feedback from the applied field to the theoretical worker could have been caused by the journal's policies. It is an area worthy of attention by journal editors and reviewers.

Category 7. Meta-rules for computations that can be generalized to similar cases

As appendix G indicates, this category has 6 meta-rules that almost exclusively are concerned with computational efficiency. Some recommendations focus on reduction of CPU or computational time (rules 98, 99), and others on general computational efficiency (rules 101, 102, 103). The extent of the advice ranges from specific computer programming methods (rule 98) to solving the problem as a network problem (rules 101 and 102). Given the major concern for solving large-scale problems, the scarcity of the meta-rules for computational aspects of large scale systems is surprising.

CONCLUSION

This paper promoted the idea of using the metaknowledge concept of AI to formalize the qualitative knowledge of applications of OR/MS methods in the field. To demonstrate this idea, the papers in integer programming applications published since 1980 were reviewed and 103 meta-rules were derived. The discussion of these meta-rules reveal a number of findings:

- The qualitative information in the reports could be formalized using the production rules of AI. This will increase our ability to store and organize those aspects of OR applications that have been considered an art.

- The concern about overemphasis on formulation and solution at the expense of problem identification, implementation and other stages of modeling is documented in the rules in that about 66 percent of the qualitative information in application papers are directed toward formulation and solution. In addition, the quantitative parts of papers are almost exclusively oriented towards formulation and solution .

- Given the focus of IP formulation and solution for large-scale problems, it is surprising to observe the lack of qualitative information about the general nature of large systems and computational aspects of large problems.

- There is little feedback from the applied field on the unresolved issues and areas to which theory workers can contribute.

It seems meta-OR approach has the potential of bringing into focus some of the neglected areas of OR practice, as well as organizing the existing wealth of practical knowledge.

REFERENCES

Bartholdi, J. J. III; Platzman, L. K.; Collins, R. L.; and Warden, W. H. III. 1983. A minimal technology routing system for meals on wheels. *Interfaces*, 13(3), June, 1-8.

Bell, W. J.; Dalberto, L. M.; Fisher, M. L.; Greenfield, A. J.; Jaikumar, R; Kedia, P.; Mack, R. G.; and Prutzman, P. J. Improving the distribution of industrial gases with an on-line computerized routing and scheduling optimizer. *Interfaces*, 13(6), 4-23.

Belardo, S.; Harrald, J.; Wallace, W. A.; and Ward, J. 1984. A partial covering approach to siting response resources for major maritime oil spills. *Management Sciences*, 30(10), 1184-1196.

Birtan, G. R.; Haas, E. A.; and Matsuo, H. 1986. Production planning of style goods with high setup costs and forecast revisions. *Operations Research*, 34(2), 226-236.

Brill, E. D.; Chang, S.; and Hopkins, L. D. 1982. Modeling to generate alternatives: The HSJ approach in land use planning. *Management Science*, 28(3), 221-235.

Brown, G. G.; Geoffrion, A. M.; and Bradley, G. H. 1981. Production and sales planning with limited shared tooling at the key operation. *Management Science*, 27(3), 247-259.

Brown, G. and Graves, G. 1981. Real-time dispatch of petroleum tank trucks. *Management Science*, 27(1), 19-32.

Cook, W. D. 1984. Goal programming and financial planning models for highway rehabilitation. *Journal of the Operational Research Society*, 35(3), 217-223.

Darby-Dowman, K. and Mitra, G. 1985. An extension of set partitioning with application to scheduling problems. *European Journal of Operational Research*, 21(2), 200-205.

Davis, R. 1980. Meta-rules: Reasoning about control. *Artificial Intelligence*, 15, 179-222.

Economides, S.; Fok, E. 1984. Warehouse relocation or modernization: Modeling the managerial dilemma. *Interfaces*, 14(3), 62-67.

Ellis, P. M. and Corn, R. W. 1984. Using bivalent integer programming to select teams for intercollegiate women's gymnastic competition. *Interfaces*, 14(3), May-June, 41-46.

Eppen, G. D.; Martin, R. K.; and Schrage, L. 1986. A mixed- integer programming model for evaluating distribution contribution in the steel industry. *Computers and Operations Research*, 13(5), 575-586.

Geoffrion, A. M. 1987. An introduction to structured modeling. *Management Science*, 33(5), 547-588.

Ghandforoush, P. and Greber, B. J. 1986. Solving allocation and scheduling problems inherent in forest resource management using mixed-integer programming. *Computers and Operations Research*, 13(5), 551-562.

Gilbert, K. C.; Holmes, D. D.; and Rosenthal, R. E. 1985. A multiobjective discrete optimization model for land allocation. *Management Science*, 31(12), 1509-1522.

Glover, F.; Hultz, J.; and Klingman, D. 1978 and 1979. Improved computer-based planning techniques, *Interfaces*, Part I 8(4), 16-25 and Part II 9(4), 12-20.

Glover, F., and McMillan, C. 1986. The general scheduling problem: An integration of MS and AI. *Computers and Operations Research*, 13(4), 563-573.

Gunther, R. E.; Johnson, G. D.; and Peterson, R. S. 1983. Currently practiced formulations for the assembly line balance problem. *Journal of Operations Management*, 3(4), 209-221.

Haessler, R. W. and Talbot, F. B. 1983. A 0-1 model for solving corrugator trim problem. *Management Science*, 29(2), 200-209.

Hayes-Roth, F.; Waterman, D. A.; and Lenat, D. B. (Eds.) 1983. *Building Expert Systems*, Addison-Wesley Publishing Co., Reading, MA.

Heiner, K.; Kupferschmid, M.; and Ecker, J. G. 1983. Maximizing restitution for erroneous medical payments when auditing samples from more than one provider. *Interfaces*, 13(5), October, 12-17.

Hillier, F. S. 1983. Heuristics: a gambler's roll. *Interfaces*, 13(3), June, 9-12.

Ignizio, J. P. 1983. An approach to the modeling and analysis of multiobjective generalized networks. *European Journal of Operational Research*, 12(4), 357-361.

Jasinska, E. and Wojtych, E. 1984. Location of depots in a sugar- beet distribution system. *European Journal of Operational Research*, 18(3), 396-402.

Jones. S. and Smithin, T. 1984. Using MS for the practice of MS. *Interfaces*, 14(3), 68-75.

Kennedy, M. 1982. Communication: On significance of the second kind. *Management Science*, 28(3), 337.

Keown, A. J.; Taylor, B. W. III; and Pinkerton, J. M. 1981. Multiple objective capital budgeting within university. *Computers and Operations Research*, 8, 59-70.

Klingman, D; Phillips, N.; Steiger, D.; Wirth, R.; Padman, R.; and Krishnan, R. 1987, An optimization based integrated short- term refined petroleum product planning system. *Management Science*, 33(7), 813-830.

Kolesar, P. and Showers, J. L. 1985. A robust credit screening model using categorical data. *Management Science*, 31(2), 123-133.

Lawless, M. W. 1987. Institutionalization of a management science innovation in police departments. *Management Science*, 33(2), 244-252.

Lee, S. M.; Green, G. I.; and Kim, C. S. 1981. A multiple criteria model for the location-allocation problem. *Computers* and *Operations Research*, 8(1), 1-8.

Little, John D. 1986. Research opportunities in the decision and management sciences. *Management Science*, 32(1), 1-13.

Martin, C. H. and Lubin, S. L. 1985. Optimization modeling for business planning at Trumbull Asphalt. *Interfaces*, 15(6), 66-72.

Marsten, R. E. and Muller, M. R. 1980. A mixed integer programming approach to air cargo fleet planning. *Management Science*, 26(11), 1096-1107.

Mulvey, J. M. 1983. Multivariate stratified sampling by optimization. *Management Science*, 29(6), 715-724.

Narula, S. and Ho, C. A. 1980. Degree-constrained minimum spanning tree. *Computers and Operations Research*, 7, 239- 249.

Nauss, R. M. 1986. True interest cost in municipal bond bidding: An integer programming approach. *Management Science*, 32(7), 870- 877.

Nauss, R. M. and Markland, R. E. 1985. Optimization of bank transit check clearing operations. *Management Science*, 31(9), 1072-1083

Rakes, T. R.; Franz, L. S.; and Sen, A. 1984. A heuristic approximation for reducing problem size in network file allocation models. *Computers and Operations Research*, 11(4), 387-395.

Sengupta, S. 1981. Goal programming approach to a type of quality control problem. *Journal of the Operational Research Society*, 32, 207-211.

Sherali, H.; Staschus, K.; and Huacuz, J. 1987. An integer programming approach and implementation for an electric utility capacity planning problem with renewable energy sources. *Management Science*, 33(7), 831-847.

Stecke, K. 1983. Formulation and solution of nonlinear integer production planning problems for flexible manufacturing systems. *Management Science*, 29(3), 273-288.

Weintraub, A.; Guitart, S.; and Kohn, V. 1986. Strategic planning in forest industries. *European Journal of Operational Research*, 24(1), 152-162.

Swersey, A. J. and Ballard, W. 1984. Scheduling school buses. *Management Science*, 30(7), 844-853.

Zahedi, F. Economics of expert systems and contribution of MS/OR. *Interfaces*, 17(5), 1987, 72-81.

Zahedi, F. A method for quantitative evaluation of expert system. Forthcoming in *European Journal of Operational Research*.

Zipkin, P. 1980. Bounds on the effect of aggregating variables in linear programs. *Operations Research*, 28(2), 403-418.

Table 1
Number of Meta-rules Derived from the IP Application Papers

Category of Meta-rules	Number of Meta-rules	% of Meta-rules
1. General system problem solving	19	18%
2. Formulation of similar class of problems	28	27%
3. Formulation of a particular attribute of a problem	16	16%
4. Solution technique for a class of formulated problems	24	23%
5. Application of an algorithm or a solution method	5	5%
6. Shortcomings and unresolved issues	5	5%
7. Computations	6	6%
Total	103	100%

APPENDIX A

1- Meta-rules for the general system problem solving

1.1. Meta-rules for automated and manual systems

1.
- If   deciding to automate part of a system
  ^ some parts heavily involve human judgment
  ^ some parts are time consuming and detailed
  ^ some parts are successfully performed manually with fairly simple rules of thumb
  then   choose parts that do not heavily involve human judgment
  ^ are not successfully performed manually with fairly simple rules of thumb
  ^ are time consuming and detailed (Brown and Graves 1981).

2.
- If   complete automation is cost-prohibitive
  then   use semi-automation (Brown and Graves 1981).

3.
- If   a sub-model and manual intuitive method are not compatible
  then   incorporate the manual intuitive method into the system
  (Bell et. al. 1983).

4.
- If   the computer generated solution does not match the habitual manual method
  ^ the machine generated solution and the habitual manual method are not reconcilable
  ^ gain in acceptability for the computerized system is needed
  then   let the user to have control over choosing between the computer generated solution and the habitual manual method
  (Bell et. al. 1983).

5.
- If   shift scheduling is manual
  ^ the system is large
  then   manually produced schedules will have overages in certain times
  ^ shortages in other times
  ^ may produce poor service
  ^ resulting schedule is likely to be substantially less than optimal
  (Glover and McMillan 1986).

6.
- If automating [dispatch] system for a [terminal]
  then larger automated systems are more successful than the small one
  (Brown and Graves 1981).

1.2- Meta-rules for large-scale systems

7.
- If deciding to automate a large system
  then determine the extent of automation
  ˆ determine the extent of reality introduced into the system.
  (Brown and Graves 1981).

8.
- If implementing a large scale automated system
  then train inexperienced users
  ˆ identify and solve unforeseen problems
  ˆ integrate the system into existing activities
  (Bell et. al. 1983).

9.
- If computerizing a large system
  then three distinct steps are needed: data extraction and database definition
  ˆ problem preprocessing and diagnosis
  ˆ optimization and report writing
  (Brown, Geofrian and Bradley 1981).

1.3. Meta-rules for gaining understanding of the system

10.
- If generating alternative solutions for decision making
  then do one of the following
  change parameters and solve the problem over generate alternatives at random
  V obtain alternate optima, if exist
  V examine alternative feasible solutions generated in the course of solution
  V generate local optima by gradient search starting from a different point
  V use different formulations for solving the problem
  V use different algorithms to obtain solutions
  V use Hop, Skip, and Jump approach

(Brill, Chang, and Hopkins 1982).

11.
- If   intend to gain an understanding of a multi- objective programming formulation
  then   maximize and minimize each objective separately
  ˆ identify the range of each objective
  ˆ identify the objectives with small range
  (Brill, Chang, and Hopkins 1982).

12.
- If   solving a multi-objective programming formulation
  ˆ have identified the maximum and minimum range for each objective separately
  ˆ want to exclude some objectives from the formulation
  then   exclude the objectives with the smallest range
  (Brill, Chang, and Hopkins 1982).

1.4. Meta-rules for increasing the usefulness of the system for users and managers

13.
- If   using an optimization model for strategic and tactical planning
  ˆ the model is in operational form
  then   make a specific staff position responsible for its use and maintenance
  (Martin and Lubin 1985).

14.
- If   implementing a large, integrated system
  then   implement the system on a prototype first
  ˆ evolve the final system gradually as management generates questions
  (Klingman et. al. 1987).

15.
- If   a large system is to be successful
  then   have the support of top management
  ˆ have adequate human, financial, and computer support
  ˆ have pictorial representation of the underlying model to enhance the ability to communicate the model to management and, the management to verify the correctness of the model and provide proper input data
  ˆ have managers dedicated to using the prototype model to help clean up and evolve the final system
  (Klingman et. al. 1987).

16.
- If applying quantitative methods to real problems
  then a considerable time should be spent to identify the question that the analyst can answer
  ^ answers to questions should be useful to management
  (Eppen et. al. 1986).

17.
- If the essence of management's request is to be identified
  then prepare an expanding list of management's "what if" questions
  (Eppen et. al. 1986).

18.
- If integrating the system into existing activities
  then design the man-machine interface into the daily work flow
  (Bell et. al. 1983).

19.
- If the problem is to institutionalize a model
  then the factor affecting long-term use should be part of the structured approach to model building
  ^ increase the flexibility of the model
  ^ increase the top management receptiveness to innovation
  (Lawless 1987).

APPENDIX B

2- Meta-rules for the formulation of a similar class of problems

2.1. Meta-rules for network design, siting and location

20.
- If   the problem is the design of a network
       ^ the problem involves transporting [objects] in [vehicles]
       ^ [objects] are small relative to [vehicles]
       ^ [objects] have numerous origins
       ^ [objects] have numerous destinations
  then   use a (single or multiple hub) spider graph
       (Marsten and Muller 1980).

21.
- If   using spider graph approach for formulating transshipment problem
  then   one may not require the exchange of cargo to occur only at the hub
       (Eppen et. al. 1986).

22.
- If   the problem is the design of a network
  then   formulate the problem in integer programming
       ^ experiment with the integer programming solution to find a satisfactory design
       (Marsten and Muller 1980).

23.
- If   the problem is a network with associated cost
       ^ the degree of node is constrained (limited number of connection to each node)
  then   formulate this degree-constrained spanning tree in 0-1 integer programming form
       ^ use one of two construction procedures (by Narula and Ho) and a branch and bound algorithm to solve the problem
       (Narula and Ho 1980).

24.
- If   the problem is the location of depots for [sugar-beet] distribution system
       ^ the problem is large-scale
  then   aggregate the suppliers into supply zones
       ^ use the mean value of distances of all suppliers in a zone from each depot

　　　　　˄ use regular mixed integer programming algorithm to solve the problem
　　　　　　(Jasinska and Wojtych 1984

25.
- If　　　the problem is the location of depots for [sugar-beet]
　　　　　distribution system
　　　　˄ the problem is large scale
　　　　˄ no aggregation is desirable
　then　　use heuristics to solve the problem
　　　　　(Jasinska and Wojtych 1984).

26.
- If　　　the problem involves the siting of [emergency service] facilities
　　　　˄ there are locations where demands happen
　　　　˄ demand should be covered within some response distance or
　　　　　time from the [emergency service] facility
　　　　˄ there are a limited number of facilities to cover all demands
　　　　˄ the goal is to minimize the number or cost of facility siting
　　　　　or maximize a positively weighted sum of demand centers
　　　　　covered by a fixed number of
　　　　　[emergency service] facilities
　then　　use integer programming to formulate the problem
　　　　　(Belardo et. al. 1984).

27.
- If　　　the problem involves the siting of [emergency service] facilities
　　　　˄ using the integer programming to formulate the problem
　　　　˄ the [emergency] event consists of multiple area types
　　　　˄ the [emergency] event consists of multiple event types
　　　　˄ the [emergency] event consists of multiple weather types
　then　　enumerate all area, event, and weather types
　　　　˄ list all the possible combination of area, event, and weather types
　　　　˄ delete the impossible combinations
　　　　˄ list the set of [emergency service] facilities needed for
　　　　　each event (Belrado et. al. 1984).

28.
- If　　　the problem involves the siting of [emergency service] facilities
　　　　˄ using the integer programming to formulate the problem
　　　　˄ maximizing the weighted sum of demand centers covered by a
　　　　　fixed number of [emergency service] facilities
　　　　˄ the probability of each [emergency] event is identifiable
　then　　use the probabilities of each [emergency] event as the weight in the
　　　　　objective function

˄ the maximization of the probability weighted covered demand center
may not be satisfactory for the low probability but
disastrous [emergency] cases
(Belrado et. al. 1984).

29.
- If   the problem involves the siting of [emergency service] facilities
    ˄ [emergency] event types have different external impacts
    ˄ using the integer programming to formulate the problem
    ˄ maximizing the weighted sum of demand centers covered by a
      fixed number of [emergency service] facilities
then   the problem has multiple objective
    ˄ identify efficient solutions for the decision maker's subsequent
      tradeoffs (Belrado et. al. 1984).

30.
- If   the problem involves the siting of [emergency service] facilities
    for [spill] events
    ˄ using the integer programming to formulate the problem
    ˄ maximizing the weighted sum of demand centers covered by
      a fixed number
      of [emergency service] facilities
then   identify the [spill] events
    ˄ specify the response time for each [spill] event
    ˄ identify the equipment bundles appropriate for the
      various combinations of [oil] types and weather types
    ˄ identify the potential response equipment locations
    ˄ determine the travel time between each equipment location and
      the centroid of each [spill] event location
    ˄ determine the covering sets for each [spill] event
    ˄ determine the relative probabilities of [spill] occurrences corres-
      ponding to each [spill] event
    ˄ assign each [spill] event to the group reflecting the external
      impact type (Belardo et. al. 1984).

31.
- If   the problem is land selection for development
    ˄ the selection is for a single land use
    ˄ one must determine the location and shape of the land use
then   use the integer multiobjective programming method by Gilbert et. al.
    (Gilbert, Holmes, and Rosenthal 1985).

32.
- If   the problem is land allocation and production planning in forest

              resource management
  then   formulate it in mixed integer programming form
             ^ apply variable reduction method
             ^ apply constraint aggregation method
             ^ solve it using ordinary mixed integer programming method
              (Ghandforoush and Greber 1986).

33.
- If   the problem is location/allocation
       ^ there are multiple, conflicting objectives
  then   formulate it as an integer goal programming problem
       ^ change priorities of the hierarchical order of goals to perform
         sensitivity analysis
       ^ perform sensitivity analysis to answer "what if" questions regarding
         the change in priorities due to changes in environmental conditions
         (Lee et. al. 1981).

34.
- If   the system is a distributed computer network
       ^ the problem is the determination of optimal number of multiple files
       ^ the problem is the determination of location of files
       ^ the variance among the network transmission costs are not high
  then   formulate the problem in integer programming form
       ^ use average costs in the objective function to reduce the
         dimnesionality of the problem (Rakes et. al. 1984).

        2.2. Meta-rules for scheduling

35.
- If   the problem is transportation scheduling
       ^ the problem is formulated as integer programming
       ^ the speed of the vehicle is of importance
  then   perform sensitivity analysis to find the impact of the change in speed of
         transportation (Swersey and Ballard 1984).

36.
- If   the problem is shift scheduling
       ^ the size of the problem is more than 1,000,000 shifts
  then   use heuristics that blend MS/AI to solve the problem
         (Glover and McMillan 1986).

37.
- If   the problem is shift scheduling
       ^ the computer is micro

then   use heuristics that blend MS/AI to solve the problem
             (Glover and McMillan 1986).

38.
- If   the problem is the determination of production schedule
           ˆ the capacity is limited
           ˆ demand is stochastic
           ˆ demand occurs only in the last season of planning
           ˆ items could be grouped into families
           ˆ families have the same setup costs
           ˆ mean of family demands is invariant in the planning period
           ˆ the item demands are forecast in each period
           ˆ production of each family is setup once
   then   aggregate the items into product families
           ˆ formulate the problem as mixed integer programming
           ˆ use Birtan et. al. algorithm for solving for families
           ˆ disaggregate family productions into its item members via
             another mixed integer programming problem
             (Birtan et. al. 1986).

   2.3. Meta-rules for screening and assignment

39.
- If   the problem is credit screening
           ˆ the attribute data are binary
           ˆ the rule should be easy to implement
           ˆ then sample data is very large
           ˆ the decision should be defendable in public forum
           ˆ in the frequently occurring profiles, the 'nesting' property holds
           ˆ external decision on the inclusion of an attribute should be
             easily implemented
   then   formulate the problem in integer programming
           (Kolesar and Showers 1985).

40.
- If   the problem is assigning coupons to different maturities in debt issues
           ˆ the criterion is the lowest true interest cost
   then   use the integer programming formulation
           ˆ apply the Lagrange relaxation developed by Nauss
             (Nauss 1986).

   2.4. Meta-rules for other types of problems

41.

- If   solving a multivariate stratification problem
        ˆ the population database is very large
        ˆ minimizing the within-cluster variance for attributes
  then   formulate the problem in integer programming
        ˆ use subgradient method to solve the formulation
          (Mulvey 1983).

42.
- If   solving system setup for flexible manufacturing systems
  then   solve sequentially part type selection problem, first
        ˆ solve machine grouping problem, second
        ˆ solve production ratio problem, third
        ˆ solve resource allocation problem, fourth
        ˆ solve loading problem, last (Stecke 1983).

43.
- If   the problem is the competition lineup in gymnastics
  then   the problem can be formulated in integer programming
         (Ellis and Corn 1984).

44.
- If   the problem is strategic planning based in a tactical model
         for [forest industries]
  then   aggregate stands, activities, and time from the tactical model for
         strategic model
        ˆ create consistency between the strategic and tactical
          models using Zipkin's
          aggregation approach (Zipkin 1980)
        ˆ the solution of the strategic model should be replicable in the tactical
          model, as a check for consistency (Weintraub, Guitart, and Kohn 1986).

45.
- If   the problem is priority planning [such as pavement maintenance]
        ˆ the problem could be partially formulated in capital budgeting form
        ˆ the capital budgeting formulation does not accommodate certain
          aspects of the problem [such as year-to-year stability, targets and
          goals, lower limits, and budget inadequacy]
        ˆ the goals could be prioritized in a hierarchical fashion
  then   formulate the problem in a two-phase process
        ˆ phase one is the planning phase to determine the minimal level of
          funding for each project as goals via linear programming formulation
        ˆ phase two is the priority phase of formulating the problem
          in a mixed integer goal programming formulation
        ˆ solve the problem via a special Lagrangian relaxation approach

(Cook 1984).

46.
- If   the problem is assembly line balancing
   then   minimize number of working areas and employees required, as a goal
        ~ make sure the sum of activity times does not exceed the (percentage of) cycle time, as a goal
        ~ make sure workload assigned to area or employees has not increased since last balance, as a goal
        ~ adhere to layout requirement of plant as a goal
        ~ combine activities for less boring assignment to worker, as a goal
        ~ avoid assigning several physically demanding activities to the same area or worker, as a goal
        ~ adhere to the technological constraints of sequencing, as a goal
        ~ assign tasks with common tooling to the same station, as a goal
        ~ assign tasks with common parts to the same station, as a goal
        ~ assign tasks requiring the same skill to the same station, as a goal
        ~ group tasks together which require simultaneity of motion as a goal
        ~ rank the goals
        ~ formulate the problem in mixed integer goal programming
        ~ solve the problem using Gunther et. al.'s branch and bound algorithm
        (Gunther et. al. 1983).

47.
- If   the problem is quality control
        ~ the output has a number of attributes
        ~ the attributes of the output should meet a number of specifications
        ~ the attributes have different priorities
        ~ all specifications may not be achieved
   then   modify the form of specification to have one- sided upper limits
        ~ use regression analysis to find the relationship output attributes in terms of input and process variables
        ~ determine the priority of output attributes
        ~ formulate the problem as in goal programming
        (Sengupta 1981).

APPENDIX C

3- Meta-rules for novel formulation of a particular attribute that one can use in any problem where that attribute is present

3.1. Meta-rules for size attribute

48.
- If   formulating a large 0-1 integer programming problem
       ^ reduction in problem size is desirable
       ^ resource allocation options are enumerated in the formulation
       ^ some undominated options could be identified in advance
then   eliminate the dominated options from the formulation
(Haessler and Talbot 1983).

49.
- If   the problem is formulated as a large integer programming problem
then   reformulate it as a binary network assignment with gains
(Brown and Graves 1981).

50.
- If   the problem is formulated as a large integer programming problem
then   develop a network factorization algorithm
(Brown and Graves 1981).

3.2. Meta-rules for time attribute

51.
- If   the problem is transportation scheduling [school buses, tankers, airlines]
then   formulate the problem as a mixed integer programming problem
       ^ approximate the problem by discretisizing time to get a tractable integer programming formulation
(Swersey and Ballard 1984).

52.
- If   the problem is formulation in one-period integer programming
       ^ there is growth in variables over time
then   carry out sensitivity analysis on the single- period model
(Economides and Fok 1984).

53.
- If   formulating [vehicle] routing schedule
       ^ time database is available

```
           ^  distance database is available
           ^  road networks are available
     then     use time and distance as decision criteria
              (Bell et. al. 1980).
```

54.
- If   the problem involves transportation from any point via a given route
    ^ the pick-up time is of importance
  then   break the time into integer time slots
    ^ formulate the problem as an integer programming problem with
              the integer variable having a subscript for time
              (Nauss and Markland 1985).

    3.3. Meta-rules for continuous variable attribute

55.
- If   the problem involves the siting of [emergency service facilities]
    ^ the location where demands happen are continuous
    ^ using the integer programming to formulate the problem
  then   divide the area into discrete homogeneous subareas
              (Belardo et. al. 1984).

56.
- If   the integer programming problem involves discretisized variables
  then   perform sensitivity analysis to evaluate the effect of changing
              discrete intervals
              (Sewersay and Ballard 1984).

57.
- If   the problem is capital budgeting
    ^ projects are indivisible
    ^ there are multiple conflicting objectives of uncommon unit to be
              satisfied
    ^ objectives could be ranked in a hierarchical order
  then   formulate the problem as mixed integer goal programming
              (Keown et. al. 1981).

58.
- If   the problem is capital budgeting
    ^ projects are indivisible
    ^ there are multiple conflicting objectives of uncommon units
    ^ objectives could be ranked in a hierarchical order
    ^ there exists uncertainty about the product demand
  then   formulate the problem as chance-constrained mixed integer

goal programming
- perform sensitivity analysis on goal priorities and chance constraints (Keown and Taylor 1980).

3.4. Meta-rules for nonexact solution attribute

59.
- If the problem is a set covering or set partitioning or set packing
- the exact solution is infeasible or too expensive
- then formulate the problem as a goal programming problem with deviational variables
- use appropriate penalty weights to control overcovering or undercovering (Darby-Dowman and Mitra 1985).

60.
- If soft constraints are desired
- then identify the constraints to which the model is sensitive
- set the constraints to which the model is sensitive to relatively loose values (Bell et. al. 1983).

3.5. Meta-rules for stochastic variable attribute

61.
- If solving an integer programming problem with probabilistic variables
- using the branch and bound method for solution method
- need a good starting solution
- then formulate the problem deterministically using expected value of stochastic variables
- solve this problem using branch and bound method
- use the solution as the starting point for the problem with stochastic variables (Sherali et. al. 1987).

3.6. Meta-rules for estimated variable attribute

62.
- If designing multivariate stratified sampling
- then include only those attributes that are uncorrelated (Mulvey 1983).

63.
- If the problem involves estimating shipping spills
- the historical data for the region is small
- then use the information on cargo throughput, shipping transits, and historical spills to estimate the probabilities (Belardo et. al. 1984).

APPENDIX D

4- Meta-rules for applying a solution technique to a formulated problem that can be generalized to the class of similar formulations

4.1. Meta-rules for solving large IP formulation

64.
- If solving large integer programming problem
  then X-system is a viable option (Brown and Graves 1981).

65.
- If solving very large mixed integer programming formulation
  ˆ very large scale is defined as the number of variables up to 800,000 and the number of constraints 200,000
  ˆ near optimal and approximate answer is acceptable
  then use Lagrangian relaxation method (Bell et. al. 1983).

66.
- If solving a very large mixed integer programming formulation
  ˆ very large scale is defined as the number of continuous variables up to 40,000, number of integer variables up to 12,000 and the number of constraints up to 26,000
  ˆ the problem has special structure
  then utilize the special structure to decouple the problem into small problems via Lagrangian relaxation method
  ˆ reformulate the small problems into network formulation
  (Brown, Geofrian and Bradley 1981).

67.
- If solving a large integer programming formulation of a network
  ˆ large is defined as the number of constraints 80, 1000 flow variables and 60 selection variables
  then use the following heuristics solve the problem by linear programming, pick an integer variable, round it, set its value, and resolve the linear programming problem till all integer variables have integer solutions (Marstern and Muller 1980).

68.
- If solving a extremely large integer programming formulation
  ˆ large is defined as 178 million integer variables
  ˆ large is defined as 183 million constraints
  then group the integer variables according to a criterion

~ utilize the concept of variable dominance to eliminate the variables that are dominated by others based on a criterion
(Nauss and Markland 1985).

69.
- If   solving very large integer programming formulation
then use the modified version of Lagrangian-based branch and bound method developed by Erlenkotter
(Nauss and Markland 1985).

70.
- If   solving a large integer programming formulation
~ speed is essential
then use heuristics
~ audit by offline optimization (Brown and Graves 1981).

4.2. Meta-rules for solving nonlinear IP formulation

71.
- If   the formulation is nonlinear objective function
~ the constraints are linear
~ the variables are integer
~ the objective function is the sum of K objective functions of K subproblems
~ each subproblem has one upper bound constraint
~ there is one resource constraint for all K subproblems
then a simple scaling of the optimum values of K subproblems give the best solution
(Heiner, Kupferschmid, and Ecker 1983).

72.
- If   solving nonlinear integer programming formulation
~ the product of nonlinear 0-1 variables is quadratic
then use Glover's linearization method (Stecke 1983).

73.
- If   solving a set of nonlinear integer programming problems
~ problems have different, high-order, nonlinear product (either in constraints or the objective function)
then do not use Glover's linearization method (Stecke 1983).

74.
- If   solving a nonlinear integer programming formulation
~ the formulation have small constraint size

then use Glover's linearization method (Stecke 1983).

75.
- If solving a nonlinear integer programming formulation
   ^ constraints have few terms in common
  then use Glover's linearization method (Stecke 1983).

76.
- If solving a mixed integer programming formulation
   ^ desire the generation of fewer variables in linearization
  then use Glover's linearization method (Stecke 1983).

77.
- If solving a mixed integer programming formulation
   ^ desire the generation of fewer constraints in linearization
  then use Glover and Woolsey's methods (Stecke 1983).

4.3. Meta-rules for solving IP problems when the most appropriate resources are not available

78.
- If solving a routing problem
   ^ the problem should be solved manually
   ^ the delivery is numerous in each route
   ^ an approximate to optimal solution is adequate
  then use Bartholdi and Platzman heuristic
   (Bartholdi et. al 1983).

79.
- If solving a 0-1 integer programming formulation
   ^ no integer programming package is available
   ^ an LP package is available
   ^ approximate answers may be sufficient
  then add a small value (e_i) to the coefficients of binary variables in the objective function
   ^ remove the binary constraints
   ^ solve the problem using an LP package
   (Ellis and Corn 1984).

80.
- If solving a 0-1 integer programming formulation
   ^ have access only to small computer systems
   ^ the computer time and memory are limited
   ^ good feasible solution is needed early in solution process

then   use an adaptation of the set-partitioning algorithm
       (Haessler and Talbot 1983).

4.4. Meta-rules for solving multi-objective IP formulations

81.
- If   generating solutions for a multi-objective programming formulation
       ˆ the alternative solutions are to be different from each other
  then use Hop, Skip and Jump method
       (Brill, Chang, and Hopkins 1982).

4.5. Meta-rules for solving IP problem with integer objective function

82.
- If   solving an integer programming formulation
       ˆ the objective function should be integral
  then solve the integer programming using the LP algorithm
       ˆ for nonintegral objective function, set a new constraint that
       the objective function should be equal to the smalles
       integer greater the objective value
       ˆ resolve the new problem using the LP algorithm
       ˆ continue the process till an integral objective is attained
       (Swersey and Ballard 1984).

4.6. Meta-rules for heuristic design

83.
- If   designing heuristic
  then identify a general region within which to search for good feasible solution
       ˆ conduct search in this region to find a feasible solution
       ˆ try to improve upon the feasible solution in small steps
       (Hillier 1983).

84.
- If   designing heuristics
       ˆ identifying a general region to find a feasible solution
  then find a good feasible solution even though it takes a longer time
       (Hillier 1983).

85.
- If   designing heuristics
       ˆ have a feasible solution
  then continue searching in the feasible region till no improvement can
       be achieved (Hillier 1983).

86.
- If   designing heuristics
        ^ locked into a well-explored neighborhood
  then allow the search to drift till it homes on a better solution in an entirely different part of the feasible region (Hillier 1983).

87.
- If   designing heuristics
  then first, search should into several different feasible regions
        ^ second, focus on the region where the feasible solutions are good (Hillier 1983).

APPENDIX E

5- **Meta-rules about the application of an algorithm within a solution method that can be generalized to cases where the solution method is applied**

88.
- If   solving a large integer programming problem
        ^ using X-system for the solution
  then tailor general X-system for the particular problem
        (Brown and Graves 1981).

89.
- If   using Hop, Skip, and Jump method to generate different solution alternatives
        ^ the number of solutions should be small and far apart
  then use relatively lax targets for the objectives
        (Brill, Chang, Hopkins 1982).

90.
- If   using Hop, Skip, and Jump method to generate different solution alternatives
        ^ the number of solutions should be numerous with relatively small differences
  then use relatively stringent targets for the objectives
        (Brill, Chang, Hopkins 1982).

91.
- If   solving an integer programming problem via heuristics that blend AI/MS
  then generate a partial or complete solution as trial solution
        ^ modify or elaborate the solution by transition rules
        (Glover and McMillan 1986).

92.
- If   using heuristics for solution
  then use bounded heuristics for their reliability
        ^ use audited error distribution
        (Brown and Graves 1981).

APPENDIX F

6- Meta-rules about shortcomings and unresolved issues

93.
- If the problem involves the siting of [emergency service] facilities
  ˆ using integer programming to formulate the problem
  ˆ maximizing the weighted sum of demand centers covered by a fixed number of [emergency service] facilities
  ˆ the probability of each [emergency] event is identifiable
  ˆ using the probabilities of each [emergency] event as the weight in the objective function
  then the maximization of the probability-weighted covered demand center may not be satisfactory for the low probability but disastrous [emergency] cases (Belrado et. al. 1984).

94.
- If using an optimization algorithm
  ˆ the problem is a combinatorial problem
  then the modanic view of optimization is a weakness (Brown and Graves 1981).

95.
- If solving large scale mixed integer problem
  then hard constraints are a shortcoming (Bell et. al. 1983).

96.
- If scheduling a vehicle routing system
  then the shortest route is not necessarily the best route (Bell et. al. 1983).

97.
- If solving the setup problem in its general form for flexible manufacturing system
  then the problem is intractable (Stecke 1983).

APPENDIX G

7- Meta-rules for computations that can be generalized to similar cases

98.
- If   reduction in CPU memory and time requirement is desired
       ˆ the computation involves a sparse matrix of density about 15%
       ˆ the elements of matrix are 0 or 1
then   replace the matrix with a matrix of row and column indices for elements with value 1
(Haessler and Talbot 1983).

99.
- If   using set-partitioning algorithm
       ˆ for time consuming solutions the algorithm should be stopped before obtaining the optimal solution
then   terminate the execution when an improved solution is not found within a time period equal to three times the cumulative CPU time needed to obtain the current best solution
(Haessler and Talbot 1983).

100.
- If   the problem is typically complex
       ˆ some important objectives cannot be captured within a mathematical programming model
then   formulate the problem in mathematical programming model
       ˆ generate a small number of solutions that are good solutions in the model and different from each other in the decision space (Brill, Chang, and Hopkins 1980).

101.
- If   the problem is an integer programming
       ˆ gain in computational efficiency is required
then   formulate the problem into an equivalent generalized network model
       ˆ solve the problem as a network problem
(Glover, Hultz, and Klingman 1978 and 1979).

102.
- If   the problem is integer goal programming
       ˆ gain in computational efficiency is required
then   formulate the problem as a generalized network
       ˆ solve it as a network problem

(Ignizio 1983).

103.
- If solving an integer programming problem with numerous constraints
      want to gain computational efficiency
  then introduce some constraints only when they are violated
      (Sherali et. al. 1987).

VI. PERFORMANCE ANALYSIS AND COMPLEXITY MANAGEMENT OF EXPERT SYSTEMS

371

There are many steps in the process of building an expert system. Particularly important tasks include verification, validation and testing. Jafar and Bahill ("Validator: A Tool for Verifying and Validating Personal Computer Based Expert Systems") have developed a general purpose tool, described in their paper, to help verify and validate knowledge bases with little human intervention.

The second paper in this section ("Measuring and Managing Complexity in Knowledge-Based Systems: A Network and Mathematical Programming Approach") by O'Leary notes that developing the knowledge base of an expert system is a complex process. The extent of that complexity is likely to impact the development time and cost of the system, quality of the system, validation and assessment efforts, and other development and maintenance issues. Thus, there is interest in analyzing the impact of complexity on those factors. One approach to that analysis is to develop some metrics for complexity of systems and compare those metrics to those factors for specific systems.

O'Leary's paper investigates the use of network representations of rule-based knowledge bases to address that broad base of complexity issues. A particular network representation of rule-based knowledge bases is presented. Complexity metrics deriving from networks are used as a basis of assessing the structural complexity of that rule-base.

The existence of complexity suggests the need for tools to manage that complexity. That same network representation is used to develop mathematical programming formulations to assist in a number of complexity management issues including ordering rules, finding rules likely to be in solutions, eliciting possible new rules, finding circular reasoning and exploiting parallel architectures. The formulation of these management problems as mathematical programming problems has the additional benefit of providing benchmarks for the computational difficulty of those problems.

The last paper in this section by Cox ("Pragmatic Information Seeking Strategies for Expert Classification Systems") describes how the costs of information gathering in classification systems can be significantly reduced. Systems for learning classification rules and for estimating "optimal" discriminant functions have played prominent roles in the theories of machine learning and statistical classification, respectively. Cox's paper addresses the converse problem of how best to apply known classification rules or discriminant functions in practical settings where the information required by the classification procedure is costly to obtain. This is one of the key extensions that is needed to win academic AI and statistical classification methods broader acceptance among practitioners for whom the expense of gathering relevant information is a primary constraint on decision-making. In addition to constructive strategies for information-gathering that will (approximately) minimize the expected costs of classification decisions, this paper presents a unifying conceptual framework for viewing certain features of expert classification systems, machine-learning algorithms for concept classification, and more conventional statistical techniques of discrete discriminant analysis. Experiments with the heuristics presented here indicate that they are extremely effective in obtaining nearly-optimal solutions to classification decision problems that are known to be NP-hard. This provides support for the often-asserted claim that in at least some settings -- here, a Boolean one with assumptions of known probability distributions and mutual statistical independence among uncertain quantities -- it is possible for ideas from AI to overcome the curse of combinatorial complexity by providing heuristics that provide excellent performance for practical purposes.

Validator, A Tool for Verifying and Validating Personal Computer Based Expert Systems

Musa Jafar
and
A. Terry Bahill
Systems and Industrial Engineering
University of Arizona
Tucson, AZ 85721
sie!musa@arizona.edu
sie!terry@arizona.edu

ABSTRACT

The most difficult tasks in expert system design are verification, validation and testing. Traditional techniques for these tasks require the knowledge engineer to work through the knowledge base and the human expert to run many test cases on the expert system. This consumes a great deal of time and does not guarantee finding all mistakes. On the other hand, brute force enumeration of all inputs is an impossible technique for most systems. Therefore, we have developed a general purpose tool to help verify and validate knowledge bases with little human intervention. Our tool, named *Validator,* has four main components: (1) a Syntactic Error Checker, (2) a Debugger, (3) a Rules and Facts Validation Module, and (4) a Chaining Thread Tracer. It was designed for knowledge bases that use the M.1[1] expert system shell; however, the principles should generalize to any rule-based, backchaining shells, i.e. MYCIN derived shells.

---

[1] Contrary to popular belief in the AI community, M.1 is still sold, supported and updated by Cimflex Teknowledge Inc.

INTRODUCTION

There are many steps in the process of making an expert system: identifying an appropriate problem domain, learning about the problem domain and the structure of the problem, specifying the input-output performance criteria, selecting a good expert, selecting an expert system shell (or perhaps selecting a language and a quantitative technique for dealing with uncertainty), extracting the knowledge from the expert, encoding this knowledge in the knowledge base, verifying the knowledge base, validating the system, testing the system, updating and maintaining the system, and finally, at the end of its life cycle, retiring and replacing the system (Lehner and Adelman, 1989). This paper discusses verification and validation.

Validation means building the right system: that is writing specifications and checking preformance to make sure that the system does what it is supposed to do. Verification means building the system right: that is ensuring that it correctly implements the specifications. Testing means running test cases on the system to see if it emulates human input-output behavior. In a typical expert system verification is done first, then validation, and finally testing.

Verification is the process of assuring completeness, consistency and correctness of the syntax of a knowledge base. There are many subtasks in verification, they must be performed in the correct order. First we proof read the knowledge base. *Validator* aids the proof reading process by displaying lists of all possible premises, all possible conclusions, all possible facts, all possible legal values for each object, and all possible goal statements. Next we run a spelling checker on it. Then we look for low level syntactic mistakes (most shells will do this function). Then we look for more subtle syntactic mistakes and finally we run our Debugger.

After verification is complete we can do validation. Validation is the process of assuring the compliance of system performance with the specified system requirements and needs. In other words, validation check the semantics of the system. However, verification and validation of a system does not mean that the system is adequate. Verified and validated systems can still exhibit unsatisfactory functional performance due to poor hardware/software design, poor system-system and human-system interfaces, poor explanation facilities or incomplete and unclear requirements specifications.

After a system is verified and validated, the human expert should run a few dozen test cases through the expert system to test it. It is important to do verification and validation before the testing, to minimize squandering the expert's time finding simple mistakes.

It seems that more mistakes would be detected if many experts tested the system. It is often possible to get an expert to devote a substantial amount of time to a project; being interviewed, verifying and validating the knowledge base, and running test cases. However, we have found that it is difficult to get other experts to devote time to testing the final product for the following reasons: their time is expensive, there are few of them in any geographical area, they might disagree on the criteria used to draw conclusions or even the conclusions themselves, finally, they do not have the personal commitment to the project to compel them to donate copious amounts of time. Therefore, it is difficult to get multiple experts to exhaustively test an expert system. So, it is important to maximize the efficiency of utilizing the domain experts.

Therefore, we have built a general purpose tool to help verify and validate knowledge bases without extensive intervention by human experts during the development stages of an expert system. We tried to make this tool generic so that it can work on any rule-based knowledge base no matter which expert system shell is used. The first two components of the system, the Syntactic Error Checker and the Debugger are a part of verification. While the last two components, The Rules and Facts Validation Module and The Chaining Thread Tracer are a part of validation.

There are other programs for verification and validation of expert system knowledge bases: *Teiresias* for the MYCIN system (Davis, 1976), a program for ONCOCIN (Suwa, Scott and Shortliffe, 1982), *Check* for programs written with Lockheed's LES shell (Nguyen, Perkins, Laffey and Pecora, 1987), ARC for production systems written with ART (Nguyen, 1987), and EVA for knowledge bases using ART and LISP (Stachowitz, Combs and Chang, 1987; Stachowitz, Chang, Stock and Combs, 1987). A sixth similar tool is ESC, a decision-table-based processor for checking completeness and consistency in rule-based expert systems (Cragun, 1987). In contrast to these programs, our system, *Validator,* is specifically designed to run on personal computers.

VALIDATOR

The Syntactic Error Checker

The first component of *Validator* is the Syntactic Error Checker. Syntactic errors are common in expert systems knowledge bases; many of these are misspellings or typographical errors. The expert systems submitted by graduate and undergraduate students as class projects for our Expert Systems course in the Fall of 1987 had an average of 12 spelling errors per knowledge base. The Syntactic Error Checker also checks for syntax that, although legal, produces unspecified behavior of the system such as (1) the use of *is known* or a negation in the conclusion of a rule, e.g.

then type(air-conditioner) is known,
then not(type(air-conditioner) = central);

(2) negations in the right hand sides of premises, e.g. *type(air-conditioner) = not(central)*, where the knowledge engineer probably wanted *not(type(air-conditioner) = central)*, and (3) instantiating objects to *known, unknown, found* or *sought* such as *type(air-conditioner) = known*, where the knowledge engineer probably wanted to use the metafact *is*, e.g. *type(air-conditioner) is known*. This component also checks each user defined object, attribute and value to make sure that it is not a reserved word such as *or, and, mod, add, is,* or *off*.

Providing legal values for an object ameliorates typing errors in response to a question. If no legal values were provided, the system would take any user's response as an answer. So typographical errors might escape detection, even if they were detected, it is hard for a user to recover without restarting the whole system. Providing legal values also allows the user to abbreviate his response, for example, he can type *y* instead of *yes* as an answer. Another type of syntactic error occurs when an illegal value is written into a knowledge base. Most shells check user responses to questions to see if they match legal values specified by the knowledge engineer. However, they do not check the rules to ensure that only legal values have been used. Legal values are associated with questions, not with rules or terms. In figure 1, rule 4 shows an example of a rule using an illegal value. Premise 2 of rules 3 and 5 will similarly be flagged, because no legal values were provided for *marking of animal*. Figure 1 also shows an unused legal value, the value *claws* for the object *extremitie*

Validator 377

of animal was never used in the rule base. In this section, we have shown examples of 7 of the 10 syntactic errors that *Validator* can detect. More examples are given in Jafar (1989), and Bahill (1990).

The Debugger

It has been estimated that testing and debugging comprised 80 percent of the cost of the NASA Apollo project (Yourdon, 1975), 44 percent of the cost of the Saturn 1 project, and 50 percent of the cost of the Naval Tactical Data System (Boehm, 1970). Manual debugging of a knowledge base is expensive and time consuming, it is also difficult, error prone and does not guarantee the finding of all bugs. Debugging means removing compiler specific mistakes that cause the system to not compile or to fail at run time. In the following rule, the variable X will be instantiated whenever the conclusion of the rule falls in the search path of a goal. However, the knowledge engineer mistakenly typed a Y instead of an X in the third premise. If this rule is encountered during a consultation the computer will halt, because the variable Y can not be instantiated.

> if air-conditioner = air-conditioner-X
> and type(air-conditioner-X) = central
> and not(maintained(air-conditioner-Y))
> then maintain(air-conditioner-X).

Validator also checks for unclosed comments. Shells and compilers can detect an unclosed comment if it is the last comment in the knowledge base, but if the unclosed comment is followed by another comment, the unclosed comment will probably escape detection. An unclosed comment warning is issued whenever the Debugger encounters a second beginning of a comment string /* without closing the first one. This type of error is typographical. It usually forces the inference engine to ignore the part of the knowledge base that lies between the two comments.

Debugging also includes the actions of a knowledge engineer running the system and watching for aberrant behavior, such as unexpected questions being asked. Interactive debugging removes many inconsistencies that are virtually impossible to detect manually by a knowledge engineer.

The Rules and Facts Validation Module

When verification is complete and all the syntactic errors are removed from the knowledge base, then validation can begin. Validation means ensuring that the system does what it was supposed to do. Rules that can never fire are typical of mistakes that can be detected by our Rules and Facts Validation Module. For example Rule 6 of figure 1 shows such a rule that can never fire. This module also checks rules and facts for validity. For example the next rule will always fail because of the declared fact.

fact: prob-cards = no.

rule: if found-tag = yes
 and prob-cards = yes
 then lost-tag = yes.

The object prob-cards will never be instantiated to the value *yes,* because it was set to the value *no* by a fact. The above example is from the Carpet Advisor, one of the student generated systems.

The Chaining Thread Tracer

The fourth and last component of *Validator* is the Chaining Thread Tracer. It was designed for backchaining shells, but was later enlarged to handle forward chaining constructs (Jafar, 1989). It checks the validity of each rule by tracing it's premises and conclusions to see if they are properly interconnected. We call this component a Chaining Thread Tracer, because it traces connectivity of the terms through the backchaining system.

Backward chaining systems start their search with a goal as the root of an inverted tree and rules as branches. For example, the goal *identity of animal* of figure 1 brings us to the conclusion of rule 3. From here the term *subtype of animal* links the first premise of rule 3 to the conclusion of rule 2. Next the term *type of animal* links the premise of rule 2 to the conclusion of rule 1. The conclusion of rule 1 is then linked to the premise of rule 1, *coat of animal = hair.* There is a question that can provide a value for this term, coat of animal, so this chain of linking is good.

Validator

To detect logical errors in backchaining systems, *Validator* checked every premise of every rule to make sure that it led to a valid end. A premise has a valid end if it appears in the conclusion of another rule or a question is provided for it by the knowledge engineer. For example, in figure 1 the first premise of rule 2 is a valid end, since it appears as a conclusion in rule 1, the second premise of rule 2 is a valid end also, since a question was provided for it. We also checked the conclusion part of every rule for valid ends. A rule's conclusion is considered to be a valid end if it is either a goal statement or it appears as a premise in another rule. For example, in figure 1 the conclusion of rule 1 appears as a premise in rule 2, in turn, the conclusion of rule 2 appears as a premise of rule 3, the conclusion of rule 3 is a goal. Hence rules 1, 2, and 3 all lead to valid ends. An added complexity in the conclusion validation checking that we have to deal with is caused by the fact that, conclusions of rules can be linked with two types of premises. A rule with a conclusion of the form *then animal = mammal*, has to be linked with the premise *if animal = mammal* and also with *if not(animal = ANY)*, where *ANY* can take any value other than mammal, since both premises are going to be in the search tree of a goal statement.

Figure 1 shows a typical knowledge base. The first three rules are correct. If you have a zebra in mind, it will correctly identify this animal. The rest of the rules illustrate various mistakes that might result when a knowledge base is expanded. Rule 6 is extraneous, it will never be reached during inferencing. Rule 5 shows a mistake that the Chaining Thread Checker will detect. There is no way to get a value for the object *feed of animal*. The object has no question and it does not appear in the conclusion of any other rule. Therefore, it will be flagged as a potential error.

infix of.
goal = identity of animal.

rule1: if coat of animal = hair
then type of animal = mammal.

rule2: if type of animal = mammal and
extremities of animal = hooves
then subtype of animal = ungulate.

rule3: if subtype of animal = ungulate and
marking of animal = black-stripes
then identity of animal = zebra.

rule4: if coat of animal = scales
then identity of animal = fish.

rule5: if feed of animal = meat and
marking of animal = black-stripes
then identity of animal = tiger.

rule6: if coat of animal = feathers and
animal-swims
then habitat of animal = antarctic.

question(coat of animal)=('What is the coat of animal?').
legalvals(coat of animal)=[hair, feathers].
question(extremities of animal) =
('What is type of the extremities of the animal?').
legalvals(extremities of animal) = [hooves, claws].
question(marking of animal) =
('What is type of the marking of the animal?').

Figure 1. A small knowledge base with typical errors. The infix operator is just a convenient way of saving space and improving readability. It allows us to say "coat of animal" instead of "coat-of-animal".

Validator

Results of Testing Validator

Over a period of three years, 50 student generated expert systems were used in the development of *Validator*. Another 14 new expert systems, which were created as final projects by students in our Fall 1988 Senior/Graduate student Expert Systems course, were then used to test it. Each of these programs took an average of 100 hours to create and an average of 50 Kbytes of disk space. Table 1 summarizes some of the potential errors detected by *Validator* in these systems. Rule 4 of figure 1 is an example of a rule using an illegal value. Figure 1 also shows an unused legal value *claws* for the object *extremities of animal*. Rule 6 in figure 1 is an example of a rule that can never fire. On the next page we show an example of a rule (rule-26) that can never succeed.

Table 1. The number of certain types of potential errors detected by *Validator* in 14 student generated expert system knowledge bases.

System Name	Rules using illegal values	Unused legal values	Use of reserved words	Unused rules
Veterinarian	0	9	0	0
Automech	0	0	0	0
Carpet	2	29	0	0
CompSel	1	10	0	1
Diet	1	4	0	2
Legal Drug	5	19	0	0
HAA	0	7	1	2
MacExpert	0	2	3	0
NAPSX	2	36	0	1
O-ring	0	60	0	0
Software	0	3	0	1
SoundFilm	0	72	0	0
Volcanic	2	10	0	0
Wire	0	3	3	0

Validator sometimes flagged items that were not mistakes, but rather were handled by the knowledge engineer outside the expert system. For example O-ring and SoundFilm had external data collection programs that caused *Validator* to think that there were errors where such errors probably did not exist. After testing *Validator* with these 14 class projects, we tested it with five more advanced expert systems. The results of these tests are shown in Table 2.

Table 2. Number of potential mistakes detected by *Validator* in five more advanced expert systems.

System Name	Rules using illegal values	Unused legal values	Use of reserved words	Unused rules
Advice	1	1	0	3
Helper	2	1	0	0
Stutter	7	5	0	6
Fund-Eye	0	2	1	0
Wine	0	2	0	1

The first three systems, Advice, Stutter and Helper, were master's theses projects in the Department of Systems and Industrial Engineering at the University of Arizona. Advice was designed to recommend a study plan for new graduate students. Helper was designed to aid students using the computer laboratory. Stutter was designed to help with the diagnoses and prognosis of children who may have begun to stutter. The fourth system, Fund-Eye, is a retinal disease diagnostic system. The fifth system, Wine, is a demonstration program provided by Teknowledge (the developer of M.1). The following set of rules, taken from the Wine Advisor, shows a rule that can never succeed.

rule-12: if has-sauce = yes
and sauce = sweet
then best-sweetness = sweet cf 90
and best-sweetness = medium cf 40.

rule-26: if best-sweetness = dry
then recommended-sweetness = dry.

rule-27: if best-sweetness = medium
then recommended-sweetness = medium.

rule-28: if best-sweetness = sweet
then recommended-sweetness = sweet.

From rule-12, the object *best-sweetness* will be instantiated either to the value *sweet* with certainty factor 90 or to the value *medium* with certainty factor 40. For rule-26 to succeed, the object *best-sweetness* would have to be instantiated to the value *dry*. But this can not happen, because the conclusion of rule-12 is the only place in the knowledge base where *best-sweetness* gets a value. Therefore rule-26 can never succeed.

Once again we note that *Validator* flags potential mistakes. It is up to the knowledge engineer to decide if it is an actual mistake or not. For example in the above example from the Wine Advisor it is obvious that the knowledge engineer wanted to include rule-26 for completeness, although it would never fire. These tables show data for only 5 out of the 17 types of potential errors that *Validator* currently detects. In previous sections we showed examples of another 8 types of potential errors that it detects.

Undetected Errors

We could continue enlarging *Validator*, but it would never be able to detect all possible errors. For example, the following constructs were meant to allow the user to change his mind about his last answer, and backup to correct it:

whencached(X=Y) = [(not(Y=oops)) and temp=X].
whencached(X=Y) = [(Y=oops) and (do(reset X), do(reset temp), temp)].

However, what resulted was circular reasoning; it put the system into an infinite loop. We did not enlarge *Validator* to detect this particular error.

SUMMARY

Ideally we want to build expert systems that pass exhaustive tests and are accepted by all experts. This goal is not reachable. We will never be able to guarantee that a system works exactly as intended or that it fully meets its requirements specifications. It is almost impossible to generate test cases that exhaust all rules. So some rules are going to escape the verification, validation and testing. Tables 1 and 2 show that *Validator* guarantees that each rule is checked for validity and consistency. It also guarantees that the rule will fire whenever enough evidence is gathered to satisfy its premises. We developed verification and validation tools to find out why a system is not working right and to increase the confidence in the level of performance of a system.

ACKNOWLEDGEMENT

Validator can be purchased from the authors. Research of this paper was partially supported by Grant Nr. AFOSR-88-0076 from the Air Force Office of Scientific Research.

REFERENCES

Bahill A.T. (1990) *Verification and Validation of Personal Computer Based Expert Systems.* Prentice-Hall Inc., Englewood Cliffs, New Jersey.

Boehm B.W. (1970) *Some information processing implications of Air Force space missions:* 1970-1980. Memorandum RM-6213-PR, RAND Corp, Santa Monica.

Cragun B.J. (1987) A decision-table-based processor for checking completeness and consistency in rule-based expert systems. *Int. J. Man-Machine Studies* **26**, 633-648.

Davis R. (1976) *Applications of Meta-Level Knowledge to the Construction, Maintenance and Use of Large Knowledge Bases,* Ph.D. dissertation, Department of Computer Science, Stanford University.

Jafar M.J. (1989) *A Tool for Interactive Verification and Validation of Rule Based Expert Systems,* Ph.D. dissertation, Department of Systems and Industrial Engineering, University of Arizona.

Lehner P.E. and Adelman L. (1989) (Eds.) Special issue on perspectives in knowledge engineering. *IEEE Trans. Syst. Man Cybern.* **SMC-19,** 433-662.

Nguyen T.A. (1987) Verifying consistency of production systems. In *Proc. of IEEE Third Conf. on AI Applications* pp. 4-7. IEEE, New York.

Nguyen T.A., Perkins W.A., Laffey T.J. and Pecora D. (1987) Knowledge base verification. *AI Magazine* 8(2), 69-75.

Stachowitz R.A., Chang C.L., Stock T.S. and Combs J.B. (1987) Building validation tools for knowledge-based systems. In *Proc. of First Annual Workshop on Space Operations Automation and Robotics* pp. 209-216. Houston, Texas.

Stachowitz R.A., Combs J.B. and Chang C.L. (1987) Validation of knowledge-based systems. In *Second AIAA/NASA/USAF Symposium on Automation, Robotics and Advanced Computing for the National Space Program,* pp. 1-9. American Institute of Aeronautics and Astronautics, New York.

Suwa M., Scott A.C. and Shortliffe E.H. (1982) An approach to verifying completeness and consistency in a rule base. *AI Magazine* 3(4), 16-21.

Yourdon E. (1975) *Techniques of Program Structure and Design.* Prentice-Hall Inc., Englewood Cliffs, New Jersey.

Measuring and Managing Complexity in Knowledge-Based Systems: A Network and Mathematical Programming Approach

Daniel E. O'Leary

Graduate School of Business
University of Southern California
Los Angeles, California 90089-1421

ABSTRACT

Developing the knowledge base of an expert system is a complex process. The extent of that complexity is likely to impact the development time and cost of the system, quality of the system, validation and assessment efforts, and other development and maintenance issues. Thus, there is interest in analyzing the impact of complexity on those factors. One approach to that analysis is to develop some metrics for complexity of systems and compare those metrics to those factors for specific systems.

This paper investigates the use of network representations of rule-based knowledge bases to address that broad base of complexity issues. A particular network representation of rule-based knowledge bases is presented. Complexity metrics deriving from networks are used as a basis of assessing the structural complexity of that rule-base.

The existence of complexity suggests the need for tools to manage that complexity. That same network representation is used to develop mathematical programming formulations to assist in a number of complexity management issues including ordering rules, finding rules likely to be in solutions, eliciting possible new rules, finding circular reasoning and exploiting parallel architectures. The formulation of these management problems as mathematical programming problems has the additional benefit of providing benchmarks for the computational difficulty of those problems.

1. INTRODUCTION

Developing a knowledge base in an expert system is a complex process. The extent of that complexity has some readily identifiable costs. First, in a more complex system we can anticipate a more lengthy development time for the knowledge base. Second, with a more complex system it is easy to speculate that the verification and validation costs of the system will be higher. Third, in more complex systems management costs likely would be higher. In addition, to the costs, the quality of the system may be impacted by the complexity. As a result of these cost and quality issues, it is important that the complexity of a knowledge base can be measured so that determinants of that complexity can be established and managed through theorical and empirical investigations.

Complexity also has a direct impact on some of the tasks associated with the management of knowledge base development and implementation. First, the more complex a system the more difficult it can be to structure the rules in its knowledge base. As a result, there is a need to develop tools to assist in the ordering of rules. Second, validating and verifying a system are complicated by the complexity of the system. Thus, there is a need to assist the validator in identifying those rules and conditions in rules for which there is a greater need to validate. Third, more complex systems generally are more difficult to validate and use, unless they can be made smaller or broken into pieces, whose solutions can be joined together to establish a solution to the overall problem.

Complexity management is complicated further because of weaknesses in human performance on complex inference tasks. Researchers such as Schum [1981] and Schum and DuCharme [1971] have established means to determine the impact of such human frailties. As a result, the purpose of this paper is to focus on these issues of measuring and managing complexity in expert systems. This is done by formalizing a rule-based knowledge base as a network and using network and mathematical programming to approach some of those specific management issues and network complexity measures to measure the complexity of the knowledge base.

1.1 Measuring Complexity

One approach to measuring the complexity of a knowledge base is analyze the structural complexity of the rule-base. This refers to the extent to which interaction between rules makes the process of representing the knowledge complex. For example, a large number of rules each of which leads directly from a set of conditions or input data to a solution is not likely to be indicative of a complex knowledge base. Alternatively, a highly integrated set of rules, where one rule leads to many subsequent alternatives is likely to indicate a complex knowledge base. Thus, there is substantial reason to suggest that the underlying structure of the knowledge base is a major component of complexity. One of the primary vechicles from which the structural nature of an entity can be assessed is network theory, alternatively referred to as graph theory.

Another approach to assessing the complexity of a knowledge base or a process associated with a knowledge base (e.g., finding structurally sensitive rules) is to formulate those problems as mathematical programming problems whose computational complexity has been established. For example, if a management or development process is formulated as a mathematical programming problem that is equivalent to a traveling salesmen problem, then that provides a measure of the complexity of that process.

Mathematical programming can be used to ensure that the rules and their corresponding conditions are ordered appropriately. This paper finds that ordering process basically is equivalent to a traveling salesman problem. This is important since it indicates that the ordering of conditions is an established computationally complex problem.

1.2 Mathematical Programming Responses to Managing Complexity

Mathematical programming can be used to formulate solutions to a number of complexity management problems. For example, mathematical programming interpretations can be used to determine various paths, such as the "most likely" inference path, through that network. Such an approach can be an efficient means of determining optimal inference strategies in those situations where a database of information is available to define the inference needs of a particular application, rather than a situation where the information must be iteratively solicited from a user. Using a similar approach, a second, third, etc. most likely path through the network can be found. Such inference paths also serve as an explanation to solutions recommended by an expert system.

The mathematical programming approach also can be applied to the validation process. A portion of the validation process of expert systems (ES) is aimed at determining what the ES knows correctly and incorrectly (O'Leary [1987]). However, there are few means of ascertaining whether the knowledge is correct without direct examination of the represented knowledge or testing the system using sample inputs. Unfortunately, it is difficult to determine correctness of rules that cascade onto other rules, through direct examination. Further, typical test data approaches inevitably are limited by resources and time so that in some cases only a small portion of the knowledge base can be investigated. As a result, it is desirable to develop other means to investigate the correctness of the rules in a knowledge base.

Accordingly, mathematical programming is
analyzed as a basis for the validation of a
rule-based knowledge base. The mathematical
program can be used to ascertain the existence
of cyclic reasoning in the knowledge base, by
analyzing the incidence matrix. If the
knowledge base contains circular reasoning then
solutions found by the system are likely to be
inappropriate. Similarly, mathematical
programming can be used to isolate those rules
on which other rules structurally depend the
most. Those isolated rules can then be used to
focus further content oriented validation
efforts. Further, mathematical programming can
be used to elicit possible new rules as a means
of testing the completeness of the knowledge
base.

In other cases, knowledge bases can be
structured as loosely coupled modules. In
these situations, mathematical programming and
decomposition analysis may provide additional
insights to assist in solution, using various
devices such as parallel architectures.
Finally, since these problems can be formulated
as mathematical programming problems, clearly
more specific and efficient network approaches
can be used to solve these problems.

1.3 Outline of this Paper

This paper proceeds as follows. Section 2
provides some definitions of characteristics of
networks that are used throughout the paper.
Section 3 summarizes network representations of
knowledge bases and investigates a particular
formulation of knowledge base rules as
networks. Then that network is used to develop
an incidence matrix representation of the
rules. Section 4 provides an example of the
approach using the representation of
ignorance. The example demonstrates the
importance of the structure inherent in the
solution space and the impact of ordering the
conditions. Section 5 continues the

investigation of the impact of structure, by providing an algorithmic approach to ordering the rules for their presentation to the user of the system. Section 6 presents a mathematical programming model for the problem of finding the best solution through the knowledge base and provides a methodology to assess the possibility that a rule should be added to the knowledge base. Section 7 presents a mathematical programming model that uses the incidence matrix for the investigation of the existence of circular reasoning, while section 8 develops a model aimed at determining a set of structurally important rules that may warrant further investigation. Section 9 discusses the potential use of mathematical programming decomposition analysis in parallel architecture. Section 10 develops measures for analyzing the complexity of a network representation of a knowledge base. Section 11 briefly summarizes the paper.

2. NETWORKS: SOME DEFINITIONS

A <u>directed network</u> or <u>network</u> is a collection of <u>nodes</u> (1, 2, ..., N) and <u>directed arcs</u> $a_{i,j}$ for i,j in (1, 2, ..., N). It is assumed that each node i and each arc $a_{i,j}$ is unique. Let M be the number of arcs. The <u>degree</u> of a node i is the number of arcs leading into (say H) and out (say J) of that node i, $D_i=(H,J)$.

An <u>undirected network</u> is a collection of nodes (1, 2, ..., N) and undirected arcs $a_{i,j}$ for i,j in (1, 2, ..., N). Each undirected $a_{i,j}$ is equivalent to directed arcs $a_{i,j}$ and $a_{j,i}$. Unless otherwise specified it is assumed that a directed network is being used. A <u>path</u> is a sequence of nodes, g, h, i, ... j, k, where $a_{g,h}$, $a_{h,i}$, ..., $a_{j,k}$ exist. A path is <u>simple</u> if it does not contain the same arc more than once.

A <u>cycle</u> is a path from node i back to node i. A path to i from i that does not go through any intermediate nodes is a <u>loop</u>. A network is <u>acyclic</u> if there are no cycles in the network. The <u>cyclomatic number</u> is equal to the maximum number of linearly independent cycles in a network. In a directed network the cyclomatic number can be measured as M-N+2 (Conte et al. [1986]).

If H=J=N-1 then the network is <u>complete</u>. If H=J=N then the network is <u>complete, with a loop</u>.

A network is <u>connected</u> if there is a path between all pairs of nodes (in an nundirected version of the network, where directed arcs are replaced with undirected arcs). A network is <u>strongly connected</u> if there is a path joining every pair of nodes, such that, if there is a path from i to j then there is a path from j to i. If a network is unconnected the unconnected parts of that network are referred to as components.

In a connected network, a node is said to be an <u>articulation point</u> if the node can be split to yeild an unconnected network. An <u>articulation set</u> is a set of nodes that can be split to yeild an unconnected network. When an articulation point (or set) is split, the connected components in the resulting network are called <u>peices</u> or modules.

A <u>cut-set</u> is a set of arcs in a network, whose removal will increase the number of pieces, whereas the removal of any proper subset will not. Accordingly, the removal of a cut set of arcs will separate the network into two pieces. A <u>minimal cut-set</u> is the cut-set with the fewest arcs.

A network is said to be <u>planar</u> if it can be mapped on a plane in such a way that two arcs meet one another only at nodes with which the arcs are incident. A <u>region</u> of a planar network is an area of the plane that is bounded by arcs and contains neither arcs nor vertices. It can be shown that a connected planar network with a arcs and n nodes has a-n+1 finite regions. <u>Euler's relation</u>, a-n+r=2, is satisfied if the network is planar, where r is the number of regions (assuming an undirected network).

Other network concepts can be explored. Berge [1966], Hu [1970], Kaufman [1967] and Liu [1968], on which this general discussion is based, provide further discussion of networks.

3. THE NETWORK AND MATRIX STRUCTURE OF "If ... Then..." RULES

Rule-based ES use rules of the form "If (Conditions) $a_1, \ldots, a_n$ Then ... (Consequence) $b_1, \ldots, b_m$." Sometimes the rules have a weight on the rule that is used to capture the "likelihood" of the rule. These weights go under a variety of names such as certainty factor (cf) and are developed under a number of assumptions (Speigelhalter [1986]).

Often one rule leads to another rule, as the knowledge "chains" together. This linking of one rule to another provides the view of a rule-based knowledge base as a network. The conditions and consequences can be viewed as nodes in a network, while the "If ... Then ..." relationship defines an arc between those nodes. The path that results from chaining together a continuous set of "If ... Then ..." rules will be referred to as an inference path or a chain of inference. An artificial node that preceeds every other node which have no other inputs, referred to as the source node, can be added to the network. While an artificial node, called a sink, which follows every node for which there are no paths out of, also can be added.

The purpose of section 3.1 is to illustrate one approach to the development of a "route" network of rules for propagating evidence. Section 3.2 summarizes some additional comments on the structure and the impact on combining probabilities. The resulting network is placed in an equation format in section 3.3. Implementation concerns are discussed in section 3.4, while sections 3.5 and 3.6 briefly review some of the previous literature on network and constraint representations of knowledge bases.

The network is developed only as a function of the existing rules, and does not attempt to elicit possible induced relationships (as noted by, e.g., Quinlan [1982]). However, this is not to say that induction of rules and their corresponding network relationships is not appropriate. Only that the network establishes a model of the current understanding of the knowledge base. In addition, the network may form the basis of an investigation of certain possible or probable relationships that should be a part of the knowledge base.

3.1 Generic Cases: Network Formulation

Rules come in either simple or compound formats. Simple rules are of the form "If a_1 then b," where "a_1" and "b" are the condition and consequence. The a_1 and b conditions and consequences can be used as either grouped sets of possible occurrences of a_1 ($a_{1,1}$, $a_{1,2}$, ..., $a_{1,k}$) or as individual occurrences. However, a grouped set would result in a more parsimonious type of structure. Compound rules employ the conditions and consequences of the rules in expert systems linked together by the use of "or" or "and." Using these types of connectors, there are at least five generic types of rules that can investigated as networks. The network that is constructed from the intersection of these cases can be used to represent the total knowledge base.

Case 1: "If a then b."

This is the simpliest and most direct case, yet still has a substantial number of applications (for example see section 3). The nodes are labeled "a" and "b." The network representation of this rule can be illustrated as follows: (a) ------> (b).

Case 2: "If a_1 or a_2 then b."

In this case either a_1 or a_2 must occur in order for b to occur. Accordingly, this relationship can be illustrated as follows:
(a_1)------------>(b)<---------------(a_2).

Case 3: "If a_1 and a_2 then b."

In this case both a_1 and a_2 must occur in order for b to occur. Since each condition is an independent entity the individual conditions must be sequenced, i.e., either a_1 or a_2 must occur first. Since sequencing a_1 or a_2 first may affect other chains of inference, two paths to b, of which at most one can be used, are required. Accordingly, this relationship can be illustrated as follows:
(a_1)--->(a_2)---->(b)<----(a_1)<----(a_2).

There are two approaches to representing this set of relationships one using only three nodes (a_1, a_2 and b) and the other using five (a_1, a'_1, a_2, a'_2 and b). The first

approach introduces loops that must be constrained so that the nodes are traversed in order, while the second approach increases the number of nodes.

Case 4: "If a then b_1 or b_2"

In this case "a" occurs before b_1 or b_2. The sequencing of b_1 and b_2 is not an issue because of the "or" relationship. However, "a" will only impact one of b_1 or b_2 in an inference chain. Accordingly, the network

representation of that rule can be formulated as follows:
$(b_1) <\text{---------}(a)\text{---------}> (b_2)$.

Case 5: "If a then b_1 and b_2"

In this case, if "a" occurs then both b_1 and b_2 occur. Accordingly, as in case 3, the network model requires sequencing both b_1 and b_2. Thus, two paths from "a" are required. This relationship can be summarized as follows:
$(b_1) <\text{----}(b_2) <\text{----}(a)\text{---->}(b_1)\text{---->}(b_2)$.

Similar to case 3 either three or five nodes can be used to represent this rule.

Other Cases

Other cases can be developed by combining these five cases. For example, rules like "If a_1 or a_2 then b_1 and b_2" can be examined.

3.2 Comments on Structure and Probabilities

Comments on Structure

A knowledge base that contains knowledge of the type in cases 1, 2 and 4, generally will be the easiest to solve. Unless some circular reasoning of the type "if a then b; if b then c; and if c then a," has been developed, an acyclic network will result. As seen in PERT and CPM (e.g., Wagner [1969]), such problems can be easily solved. AI researchers apparently have recognized the impact of such structure on ease of solution (e.g., Kim and Pearl [1983]).

However, cases 3 and 5, with a three node approach, introduce the possibility of cycles. Thus, with these cases the acyclic structure may be lost. In the three or five node representations either the computational complexity is increased or the size of the network is increased. Typically, commerical expert system shells, such as EXSYS, get around this problem by forcing the user to develop the equivalent of a smaller acyclic network by establishing partial orderings between the rules, conditions and consequences.

Comments on Combining Probabilities

The primary focus of this paper is on the structural representation of rules and the use of those representations to manage complexity in a knowledge base. Thus, each of the applications discussed in this paper, with one exception application and the example, are independent of the process used to combine probabilities. In that application (determinining the most likely path), a simplifing assumption is made to circumvent the problem of combining probabilities.

However, a related issue is the manner in which probabilities are combined as we move through the structure implicit in a set of rules. A number of approaches to combining the probabilities (and beliefs) have been proposed. Buchanan and Shortliffe [1985] discuss the approach used in EMYCIN, while, Duda et al. [1976] and Duda et al. [1979] discuss the approach used in AL/X. A summary of these and others appears in Speigelhalter [1986]. Typically, various heuristics, such as conditional independence, are assumed so that we can avoid problems such as knowing how to update the probability of b given a_1 or a_2, separatedly, rather than the probability of b, given a_1 and a_2 (Quinlan [1982]). In this paper it is assumed that probabilities are combined in a consistent manner as outlined in any of those approaches.

3.3 Incidence Matrix Formulations

The networks for these rules can be used as the basis for incidence matrix formulations of the rules. Let $I_{i,j}$ be associated with each

$x_{i,j}$ that exists, else let the incidence

matrix component be 0.

Case 1: "If a then b."

This rule can be formulated using a single constraint in an incidence matrix. Let $x_{a,b}$ represent the rule. If that rule is invoked then

(1) $\quad x_{a,b} \leq 1$,

where, in addition, $x_{a,b} = 1$ if the rule is used as part of the inference process and 0 if the rule is not used as part of the inference process. For each of the following rules the variables $x_{i,j}$ are interpreted similarly as 0/1 integer variables. Since the $x_{i,j}$ are less than or equal to 1, the constraints (1) are redundant.

Case 2: "If a_1 or a_2 then b."

This rule also can be formulated as a single constraint in an incidence matrix. Let $x_{a1,b} + x_{a2,b} \leq 1$. This allows only one of a_1 or a_2 to lead to b in a chain of inference.

<u>Case 3:</u> "If a_1 and a_2 then b."

Because there are two paths to b this type of rule requires multiple constraints in the incidence matrix for the three node approach. Let

(2) $\quad d_1 * 2 \leq x_{a1,a2} + x_{a2,b} \leq d_1 * 2$

(3) $\quad d_2 * 2 \leq x_{a2,a1} + x_{a1,b} \leq d_2 * 2$

(4) $\quad d_1 + d_2 \leq 1$, where $d_1 = 1$ if the path $(a_1)\text{-----}>(a_2)\text{-----}>(b)$ is used

and 0 if is not used and similarly for d_2.

These constraints ensure that at most one of the two paths is included. They also ensure that if a path is taken then the entire path is included not just a portion of the "and" relationship is used.

An alternative approach is to use the variables as upper bounds on the other variables. Rather than (2) and (3) the following bounds and constraint could be used:

(2') $x_{a1,a2} \leq x_{a2,b}$ and

(3') $x_{a2,a1} \leq x_{a1,b}$ could be used in conjunction with

(4') $x_{a2,b} + x_{a1,b} \leq 1$.

For the five node case no additional constraints over the incidence matrix are required.

Case 4: "If a then b_1 or b_2"

This rule also can be formulated as a single constraint in an incidence matrix. Let
(5) $x_{a,b1} + x_{a,b2} \leq 1$.

This allows "a" to lead to only one of b_1 or b_2 in a chain of inference that includes this rule.

Case 5: "If a then b_1 and b_2"

Because there are multiple paths from "a," this type of rule requires multiple constraints in the incidence matrix. These constraints ensure that at most one of the two paths can be taken, for the three node approach. In addition, they ensure that if a path is taken then the entire path is taken. Let
(6) $d_1*2 \leq x_{a,b1} + x_{b1,b2} \leq d_1*2$

(7) $d_2*2 \leq x_{a,b2} + x_{b2,b1} \leq d_2*2$

(8) $d_1 + d_2 \leq 1$, where $d_1 = 1$ if the

path $(b_2) \texttt{<----} (b_1) \texttt{<------} (a)$ is used and 0 otherwise, and similarly for d_2.

Alternatively, the same bounding approach as in Case 2 can be used. In particular, we have the following bounds and constraint:

(6') $x_{a,b1} \leq x_{b1,b2}$

(7') $x_{a,b2} \leq x_{b2,b1}$

(8') $x_{b1,b2} + x_{b2,b1} \leq 1.$

For the five node approach no additional constraints over the incidence matrix are required.

3.4 Implementation Considerations

In many cases, there will be a partial ordering between either the conditions or consequences investigated in cases 3 and 5. That partial ordering would eliminate the problem of ordering those elements. Further, cases 3 and 5 generally are reduced to a single path when the knowledge base is structured for most expert system shells. This is because those shells generally require establishing an order between each of the conditions, a priori. By ordering the conditions in case 5, either one of the bounds (6') or (7') would be eliminated and the constraint (8') also would be eliminated. A similar reduction in constraint sets would be incurred with case 3.

3.5 Previous Research on Network Formulations

This approach is different than Chomsky's [1957] (see also Sowa [1984]) grammar analysis, where the arcs represent a series of rules that determine the grammatical combinations for a language. This approach also is different than that used in Buchanan and Shortliffe [1985], where explicit use of a Boolean "and" eliminates the explicit enumeration of the two sets of paths in cases 3 and 5. Each of these approaches allows multiple consequences to be at the same node, e.g., i --> (j,k).

The network approach discussed in this paper appears consistent with that used in the discussion of AL/X (Duda et al. [1976] and Duda et al. [1979] and Quinlan [1982]) and with discussions of the structure of probability-based systems. Pearl [1988] has discussed many other approaches, including Bayesian Networks, Markov Networks, Singly Connected Causal Polytree Networks and Influence diagrams. Throughout, the purpose of the network is critical to its design. For example, in singly connected causal polytrees no more than one path exists between any two nodes. Thus, that analysis is useful with, e.g., decision trees.

3.6 Previous Research on Constraint Representations of Knowledge Bases

However, there apparently have been few mappings of knowledge bases into a mathematical programming framework. In that Konolige [1979], constraints were placed on the probabilities to insure their consistency, as in the following example:

$P(U) = c$ => the sum of $P_i = c$, for all i, such that P_i is part of the marginal probability $P(\tilde{U})$.

Konolige also noted that there may be other constraints imposed, such as the constraint that one conditional probability is always greater than another.

4. EXAMPLE: REPRESENTATION OF IGNORANCE

In order to illustrate the development of a sample network, an example that Shafer [1976, pp. 23-24] used to illustrate the "ineffectiveness" of Bayesian probability theory in representing ignorance is used.[1] Probability theory has been criticized by for its inability to represent ignorance. As noted by Shafer [1975, p. 23], "the basic difficulty is that the theory cannot distinguish between a lack of belief and disbelief."

> Life Near Sirius? Are there or are there not living beings in orbit around the star Sirius? Some scientists may have evidence on this question, but most of us will profess complete ignorance about it. (Suppose) ... t_1 denotes the possibility that there is such life and t_2 denotes the possibility that there is not...
>
> We can also consider the question in the context of a more refined set of possibilities. We might, for example, raise the question of whether there even exist planets around Sirius. We would then have a set of possibilities $M=\{m_1,m_2,m_3\}$, say, where m_1 corresponds to the possibility that there is life around Sirius, m_2 corresponds to the possibility that there are planets, but no life, and m_3 corresponds to the possibility that there are not even planets. (The set M is related to the set T in that m_1 corresponds to t_1 and $\{m_2,m_3\}$ corresponds to t_2.)...
>
> The Bayesian will find it difficult to specify consistent ...(probability measures) over T and M that he can defend as representing ignorance. Focusing on T, he might claim that ignorance is represented by
> $Pr(t_1) = Pr(t_2) = 1/2$
>
> But when he turns to M, he has to satisfy
> $Pr(m_1) + Pr(m_2) + Pr(m_3) = 1$,

the best he can do to represent
ignorance among the three
alternatives is to set
$Pr(m_1)=Pr(m_2)=Pr(m_3)= 1/3$.

But this yeilds
$Pr(m_1) = 1/3$, $Pr(m_2,m_3) = 2/3$.

And since $\{t_1\}$ has the same meaning
as $\{m_1\}$ and $\{t_2\}$ has the same
meaning as $\{m_2,m_3\}$, (1) and (2)
are inconsistent.

The space of outcomes serves as a source model for information about the problem, and the case with three outcomes involves greater source complexities, thus causing the difficulties. The meaning of "ignorance" is modified between the two models in the example because of the difference in source complexity.

Using a network approach to this problem, such as the one discussed in section 2, we still end up with arbitrary probability measures, however, we do not encounter the inconsistencies outlined for the Bayesians. Instead we develop the following structural rule-based model, using probabilities as certainty factors. The network developed from these rules (see figure 1) can allow us to investigate alternative configurations of probabilities on the weights. In terms of the example, the approach of section 2 yields $t_1=.25$ and $t_2=.75$, while, $m_1=.25$, $m_2=.25$ and $m_3=.5$.

Life Near Sirius--With Hypothetical Probabilities

If there is Sirius then there is a Planet (.5)
If there is Sirius then there is no Planet (.5)

Figure 1 -- Continued on next page

If there is a planet then there is life (.5)
If there is a planet then there is no life (.5)

If there is no planet then there is life (0.0)
If there is no planet then there is no life(1.0)

Network Model

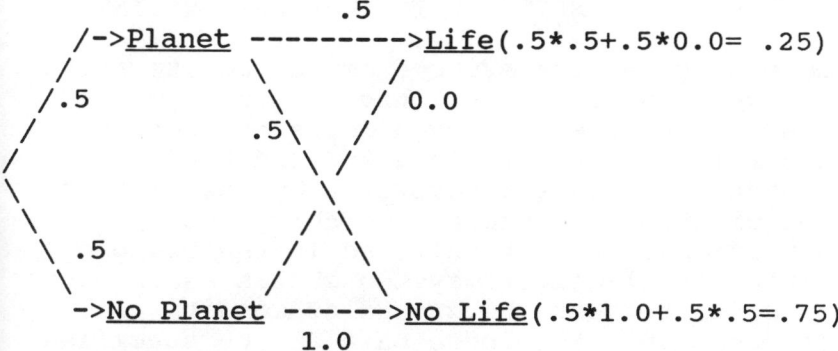

Figure 1 -- Completed

The network approach appears particularly useful for "multistage" reasoning problems, like those in the Sirius problem. The outcome space was partitioned into two sequential sets of outcomes that can be exploited in the analysis of the problem: (Planet, No Planet) and (Life, No Life). Accordingly, the network approach discussed here adopts a finer partitioning and establishes a relationship between the partitionings.

4.1 Impact of the Network Approach

As noted in Baim [1988], since the task of looking for patterns of interdependency and interaction within data is so complex, and because the amount of data that needs to be considered should be large if the results are to be reliable, data reduction is necessary to establish an efficient system. Networks such as that in section 2 provide one such basis for reduction. For example, researchers, such as Pearl et al. [1982] have developed systems to support decision making that assist the user in the development of decision trees in order to understand the particular decision at hand. The inherent structure of the outcome spaces provides additional insights into understanding the domain that may be accessible using approaches such as those developed in this paper.

5. ORDERING CONDITIONS IN AN EXPERT SYSTEM

Generally, expert systems employ an iterative approach to soliciting information from a user. The order in which different conditions and rules are stored in a knowledge base impacts the order encountered by the user for solicitation of information about the conditions. As a result, as in the example, virtually all expert system shells assume that an order between the groups of conditions inherent in a knowledge base can be specified.

However, in some situations there is considerable flexibility because there are few partial orderings of the conditions and consequences. For example, the "and" conditions of rule types 3 and 5 each resulted in the need to establish a partial ordering. In addition, many of the "or" or simple rules have no ordered established, a priori. Thus, there is a need to be able to establish an order of the conditions.

5.1 Impact of Ordering Information on Decisions

The importance of the ordering of the information increases when we note that there also is evidence that the order in which information is presented has a substantial impact on understanding and decision making. For example, as noted by Hogarth [1985, p. 46], "... the order in which information is presented can produce so-called 'primacy' or 'recency' effects. That is, when presented with a number of items of information, sometimes the earlier items dominate the individual's final opinion (a primacy effect) and sometimes the latter (a recency effect)."

Measuring and Managing Complexity ...   407

Hogarth [1985, p. 46] also notes that "... much
evidence indicates that the middle of a series
of informational inputs receives less attention
than either the beginning or the end.
Consequently, if you want to make an
impression, avoid the middle of a series!"

Similarly, Hogarth [1985, p. 45] indicates that
"... the 'availability' of information is an
important clue that people use in making
judgments. Specifically, if you can think of
or see several instances of one kind of event
as opposed to another, you can be led to
believe that the former is more frequent than
the latter." (p. 45)

While Glass and Holyoak [1986, p. 396] also
indicate that the order of the information can
impact problem solving capabilities, "...
functional fixedness: If an object has one
established use in a situation, subjects have
difficulty in using the object another way."

As a result, the order in which information is
presented to the user in an iterative expert
system, can impact the solutions found and
decisions made. Thus, the ordering of those
rules in a knowledge base can be a critical
issue.

5.2 Ordering Conditions

Given the need to establish the order of a set
of conditions, the purpose of this section is
to investigate ordering the conditions.
Without loss of generality assume that each of
the rules are of the type "If $a_1, \ldots, a_n$
then b_j," where each rule uses the same set
of n different conditions ($a_1, \ldots, a_n$) and
results in a consequence chosen from a single
set of consequences ($b_1, \ldots, b_p$). Suppose
further that each of the conditions (a_i) can
result in any of $m_i \geq 0$ different alternatives.

In terms of the Sirius problem, there were two sets of conditions "Sirius" and "Planet/No Planet." While there was a single set of consequences "Life/No Life." In the first condition set there was a single alternative, while in the last set there were two different alternatives. For this example, the rules would be structured as "If (Sirius, Planet/No Planet) then (Life/No Life)." In particular, the rules could be summarized as

"If Sirius exists and there is a planet then there is life."
"If Sirius exists and there is a planet then there is no life."
"If Sirius exists and there is no planet then there is no life."
"If Sirius exists and there is no planet then there is life."

There are at least two ordering concerns. First, there is a matter of ordering the conditions (for example, Sirius before Planet/No Planet). Second, there is the problem of ordering the alternatives within a condition (Planet/No Planet or No Planet/Planet).

There also is a need to establish a goal to optimize. There are a number of options. One approach is to use the ordering that leads to the largest value associated with an inference path. The primary limitation associated with this approach is that the conditions would only be ordered to meet the needs of that inference path. An alternative approach is to order the conditions and alternatives based on a need to minimize the inefficiencies in cognitive processing outlined above. Weights could be developed for this approach using any of a number of psychometric type tools (e.g., Saaty [1974]).

If the ordering of the alternatives is independent of the order of the conditions then ordering the alternatives for each condition can be done independently, as can the process

of ordering the conditions. After that has
been established the order of the conditions
can be established. In each case the two
subproblems are equivalent to traveling
salesman problems. Such problems are tractable
since an expert system typically might have
10-20 conditions, each with 2-15 alternatives.
Thus, this approach would result in a number of
different traveling salesman problems.

If ordering the alternatives is not independent
of the conditions then the problem could take
on substantial difficulty. Based on the
parameter estimates for conditions and
alternatives for typical ES, the problem would
be comparable to a traveling salesman problem
of 20 to 300 cities.

6. FINDING THE MOST LIKELY INFERENCE PATH AND ELICITING POSSIBLE NEW RULES FROM THE STRUCTURE

Expert systems function by soliciting
information from the user or a database in
order to determine which of various rules and
resulting consequences should be invoked to
solve the particular problem, for which the
user is addressing the expert system. In a
large network of rules the most probable
conclusions are supported by multiple chains or
paths of inference. However, it is typical of
expert systems to present a "trace" or a path
of the rules through which the system goes in
order to mimic the system's reasoning through
the rules. Similarly, in order to "understand"
recommendations generated by the system, users
in an investigation of "why" a recommendation
was made by the system may not be interested in
how all of the probabilities cascaded to result
in the preference ranking developed by the
system. Instead, the user may be interested in
the most likely inference path or such
understanding may then be generalized to
finding the k most likely inference paths.

This information also can be useful to the
validator/ verifier of a knowledge base. Since
the rules in the k most likely inference paths
are likely to appear in a solution provided by
the system, the validator might focus on those
rules to ensure that they are correct.

The incidence matrix can be used in conjunction
with a shortest path-like formulation to find
the most likely inferred consequence. In
addition, clearly the ability to formulate it
as a mathematical program indicates that more
efficient network methods can be used. First,
database information can be used to set the
values of $I_{i,j}$ to 1 if the relationship
between i and j is true for the particular
situation and 0 otherwise. In that
formulation, in addition to the above
constraints, path conservation constraints also
would be used (for example, Wagner [1969]).

Those constraints would be formulated as
follows:

$$x_{k,j} - x_{i,k} = \begin{cases} 1 & \text{for k=s (source)} \\ 0 & \text{for all other k} \\ -1 & \text{for k=r (sink),} \end{cases}$$

$x_{i,j}$ = 0 or 1, where, $x_{i,j}$ exists when
$I_{i,j}$ =1.

However, instead of minimizing the distance, as
in a shortest path problem, the formulation
would be aimed at maximizing an objective
function that employs information about the
cf's so that the inference path with the larger
probability of occurence would be found. A
multiplicative combining relationship among the
cf's can be constructed by using the logs of
the cf's. Assuming that the cf's are between 0
and 1, the logs would be negative. Thus, there
would be no concern with finding the longest
path with that formulation, even if the
incidence matrix is not acyclic, since each
weight would be adding negative distance to the
objective function, thus, there would be
negative incentive by the program to add
additional arcs.

6.1 Eliciting Possible New Rules

This same approach can be used to examine the possibility of new rules outside the structure provided by the knowledge base. For example, consider some pair of nodes (n,m), for which no $a_{n,m}$ exists in the knowledge base. The same math program with n=s and m=r can be constructed to determine the existence of a path from n to m. If a path exists but no $a_{n,m}$ exists, then that means that n leads to m only through some intermediate set of nodes. This suggests that possibly an arc $a_{n,m}$ should exist. Based on this inquiry the knowledge acquisition process may include going back to the expert and ascertaining the possibility of a direct relationship between n and m.

6.2 Total Number of Paths

Conceptually, this approach can be used to establish the number of paths through a knowledge base (cyclomatic number). Although this may be an infeasible task, it may be a useful complexity measure for a knowledge base. This is discussed in more detail in section 10.

7. DETECTING CIRCULAR REASONING

Circular reasoning occurs when it is possible to go from some condition to other conditions/consequences and then back to the original condition. For example, the rules

"If a then b / If b then c / If c then a"

allow us to start with condition a and end up back at condition a by following a chain of reasoning from a -> b -> c -> a. The approach of section 2 does not introduce any cycles that are not already in the knowledge base being represented. This is because either constraints limit the choice to at most one of the alternatives for cases 3 and 5 or an increase in the number of nodes eliminates the introduction of any new cycles.

One of the goals of verification is to locate circular reasonings. The concern is that the inference engine will view each new trip through a cycle as a new set of inferences to evaluate. However, it may be inappropriate to say that such cycles are "unwanted." For example, $P(a|c)$ cannot be uniquely derived from $P(b|a)$ and $P(a|c)$, so the information in circular reasoning is not redundant -- cycles just get in the way of our algorithms to process rules.

Mathematical programming also can be used to ascertain the existence of cycles in the incidence matrix. Rather than using an objective function with a set of weights based on the cf's, a weight of $c \geq 1$ on each arc for which $x_{i,j}$ exists can be used to determine the existence of a cycle. If there is a cycle then there would be a solution longer than the longest (in terms of total number of arcs) feasible loopless path. Thus, the number of arcs in the path would be larger than would occur in that loopless path and that would indicate that there was a solution with a cycle or circular reasoning.

8. ASCERTAINING THE IMPORTANCE OF RULES

Given a large knowledge base of rules, it may be feasible to only examine in detail a subset of the total set of rules. Researchers (e.g., O'Leary [1988] have suggested statistical approaches to identify that subset of rules. However, it may be important to examine those rules on which the inferred solutions depend the "most" because of the "structure" of the network. This identification process is independent of content of the rules and dependent only on the structural relationship between the rules.

One critical aspect of the structural nature of the knowledge base occurs with that minimal set of "if ... then ..." relationships through which any of the paths of inference leading to solutions must pass. In terms of the network of those relationships, this is referred the minimal cutset. This set is the minimal set of relationships whose removal makes it impossible to reason further than the cutset. Thus, those "if ... then ..." conditions in those sets are likely to deserve additional attention because of their critical relationship with the rest of the knowledge base.

The cutset can be identified using a mathematical program that consists of the incidence matrix in section 2, an artificial source and sink and the following set of constraints and objective function (Wagner [1969, p. A3]):

Maximize F

$$x_{k,j} = F \text{ for } k=s \text{ (source)}$$
$$x_{k,j} - x_{i,k} = 0 \text{ for all other } k$$
$$- x_{i,k} = -F \text{ for } k = r \text{ (sink)}$$

(for all (k,j) and (i,k) in the network) and $x_{i,j} = 0$ or 1, where each arc has a capacity of one.

The general max flow-min cut algorithm (e.g., Wagner [1969]) can be used if there are no case 3 or case 5 arcs. Alternatively, the multiple path nature associated with cases 3 and 5 could make the cutset meaningless or at least distort the number in the cutset. Thus, unless an a priori decision is made on the order of the conditions, that approach could be inappropriate.

Although this section has focused on cutsets of arcs, a similar approach could be used to find node articulation sets rather than just arc cutsets. In that approach, the conditions that are the most structurally critical could be found. Clearly, articulation points are a special case where the node cutset is a single node.

9. PARALLEL ARCHITECTURES

Oftentimes network representations of expert systems lead to subnetworks where those subnetworks are relatively independent. For example, in Dungan and Chandler [1985] the modular nature of individual subnetworks in the knowledge base is readily apparent in a diagram that shows how the rules are connected. Subnetworks could be constructed using using either a cutset or articulation set. By using such an approach the resulting total network would have only a few "dangling" arcs in the cutset that connect the subnetwork to the rest of the network of rules. Such structures appear particularly amenable to processing with parallel processors. Each of the subnetwork problems could be solved and put together, in order to assist in the generation of a solution to the total problem.

Dantzig [1963, Chapter 23] addressed such an issue with the "decomposition principle." Linear programs of the following general form, were investigated.

$$
\begin{aligned}
A_1 x_1 & & & & = B_1 \\
 & A_2 x_2 & & & = B_2 \\
 & & \cdots & & = \cdots \\
 & & & A_n x_n & = B_n \\
D_1 x_1 + & \cdots & & D_n x_n & = B_{n+1}
\end{aligned}
$$

Dantzig [1963] developed an algorithm to solve problems of this type, by exploiting the structure. The approach was to decompose the problem into each set of "almost" independent subproblems, $A_i x_i = B_i$, for i=1, ..., n, and a master program, with the constraint matrices D, which tied the programs together. Using this approach, both the master program and the subprograms would have to be solved many times. However, the decomposed problems were substantially smaller than the overall problem, and thus, easier to solve. From a parallel computing perspective, the master program would be solved, then n processors could solve the independent problems, a revised master problem would be solved, etc., until the process generated a solution to the overall problem.

In addition, Dantzig [1963, pp. 455-466] develops a "story" that is useful in understanding the relationship between the agents assigned to solve subproblems and the agent responsible for solution of the overall problem of such loosely connected networks. In general, the master agent, responsible for the overall problem, would establish prices on the scarce resources, the other agents would then solve their individual problems, turn in their solutions to the problems, and then receive new prices. Such a "story" can provide insight into parallel processing, multiagent schemes, so that the parallel processors have the ability to function virtually independently.

10. MEASURING COMPLEXITY OF CONSTRUCTED KNOWLEDGE BASES

Since the management of complexity is such a critical issue, so is the measurement of that complexity. The same network approach that proved useful in identifying various approaches to use mathematical programming to measure that complexity, can be used to measure complexity. Such measures can be useful as a basis to allocate resources to management of the resulting complexity, particularly in the areas of validation of the resulting expert systems and in anticipating potential problems from the complexity, such as errors. In addition, complexity metrics can be used to study the impact of complexity on such concerns as cost, necessary validation efforts, quality and other issues.

10.1 Measuring Complexity in Traditional Computer Programs

The importance of complexity in traditional computer programs has led software engineering researchers to develop measures of the complexity of computer programs. Myers [1979, p. 166] suggests that, "most complexity measures are derived from the program's control flow (e.g., number of paths)." Both Myers [1979] and Shooman [1983] discuss the notion that network-based measures such as the cyclomatic number provide easy-to-compute measures for control flow complexity. Beizer [1984] discussed a number of other approaches to measuring complexity. These include lines of code, number of decision statements in a program, extent of modularity of a program and the extent of code masquerading as data. Conte et al. [1986] summarize much of the metric literature, including a chapter on software metrics.

10.2 Different Approaches are Required for Expert Systems

However, measuring complexity in traditional computer programs is not the same as measuring complexity in expert systems. Complexity in traditional computer programs comes from the instructions in a program, the interaction of the instructions, because the operations to data or because of the data the program is processing. In addition, traditional computer programs have a broad base of operators that are used to investigate that data, e.g., "if," "do," etc., that are not used in expert systems.

The "classic" expert system set of components (user interface, database interface, inference engine and knowledge base) allows for the same set of interfaces and inference engine to be used in many different situations. Thus, complexity in expert systems comes from primarily from constructing and processing the knowledge base. In this case the program provides little of the complexity. Further, typically there is a single operator in a knowledge base: the "if ... then ..." statement. In addition, consistently the bottleneck in the development of expert systems is in the development of the knowledge base.

10.3 Complexity Measures and Other Metrics of Network Representations of Knowledge Bases

Since complexity in knowledge bases is not the same as complexity in computer programs there is a need to develop specific metrics of size and complexity for knowledge bases. Intuition suggests that complexity should reflect the interaction of the rules in the knowledge base and the more interactions the more complex the knowledge base. Chomsky's [1957] notion of "deep structure" substantiates that intuition.

This section of the paper provides a number of different metrics that can be used to develop measures of complexity and the closely related notion of size of a knowledge base. Clearly, as with measuring the complexity of computer programs, each of these metrics has their own strenths and weaknesses.

Number of Rules

Typically, in the discussion of an expert system the number of rules is provided either as a measure of the size of the knowledge base or its complexity. However, such a measure ignores the interaction between different rules, the manner in which such rules are written, the extent to which a rule includes branching and other concerns. As a substitute for the number of rules, the network provides at least three other size measures: number of nodes, number of arcs, their sum or the difference between the number of arcs and nodes. the first three are measures of size. While, the last measure is more a measure of complexity. It is closely related to the cyclomatic number discussed earlier.

Degree

The degree of a node is a measure of complexity because if the arcs represent information flows or conditions then if the degree of one node is

greater than the degree of another node then there are more conditions associated with the first node.
If the network is complete or complete with loops then that is the most complex in terms of degree. However, degree also can be used to establish intermediate degrees of complexity. One straightforward approach is to compute the total of the degrees on all of the nodes in the network. This measures the potential number of decisions facing the inference engine.

Unfortunately, degree provides an "uneven" measure of complexity. One node may be very complex, while others could have limited degrees of 1 or 2.

This limitation can be mitigated to a certain extent by using Pareto optimality define, in terms of degree on the nodes, an order to rank the complexity of alternative knowledge bases. A matrix of degree numbers, $S_k(i,j)$, can be associated with each network, k, where $S_k(i,j)$ is the number of nodes with incoming degree greater than or equal to i and the number of outgoing degree greater than or equal to j. Then network, A, can be said to be pareto complex over network B if $S_A(i,j)>=S_B(i,j)$ for all i and j. Other partial pareto definitions can also be established.

Planarity

If a network representation of a knowledge base is nonplanar then that suggests that the network is more complex than a planar network. Euler's relation allows for an easy-to-use measure of planarity.

If a network is planar then the number of regions in the network may be a good measure of complexity. If two networks are planar then the network with more regions likely is to be more complex than the one with fewer regions.

Modules

Another approach to measuring the complexity of a knowledge base is to investigate the number of relatively independent modules or subnetworks in a knowledge base. The number of modules can be used to establish metrics for both size and complexity. There are at least two concerns with turning the notion of module into an operational concept: determining what is a module and the relationship between the number of modules and complexity.

There are a number of approaches to establishing what distinguishes one module from another. One approach is to establish two modules if there is an articulation point or a single arc cut-set. Clearly each of these approaches can be generalized to the case where there is an articulation set of say k nodes or cutset of say k arcs. Another approach is to chose some subnetwork where the arcs lead only from that subnetwork. Further, prior to analysis of network representation, it may be possible to a priori establish logically distinct modules.

Once the number of modules has been established there are various approaches to establishing metrics of size and complexity. The number of modules is likely to be a measure of size, because it is likely to reflect different content for each of those modules.

However, if there are large number of arcs and nodes then it is likely to be very helpful to be able to break that problem into a number of modules. As a result, any of the size meaures (say, number of nodes plus number of arcs) divided by the number of modules is likely to be an important measure of complexity. The higher the average number of rules in a module, the more difficult it is to break the problem down into different components.

Alternatively, rather than the average size of a module, the extreme points or standard deviation may be informative. For example, if one of the modules is substantially larger than the others then an average measure may not that useful.

Finally, the extent to which the modules themselves interact may be a measure of complexity. If the modules are only loosely coupled to each other then that may indicate that knowledge base is less complex than another knowledge base with similar sized components and number of components, with many interactions between the modules.

Connectedness

If a network representation of a knowledge base is connected then that means that there are not isolated and unrelated pieces of knowledge in the knowledge base. Generally, this suggests that similar sized connected networks are more complex then a network with multiple pieces. An unconnected knowledge base can be treated as multiple small knowledge bases rather than one large knowledge base, decreasing the complexity.

In addition, as with the occurence of modules, the average component size and extreme values of the size of the components also may provide valuable insights in the area of size and complexity.

Cyclomatic Number

As with traditional computer programs, if there are cycles in a knowledge base, then there is greater complexity. It is much more difficult to traverse a network with cycles than an acyclic one. Thus, the number of cycles in a knowledge base also provides a measure of the complexity in that knowledge base. However, as noted above, in some cases cycles in the knowledge base may be indicative of an error in the knowledge base. Thus, for this purpose, this approach may be of limited use.

However, determining the number of cycles in a computer program is not the only use. Typically, software metrics (e.g., Conte et al. [1986]) have used this approach to determine the number of paths through a computer program. That same approach can be used here on the network generated above, using the cyclomatic number, to determine the number of paths through a knowledge base.

11. SUMMARY

This paper has discussed an approach for developing network representations of rule-based knowledge bases that then allowed for the formulation of the knowledge base as a set of linear equations. Those equations could then be used as the basis to structure a number of complexity management problems as mathematical programs. Further, once those knowledge base problems have been formulated as mathematical programming problems, specific network solution methodologies may be used to establish more computationally efficient approaches, as has been done in various operations research problems.

A number of different problems were approached using a network and mathematical programming approach. First, the network approach yielded a consistent representation of ignorance that exploited the implicit underlying structure of the outcome space. Second, this paper argued that ordering the conditions in an expert system can be investigated as either a set of traveling salesman problems or a single large traveling salesman problem. Third, the incidence matrix of the corresponding network was used for developing a mathematical programming formulation that could be used to determine the best inference path through a knowledge base, assuming that appropriate information had been gathered. A similar approach can be used to find, for example, the second or third best inference path. Fourth, the mathematical programming approach also was extended to test for the existence of cycles and determining important relationships that may deserve further investigation. Fifth, a max-flow and min-cut approach was used to isolate those rules that were critical to the inference process. Sixth, exploiting the underlying structure associated with knowledge bases can be investigated for parallel processing environment by using a decomposition approach.

In addition to using the network and mathematical programming approach as a basis to manage complexity, the networks can be used to establish some size and complexity measures for knowledge bases. A number of network approaches were proposed to measure that complexity, including cyclomatic number, total degree on the nodes, and the number of rules (arcs). The individual modules and components in a knowledge base were also used as metrics.

Footnote

In order to minimize the need to fully discuss the work of Schafer (which would require more space than this paper could allocate to it) I have replaced Shafer's use of "belief functions" with probability functions, which I assume have a larger group of understanding readers.

Acknowledgement

The author would like to acknowledge the extensive comments of anonymous referees on two earlier versions of this paper.

References

Baim, P., "A Method for Attribute Selection in Inductive Learning Systems," *IEEE Transactions on Pattern Analysis and Machine Learning*, Vol 10, No. 6, November 1988, pp. 888-896.

Barlow, R. and Singpurwalla, N., "Assessing the Reliability of Computer Software and Computer Networks: An Opportunity for Partnership with Computer Scientists," *The American Statistician*, May 1985, Vol. 39, No. 2, pp. 88-94.

Berge, C., *The Theory of Graphs and Its Application*, Wiley, New York, 1966.

Breizer, B., *Software System Testing and Quality Assurance*, Van Nostrand Reinhold, New York, 1984.

Buchanon, B. and Shortliffe, E.H., *Rule-Based Expert Systems*, Addison-Wesley, Reading, Ma., 1985

Chomsky, N., *Syntactic Structures*, Mouton, The Hague, 1957.

Conte, S., Dunsmore, H. and Shen, V., *Software Engineering Metrics and Models*, Benjamin/Cummings Publishing Company, Menlo Park, California, 1986.

Dantzig, G., *Linear Programming and Extensions*, Princeton University Press, 1963.

Duda, R., Gaschnig, J., and Hart, P., "Model Design in the Prospector Consultant System for Mineral Exploration," in D. Mitchie, (ed.) *Expert Systems for the Micro Electronic Age*, Edinburgh, Edinburgh Press, 1979.

Duda, R., Hart, P., Konolige, K. and Reboh, R., *A Computer-Based Consultant for Mineral Exploration*, SRI International, Menlo Park, CA, US Department of Commerce, National Information Service, PB80-106347, September 1979.

Duda, R., Hart, P., and Nilsson, N., "Subjective Bayesian Methods for Rule-Based Inference Systems," *National Computer Conference,* pp. 1075-1082, 1976.

Dungan, C. and Chandler, J., "Auditor: A Microcomputer-based Expert System to Support Auditors in the Field," *Expert Systems*, October 1985, pp. 210-225.

Glass, A. and Holyoak, K., *Cognition*, Random House, New York, 1986.

Hogarth, R., *Judgment and Choice*, John Wiley & Sons, Chichester, 1985.

Hu, T.C., *Integer Programming and Network Flows*, Addison Wesley, Reading, Ma., 1970.

Kaufman, A., *Graphs, Dynamic Programming and Finite Games*, Academic Press, New York, 1967.

Kim, J. and Pearl, J., "A Computational Model for Causal and Diagnostic Reasoning in Inference Systems," *International Joint Conference on Artificial Intelligence*, 1983, pp. 190-193.

Konolige, K., "Bayesian Methods for Updating Probabilities," Appendix D, in Duda et al. [1979]

Liu, C.L., *Introduction to Combinatorial Mathematics*, McGraw-Hill, New York, 1968.

Myers, G., *The Art of Software Testing*, John Wiley, New York, 1979.

O'Leary, D., "Validation of Expert Systems," *Decision Sciences*, Volume 18, Number 3, Summer, 1987, pp. 468-486.

O'Leary, D., "Methods of Validating Expert Systems," *Interfaces*, Volume 18, Number 6, November-December 1988, pp. 72-79.

Pearl, J., <u>Probabilistic Reasoning in Intelligent Systems</u>, Morgan Kaufmann, San Mateo, CA, 1988.

Pearl, J., Leal, A. and Saleh, J., "GODDESS: A Goal Directed Decision Structuring System," <u>IEEE Transactions on Pattern Analysis and Machine Intelligence</u>, Vol. PMAI-4, No. 3, May 1982.

Quinlan, J., "Inferno: A Cautious Approach to Uncertainty Inference," <u>Rand Note N-1899-RC</u>, Rand, Santa Monica, Ca, 90406, September, 1982.

Saaty, T. L., Measuring the Fuzziness of Sets," <u>Journal of Cybernetics</u>, Vol. 4, No. 4, pp. 1650-1657.

Schum, D. and DuCharme, W., "Comments on the Relationship Between the Impact and Reliability of Evidence," <u>Organizational Behavior and Human Performance</u>, Vol. 6, pp. 111-131, 1971.

Schum, D., "Sorting Out the Effects of Witness Sensitivity and Response Criterion Placement Upon the Inferential Value of Testimonial Evidence," <u>Organizational Behavior and Human Performance</u>, Vol. 27, 1981, pp. 153-196.

Shafer, G., <u>A Mathematical Theory of Evidence</u>, Princeton University Press, Princeton, New Jersey, 1976.

Shooman, M., <u>Software Engineering</u>, McGraw-Hill, New York, 1983.

Sowa, J.F., <u>Conceptual Structures</u>, Addison-Wesley, Reading, MA, 1984.

Speigelhalter, D., "A Statistical View of Uncertainty in Expert Systems," in Gale, W. (ed), <u>Artificial Intelligence and Statistics</u>, Addison-Wesley, Reading, Massachusetts, 1986.

Wagner, H., <u>Principles of Operations Research</u>, Prentice-Hall, Englewood Cliffs, New Jersey, 1969.

Pragmatic Information-Seeking Strategies for Expert Classification Systems

Louis Anthony Cox, Jr.
U S WEST Advanced Technologies
6200 South Quebec Street
Englewood, CO, 80111

ABSTRACT

In the artificial intelligence and statistical theories of supervised learning for pattern recognition systems, it is usually assumed that a classifier system has access to a "training set" of pattern vectors with known correct classifications. New pattern vectors are observed for free and classified so as to minimize the estimated probability od misclassification. This paper describes heuristics for minimizing the expected costs of classifying new pattern vectors when it is expensive to observe their components and when different types of misclassification carry different penalties.

INTRODUCTION

Interactive modeling systems seek relevant information from a user in order to answer a query or to solve a problem that the user has posed. A fundamental design issue for such a system is therefore its *information-seeking strategy*, which determines the order in which it asks questions to gain the information that it needs to respond to the user. This paper examines algorithms for sequencing questions so as to acquire the information needed to accomplish an important type of task -- object classification -- at minimum expected cost, assuming that it is expensive to obtain answers to questions about an object. The general classification problem addressed in this paper is first formulated using a simple abstract model of an expert classification system, considered as a set of logical classification rules supplemented by some prior statistical knowledge about attribute frequencies. This model is used to characterize the complexity of the classification task and to present algorithms for doing probabilistic interactive (question-based) classification. Next, a new class of *sorting heuristics* is presented for adaptively generating and scheduling questions in such systems during a knowledge acquisition dialogue. These heuristics are contrasted with the more usual "recursive partitioning" algorithms now being used in artificial intelligence and computational statistics to learn classification rules from empirical data. Finally, theoretical optimality and empirical performance results are reported for the sorting heuristics that indicate their practical utility in contexts where information costs can not be ignored.

This paper describes how to use a knowledge base of classification rules to efficiently classify objects in situations where each object's class depends only on the values of its attributes. It is assumed that prior statistical information about attribute frequencies supplements the logical classification rules in the knowledge base and that attribute values for an object can be observed for some cost before drawing final conclusions about the object's class. This work is directed toward the design of practical interactive expert classification systems for contexts where acquisition of information is expensive, e.g., because it requires costly experiments or diagnostic tests. The trade-offs involved in "optimal" rule-based classification are treated in a standard statistical decision theory framework. It is assumed that the classification rules are already known (e.g., through application of a rule-learning algorithm like ID3 to a large training set) but that the order in which to apply them during a dialogue with the user needs to be (adaptively) determined.

1. The Minimum Expected-Cost Classification Problem

The problem of efficient classification of an object or "case" based on sequential inspection of its attributes can be approached from either an artificial intelligence (A.I.) or a traditional statistical decision theory/pattern recognition perspective. An A.I. formulation is as follows. Suppose we have a set of logical rules for classifying cases based on the values of each of n descriptive attributes. Such rules map n-tuples of attribute values into corresponding assigned classes. In the simplest situation, to which a substantial fraction of current research on machine learning is devoted, all attributes are binary: each one is either present or absent in an object (14-16). Binary attributes will be assumed throughout this paper. If the set of rules is deterministic, then it can be expressed without loss of generality in the following canonical form:

$$C_1 \text{ if } [a_{11} \And a_{12} \And ... \And a_{1n}]$$
$$C_2 \text{ if } [a_{21} \And a_{22} \And ... \And a_{2n}] \quad\quad\quad (1)$$
$$\vdots$$
$$C_m \text{ if } [a_{m1} \And a_{m2} \And ... \And a_{mn}]$$

This array is called a *rule base*, or an expert classification knowledge base. Each of the a_{ij}'s is a *proposition* stating either that attribute j is present ($a_{ij} = 1$) or that it is absent ($a_{ij} = 0$) or that it does not matter for the classification conclusion expressed on the left hand side of rule i whether attribute j is present ($a_{ij} = *$). (In the latter case, the corresponding a_{ij} automatically evaluates to "true." Such "don't care" conditions can be omitted, but it is useful for uniformity of notation to include them.) The propositions a_{ij} have truth values for any object that is being classified. Nondeterministic rules of the form

Pragmatic Information-Seeking Strategies for Expert Classification Systems 429

(C_j with confidence factor p) if [a_{i1} & a_{i2} & ... & a_{in}]

can be included within this formalism, if desired, by treating the value of one or more attributes j as unobservable, so that only the probability that a_{ij} is true is known.

Each row in (1) is a *rule* (a Horn clause in logic programming terminology.) The right-hand sides in square brackets are called *clauses*. Each of the labels $C_1,...,C_m$ is the name of a class; C_i asserts that that the class of the object is C_i. If all of the propositions in a clause evaluate to "true" for an object then the clause is said to be *satisfied* and the rule asserts that the object belongs the class named on the left-hand side of the rule. If C_i and C_j are the same, then rule base (1) asserts that, if either of the clauses [a_{i1} & a_{i2} & ... & a_{in}] or [a_{j1} & a_{j2} & ... & a_{jn}] is satisfied, then the object being classified belongs to class $C_i = C_j$. [Thus, any logical rule that is expressed in disjunctive normal form (DNF) -- a canonical form that has been widely used in the machine-learning literature (14-16) -- can be reexpressed in the canonical form (1) by distributing disjuncts across rules.] In practical applications, class labels might represent predictions, diagnoses, or prescriptions for control actions to be applied to the object being classified.

Previous work emphasizing induction of logical classification rules from data has often assumed that a set of objects is available for which both the values of each object's attributes and the correct classification of each object are known (1-8.) From the data in this "training set," the induction algorithm seeks to learn a general decision rule, i.e., a classification rule, mapping logical combinations of attribute values into corresponding classes. This general classification rule can then be reduced to a set of Horn clauses as in rule base (1), post-edited to remove redundant or uninformative attributes from the right-hand side clauses, and implemented in an expert system (7). Induction algorithms that have been used successfully to learn classification rules from data in a wide variety of commercial applications include Quinlan's ID3 algorithm for objects characterized by the values of a few discrete attributes (6-8), and the CART system for classification of objects with either discrete or continuous attributes (1-4). An important special case occurs when there are only two possible classes. Then the correct decision rule may be interpreted as a logical *concept* to be learned, and the learning mechanism or algorithm is trained to classify each object as either an instance (class 1) or as not an instance (class 0) of the target concept. This paper addresses the more general case in which there may be any finite number of classes.

Instead of a training set, this paper assumes that some prior statistical knowledge about attribute frequencies is available to guide classification. Specifically, the prior probability p_j that each attribute j will be present in any object is known, and all attributes are assumed to be mutually statistically independent: whether one attribute is present does not change the probability of any other attribute being present. Rather than trying to learn new classification rules from a training set of data, we assume that a set of logical classification rules having the canonical form (1) is given (perhaps after applying a learning algorithm such as ID3 to a large training set.) The problem addressed here is not how to learn such rules during a training phase, but how to apply them efficiently to classify new cases in application. This is difficult because, in contrast to most previous work on machine learning, we assume that it costs an amount c_j to learn the value of attribute j for an object. It is therefore not enough to know the logical rules mapping attribute values into classes. Rules are applied to new objects by asking questions in an effort to show that a right-hand side clause is satisfied, so that a left-hand side conclusion can be drawn. When questions are expensive to answer, a strategy for applying these rules so as to minimize the average cost of reaching a conclusion must be developed. Developing such strategies is the principal goal of this paper.

The ultimate purpose of classification is assumed to be to help decide what action to apply to each object or case being classified. A set A of possible *actions* is assumed to be available, exactly one of which must be selected for application to each object. Each object has some true but unknown *state* (e.g., its true class) belonging to a set X of possible states. If action a from A is applied to an object whose true state is x, then a loss $L(a,x)$ results. This conventional statistical decision theory formulation generalizes the "pure" classification problem in which A = X and in which the loss function $L(a,x)$ is $L(a,x) = 0$ if $a = x$ and $L(a,x) = 1$ otherwise. Allowing there to be a different number of actions than classes, and concentrating on the mapping from attribute values to actions, substantially extends the range of practical applications described by this model. The "classes" C_1 through C_m in (1) should therefore be interpreted henceforth as actions or decisions. Problems with this structure will be called *generalized classification problems* (GCPs.) Standard example interpretations of this mathematical formulation include (i) X is a set of possible diseases and A is a set of possible prescriptions; (ii) X is a set of possible signals and A is a set of possible responses (e.g., fire a torpedo or don't, in a military application); (iii) X is a set of possible symptoms and A is a set of possible diagnoses. The *minimum expected-cost generalized classification problem* (MEGCP) is to construct a question-asking and action-selection strategy that balances the costs of inspecting (i.e., discovering) attribute values against the expected losses from misclassification (i.e., from incorrect action) so as to minimize total expected loss, defined as the sum of inspection costs and the "terminal loss" $L(a,x)$.

The following sections examine the computational complexity of MEGCP and present several versions of a new, computationally efficient heuristic algorithm for solving it. The new algorithm is contrasted with recursive partitioning or "tree-structured" algorithms, which provide an alternative heuristic approach to sequential inspection and classification when observation costs can be ignored. The context for the analysis is the problem of designing an interactive expert classification system that will adaptively generate a sequence of questions to minimize the expected cost of reaching a conclusion, including the costs of misclassification error.

2. Formulation as an Expert System Design Problem

Many expert systems can be looked at abstractly as pattern-recognizers or classifiers. The expert system paradigm is roughly as follows. A builder or user supplies the machine with some generic knowledge about an application domain, formulated as a system of logical rules as in the canonical rule base (1). These items of generic knowledge are stored in a rule base K. A user then provides the machine with some empirical assertions, contained in an empirical fact base I(t). I(t) denotes the set of all empirical observations made (and supplied to the machine) through time t. Finally, the user submits a query, meaning a question to be answered, to the machine.

Typically, a query is a conclusion whose truth is in doubt. The task of the machine is to try to deduce or refute the conclusion by applying rules in K to known facts in I(t). In a *passive* expert system, the result of this effort will be either a "proof" [meaning a sequence of applications of rules from K leading from the given facts in I(t) to the desired conclusion] that the conjecture in the query is correct; or a proof that it is incorrect, e.g., because it contradicts an assertion in I(t); or a statement that the machine was unable to either prove or disprove the conclusion. (In many expert systems and in Prolog, the latter two responses are not distinguished; theoretical trade-offs between soundness and completeness and between tractability and representational power impose some fundamental limitations on the performance of any system of the type described here.) In an *interactive* expert system, the machine may also generate relevant questions asking the user for empirical information that will help it to answer the user's query. (In our terminology, a query is directed from the user to the machine; a question is directed from the machine to the user.) If the user answers a question, the answer is formulated as a new empirical assertion and added to I(t).

As previously stated, our concern is with questions that are expensive to answer, e.g., because they require resource-consuming tests to be made. When the user can not answer questions cheaply (and may not be able to answer some questions at all, corresponding to prohibitive costs of answering), the questioning strategy that the machine uses in attempting to establish or refute a conclusion becomes critical in determining the average cost of applying it. Simple systematic patterns, e.g., based on resolution-refutation or other backtracking strategies, can be prohibitively expensive. The problem of how to adaptively generate a sequence of questions that minimizes total expected loss, assuming that an expected-loss-minimizing terminal action is taken when questioning ends, must be solved in order to evaluate the overall practical usefulness of the classification knowledge base.

To make this concrete, we adopt the following simple model of an interactive expert classification system:

- *Empirical facts* consist of assertions of the form $T(a_{ij})$ or $-(a_{ij})$, meaning that attribute a_{ij} is or is not present, respectively, in the object being classified. These assertions reflect (usually costly) empirical observations. (Because the prior probabilities p_j are known and stationary, there is no loss in assuming that the fact base $I(t)$ consisting of all such empirical assertions is cleared between successive objects.)

- *Rules* are of the form shown in (1). Each rule maps a conjunction of attribute values into a conclusion that the object with that attribute value pattern belongs to a specified class.

- A *conclusion* is a probability distribution over possible classes, giving the posterior probability that the object being classified belongs to each class.

The expert system also has some <u>statistical knowledge</u> in the form of a prior probability vector $p = (p_1,...,p_n)$ for the presence of each of the n attributes independently. It is assumed that p is an accurate "well-calibrated") probability model for the data-generating process.

In this model, there is only one kind of query: the user always wants to know what class an object belongs to. A question from the machine to the user is the name of an attribute whose presence or absence is to be determined and reported to the machine. The cost of testing whether attribute j is present is c_j. The performance of the expert classification system is measured by the average cost per object classified – the sum of observation and misclassification (i.e., action error) costs. It is assumed that the expert system's conclusion (posterior probability distribution for class membership) is mapped into an action in A that minimizes conditional expected loss with respect to it. We now design a questioning strategy to minimize the expected combined loss from reaching and applying a conclusion.

3. Classification Trees and Computational Complexity of MEGCP

A minimum expected-cost generalized classification problem (MEGCP) can be specified by a quadruple [K,p,c,L], where K is a logical rule base of the form (1); p is the n-vector of probabilities that each attribute j has the value "present"; c is the n-vector of attribute inspection-costs; and $L = L(a,x)$ is the loss function. Let N be the set of n attributes that objects can have. Let N* denote the set of partial sequences of attribute values that can be generated by sequentially inspecting some of the values of attributes in N. Any sequence s* in N* has the form $(a_1,a_2,...,a_j)$, for some $j \le n$, where each a_i in s* is a value (either 0 = absent or 1 = present) of an attribute in N, and where no attribute is represented more than once in s* (since its value can be inspected at most once.) A *strategy* or *sequential decision rule* for solving a MEGCP is a function d from N* to N ∪ A, the union of N and A. For any partial sequence s* of length s in N*, d(s*) is either an attribute in N whose value has not yet been recorded in s*; or an action in A. If d(s*) is an attribute in N, then the strategy requires that the value of that attribute be inspected. This produces a new partial sequence of length s + 1, to which the decision function d again applies. If d(s*) is an action *a* in A, then d prescribes that inspection of attribute values stop and that terminal action *a* be taken.

Any decision rule d can be represented as a rooted tree, called a *classification tree*. (We use "classification tree" instead of the more common "decision tree" to avoid confusion with the decision trees of decision analysis.) Each internal node of the tree represents an attribute to be inspected. Assuming that there is more than one node, the root node represents the first attribute to be inspected. Two arcs emanate from each internal node: one corresponding to an observed attribute value of 1 and the other to an observed attribute value of 0. Each arc leads to another node, representing the next observation or decision to be taken if the corresponding attribute value is observed. Each terminal node represents a terminal action in A to be taken if the sequence of nodes leading to it has been traversed. There is a one-to-one correspondence between classification trees and sequential decision rules d: N* --> N ∪ A.

The total number of decision rules, or classification trees, is enormous: it is greater than 10^{24} for n = 7. (If G(n) is the number of classification trees for n attributes, then G(n) grows roughly according to the recursion $G(n) = n[G(n-1)]^2$; see [17] for further discussion.) Each tree determines a corresponding expected loss. When a new object is presented to a classification tree, it can be thought of as being "dropped" down through the tree, starting from the top (root) node and following some sequence of arcs according to the observed values of its attributes [1]. Since the prior probabilities of different attribute values are known, the probability that a new object will follow any particular path (uniquely identified by its terminal node) through the tree is just the product of the attribute value probabilities along that path. The cost of a path is the sum of the attribute inspection costs along it, plus the conditional expected terminal loss at its terminal node [found, for example, by summing $L(a,x)p(x;s)$ over all states x that are compatible with the set of observed attribute values in the sequence s leading to the terminal node. Here, p(x;s) is the conditional probability of state or attribute combination x given the observed attribute values in s.] The expected cost for the classification tree as a whole is the sum of the expected costs for each path to a terminal node, weighted by the corresponding path probabilities. Thus, the expected cost of any given classification tree can be found quite easily. The problem of identifying a minimum expected cost strategy (tree), however, is NP-hard except in special cases.

More precisely, given n binary random variables, the jth one having probability p_j of being b_j and probability $(1 - p_j)$ of being 0, where the b_j are arbitrary known positive constants; and given that it costs c_j to inspect the value of the jth variable, what inspection strategy will minimize the expected cost of determining with certainty whether the sum of these random variables is greater than some threshold k? It is straightforward to show that this problem, while trivial if k is less than any of the b_j's (examine the variables in decreasing order of the ratio p_j/c_j) or if k is greater than the sum of any (n - 1) of the b_j's, but less than the sum of all n of them [examine the variables in order of decreasing $(1 - p_j)/c_j$] is not in general any easier than the notorious "knapsack problem." (Take all p_j's very close to 1 and consider the recognition form of the knapsack problem; see [17] for details.) Therefore, like the knapsack problem, it is NP-complete. Moreover, this problem is a special case of MEGCP with only two classes (corresponding to the sum of the random variables being less than k and the sum being greater than or equal to k, respectively), except that all the subsets having at least k attributes present in them have not been explicitly listed in a rule base K. The implication is that *the combined task of formulating and solving the corresponding MEGCP is NP-complete.*

4. Algorithms for MEGCP

This section presents three types of algorithms for the MEGCP [K,p,c,L]. The first calculates a conclusion (i.e., a set of class probabilities) from the statistical knowledge p and the logical classification rules in K, assuming that K is in the canonical form (1). The second generates classification trees when observation costs can be ignored: this is the class of "recursive partitioning" algorithms that has come to dominate much of the literature on machine-learning and computational statistics approaches to classification. Finally, we present a new, computationally efficient algorithm for the case of costly observations.

4.1 Drawing Conclusions: The Inclusion-Exclusion Algorithm for Calculating Class Probabilities

Given a set of logical classification rules in the canonical form (1) and a prior probability vector p for the n descriptive attributes, what is the probability of each of the classes on the left-hand side of (1)? We call this the *conclusion problem*: it is specified by (K,p). Note that after some empirical observations have been made on a new object, the classification rules in K can be updated and simplified to show the logical dependence of class membership on only the remaining (as yet uninspected) variables. Any rules that have become irrelevant because the empirically observed attribute values for the object fail to match the attribute values specified in their right-hand sides can be eliminated from further consideration. The problem that remains, of calculating posterior class probabilities, is the same as the original one, but with a reduced set of variables and rules. The same algorithm solves both.

The most important facts about the conclusion problem are that (i) It can only be solved at all if K is logically "complete," in a sense specified below. (ii) Even if K is complete, there is in general no polynomial-time algorithm for solving it exactly. (iii) For any pair (K,p), where K is complete, the solution to the conclusion problem can be computed by an algorithm that gives monotonically improving approximations to the correct answer. Although the exact answer can not be computed in polynomial time, close approximations can be computed with much less work.

As an example of an "incomplete" knowledge base, consider the two rules (i) C_1 if [a & -b]; and (ii) C_2 if [b & -a], where the class labels C_1 and C_2 are distinct. Assume that attributes a and b each have probability 0.5 of being present in a randomly selected object to be classified. Here, the notation is that "a" means "attribute a is present" while "-a" means "attribute a is absent." Then what is the probability of C_1? There is not enough information in this rule base to answer the question. Unless we know what happens (i.e., which class an object belongs to) if [a & b] or [-a & -b] occurs, and are willing to make the closed-world assumption that no other rules are relevant to C_1 (so that, for example, there is not another rule "C_1 if e") there is no way to compute the probability of C_1. In such cases, it is still possible to draw useful inferences about which classes are *possible*: numerical probabilities, however, require more knowledge than is available.

A rule base K is *complete*, for the purposes of probabilistic classification, if and only if it determines a unique class corresponding to any n-vector of attribute values. For example, the two rules {(i) C_1 if a; (ii) C_2 if -a} form a complete rule base in this sense, even though they only mention one attribute. Completeness entails a form of consistency. An example of an inconsistent rule base is {(i) C_1 if [a,b]; (ii) C_2 if [b,c]}, where C_1 and C_2 are distinct (and where literals in square brackets are henceforth understood to be conjoined.) Suppose, for example, that a, b, and c each has prior probability 0.5. Then mutual independence implies that [a,b,c] occurs with probability 0.125. But according to this rule base, [a,b,c] implies both C_1 and C_2 -- an impossibility. Completeness rules out such inconsistencies. In the remainder of this section, we will assume that K is complete, i.e., that it assigns exactly one class label to each of the 2^n possible vectors of attribute values.

Given a pair (K,p), where K is complete, what is the probability that an object belongs to class C_j? This question can be answered by the following *inclusion-exclusion algorithm* from systems reliability theory.

1. Identify all rules having class C_j on the left-hand side. Suppose there are M such clauses.

2. Form the set of all the clauses on the right-hand sides of these rules. For example, suppose this set is {[a,b],[b,c],[a,c]}, so that an object belongs to class C_j if and only if at least two out of the three attributes in {a,b,c} are present.

3. Construct the unions of these clauses taken i at a time, for i = 1 to M. For example, the unions of the three clauses in our example taken one at a time are just the clauses themselves, [a,b], [b,c], and [a,c]. The unions of these clauses taken two at a time are [a,b] ∪ [b,c] = [a,b,c]; [a,b] ∪ [a,c] = [a,b,c]; and [b,c] ∪ [a,c] = [a,b,c]. The union of these clauses taken three at a time is [a,b] ∪ [b,c] ∪ [a,c] = [a,b,c]. Form the inclusion-exclusion series $S_1 - S_2 + S_3 - S_4 + ... + S_n$, where S_i is the sum, over all unions of the clauses taken i at a time, of the products of the probabilities of the attribute values in these unions. In our example, this series would be AB + BC + AC - ABC - ABC - ABC + ABC = AB + BC + AC - 2 ABC, where A denotes the probability of a (i.e., the probability that attribute a is present); B is the probability of b; and C is the probability of c.

4. The sum of this series is the probability of class C_j.

This algorithm follows from the inclusion-exclusion formula for unions of events in elementary probability theory. Its correctness can be proved by induction on the number of clauses, M.

Although the inclusion-exclusion series in principle requires the summation of $2^M - 1$ terms, where M is the number of clauses in K, in fact the error introduced by truncating the series after all ith-order terms have been accounted for (i.e., after S_i) is no greater than S_i, and is monotonically decreasing in i. Very close approximations to the correct probability can therefore be obtained by including only the lower-order terms of the inclusion-exclusion series, thereby saving most of the computational effort that is theoretically required for an exact answer. In essence, the bulk of the computational effort that would be required for an exact answer is concentrated where it can safely be ignored. Thus, computation of a conclusion from (K,p) can be performed effectively.

4.2 Expected-Cost Recursive Partitioning Algorithms for Growing Classification Trees

Given a pair (K,p) with K complete, the inclusion-exclusion algorithm applied m times (once for each possible class) will compute the corresponding posterior m-vector, call it q, of class probabilities. The expected loss from each action in A with respect to q can easily be calculated, and the "optimal" action with respect to (K,p), or, equivalently, with respect to q, can be found. Let L(K,p) denote the minimized expected loss resulting from this optimal action in the absence of any additional information.

Now suppose that some empirical observations are introduced. The initial statistical knowledge base (K,p) must be *updated* by eliminating from further consideration all rules whose right-hand side clauses do not match the observed attribute values and by replacing the probabilities p_j for observed attributes with ones or zeros. Let [K(t),q(t),I(t),p(t)] denote the updated statistical knowledge base, where I(t) is the set of empirical observations made (here, the attribute values observed) up through time t; p(t) denotes the updated vector of *attribute* probabilities [in which some of the original components of the initial p have been replaced with zeros or ones to record observed attribute values in I(t)]; q(t) is the posterior probability vector (a "conclusion") for *class* probabilities, found by propagating p(t) through rule base K(t) using the inclusion-exclusion algorithm; and K(t) is the updated rule base in which all rules whose clauses fail to match the attribute values observed so far have been deleted. Now let $L[I(t)]$ denote the corresponding minimized expected loss when the "optimal" action with respect to I(t) is taken. $L[I(t)]$ is found by updating (K,p) as just described with the information in I(t) to obtain the posterior conclusion, say q(t), corresponding to this updated knowledge base. Once q(t) has been found, the act that minimizes expected loss with respect to it (i.e., that minimizes over all a in A the sum of the $q_i L(a,i)$ summed over all classes i in X) can easily be found.

One possible decision rule for deciding which attribute to observe next, or which terminal action to take, given the current information set I(t), is as follows:

1. For each attribute, a_j, that has not yet been inspected [i.e., whose value is not recorded in I(t)], compute the *expected net value of information* from inspecting a_j by the formula

   $V_j = L[I(t)] - p_j L[I(t) + \{a_j = 1\}] - (1-p_j)L[I(t) + \{a_j = 0\}] - c_j$,

   where I(t)+{a_j = 1}, for example, denotes the information set consisting of all the observations in information set I(t) plus the additional observation that the value of a_j is 1.

2. If $V_j > 0$ for at least one attribute, then inspect next the attribute for which V_j is greatest. This will produce information set I(t+1). Update [K(t),q(t),I(t)] to get [K(t+1),q(t+1),I(t+1)]. Set t = t+1 and return to Step 1.

3. If $V_j < 0$ for all remaining uninspected attributes, then stop inspection and choose that act in A that is optimal with respect to the current information set, i.e., that minimizes expected terminal loss, L(a,x).

This is an example of a *recursive partitioning algorithm*. It generates a classification tree by successive (stepwise) myopic optimization of some criterion -- in this case, reduction in expected loss. Notice that the update in Step 2 requires use of the inclusion-exclusion algorithm [to compute q(t+1) from p and K(t+1)], making this a potentially computationally expensive algorithm.

Different recursive partitioning algorithms that are less computationally demanding can be obtained by varying the *node selection criterion* (here, reduction in total expected cost) in steps 1 and 2. Easily calculated criteria that ignore observation costs have been implemented and widely used in past machine learning and statistical classification systems. For example, the attribute selection score V_j can be defined as expected reduction in *classification entropy* of the set of objects being classified at a node in the classification tree -- the criterion used for objects characterized by discrete attributes in ID3 [6]. Or it might be chosen to reflect the expected increase in the *"separation"* of subsets that have different values for the attribute corresponding to that node -- one of the strategies implemented for both continuous and discrete attributes in CART [1-5]. Or V_j may be defined as decrease in total probability (or, more generally, in expected terminal cost) of misclassification, which is one of the criteria considered in [1]. Similarly, alternative *stopping rules* can be used to end inspection in Step 3. One possibility is to stop inspection when the best of the remaining attributes is not expected to produce a "statistically significant" difference (as defined by any of various statistical tests) between the distribution of class probabilities if that attribute has a value of 0 and the distribution of class probabilities if it has a value of 1 [6,7.] When the probability vector p is not known, but must be estimated from data, it may be desirable to reject the idea of a stopping rule altogether, and instead to first grow a classification tree of depth n (which would call for all attributes to be inspected) and then "prune" this tree back to a subtree of a size that is expected to minimize the true (as opposed to the estimated) error rate; see [1] and [5]. In general, any algorithm that uses a myopic node-selection criterion to determine which attribute to inspect next may be referred to as a recursive partitioning algorithm; there are many such algorithms and they vary considerably in their node selection criteria and implementation details.

From a decision-analytic perspective, all of these strategies can be viewed as heuristics for obtaining computationally tractable approximations to the "optimal" decision rule/classification tree. By the criterion of minimizing total expected cost, including inspection costs, all of these strategies are suboptimal -- not only because they ignore observation costs, but also because even if observation costs were all zero, the myopic optimization called for in recursive partitioning does not in general lead to globally optimal choices. Recursive partitioning continually has the expert system ask "If I could only choose one more attribute to inspect, which one would be the most valuable one to look at?" (where "most valuable" is defined by the node selection criterion.) It then prescribes that the attribute that answers this question be inspected, unless some stopping criterion is met. But choosing the "best" attribute assuming that only one can be selected does not, in general, produce the same selection as would choosing the "best" attribute, bearing in mind that other attributes may subsequently be inspected. Nonetheless, recursive partitioning algorithms are currently the dominant way of generating classification trees in practical applications.

At least two systems -- the EXPERT EASE system that implements ID3 and the CART statistical software package -- have proved that recursive partitioning is a commercially viable computational technology for many applications where large training sets and negligible inspection costs can be assumed, at least as a first approximation.

5. Growing Optimal Classification Trees when Information is Costly: A New Heuristic

We now return to the MEGCP [K,p,c,L], where K is the rule base

C_1 if $[a_{11},...,a_{1n}]$
C_2 if $[a_{21},...,a_{2n}]$
:
C_m if $[a_{m1},...,a_{mn}]$

(and the literals a_{ij} within clauses are understood to be conjoined.) The following heuristic algorithm differs fundamentally from recursive partitioning algorithms. It examines the (heuristically assessed) expected value per unit inspection cost of each attribute in the context of each classification rule before deciding which one should be inspected next. It is also explicitly designed to take into account the costs of discovering attribute values.

Algorithm A: The Sorting Algorithm

1. Order the literals (the a_{ij}) in each right-hand side clause as follows: if a_{ij} asserts that attribute j is present, then define $r_{ij} = (1 - p_j)/c_j$. If a_{ij} asserts that attribute j is not present, then let $r_{ij} = p_j/c_j$. If a_{ij} asserts that the value of j (present or absent) is irrelevant, then set $r_{ij} = 0$ and delete a_{ij} from the clause. Sort the remaining literals from left to right within the clause in order of decreasing r_{ij}. We call r_{ij} the *failure probability per unit cost* for attribute j in rule i.

2. Compute the *expected inspection cost* for each right-hand side clause, assuming that the attributes in it are inspected in the order determined by Step 1 until the clause is completed (satisfied) or until an empirically observed attribute value fails to match the value (present or absent) prescribed by the corresponding a_{ij}. If $[a_1,...,a_n]$ is an ordered sequence of attributes to be inspected, and if c_j is the cost of inspecting the value of a_j and w_{ij} is the *match probability* that the true value of a_j will match the value prescribed by the corresponding a_{ij} (i.e., $w_{ij} = p_j$ or $1 - p_j$, depending on whether a_{ij} calls for a_j to be present or absent, respectively) then the expected inspection cost for $[a_1,...,a_n]$ can be put in the form $c_1 + w_{i1}c_2 +$

$w_{i1}w_{i2}c_3 +...+ w_{i1}w_{i2}...w_{i,n-1}c_n$. (Note that only attributes that have not yet been inspected are included in this sum.) In this form it can be computed extremely quickly by a single pass through the chain of successive (c_j, w_{ij}) values. Let E_i denote the expected inspection cost for the clause on the right-hand side of rule i.

3. Let P_i denote the product of the match probabilities, $P_i = w_{i1}w_{i2}...w_{in}$. Thus, P_i is the probability that the clause of rule i is empirically satisfied. Find the rule with the greatest P_i/E_i ratio. Let j^* denote the left-most attribute (according to the ordering determined in Step 1) in the clause of this rule.

4. Let V_{j^*} be the expected reduction in cost from inspecting the value of attribute j^*, as given by the formula in Section 4.2. If $V_{j^*} < c_{j^*}$, or if all attribute values have been inspected, then stop inspection and choose that action in A that minimizes conditional expected loss. Otherwise, ask for (or inspect) the value of attribute j^* next.

5. Once the value of attribute j^* has been determined, delete from further consideration all rules with right-hand side clauses containing literals a_{ij^*} that do *not* match the revealed empirical value (present or absent) of attribute j^*. Update the P_i/E_i ratio for all right-hand side clauses containing literals a_{ij^*} that *do* match the empirically observed value of j^* by replacing w_{ij^*} with 1 in the formulas for P_i and E_i and deleting sunk costs from the expression for E_i. (Thus, in effect, let the new $P_i = P_i/w_{ij^*}$ and set $c_{j^*} = 0$.) Return to Step 2.

As Algorithm A is executed, it produces a sequence of questions (in Step 4) that ask the user for the values of different attributes. When enough questions have been answered so that the expected costs of answering further questions exceed the expected benefits of the additional information that will be gained from answering them, then Algorithm A stops and recommends an action that minimizes expected loss with respect to what it currently knows.

The performance of Algorithm A depends on both the logical structure of the rule base K and on the cost vector c. The following result show that algorithm A yields "good" results under some interesting conditions (for a proof, see [17]):

Theorem: Suppose that an object belongs to class 1 if at least k out of its n attributes have the value "present," and that it belongs to class 0 otherwise. Then the classification tree constructed by algorithm A is optimal, i.e., it determines which class the object belongs to at minimum expected cost, thus solving MEGCP exactly for this special case.

However, in general, Algorithm A is only a heuristic: it is possible to construct small examples in which the sequence of questions generated by the procedure does not solve MEGCP exactly.

The intuitive motivation for Algorithm A is that the expert system should try to satisfy the easiest-to-satisfy clauses first, where "easiest" is measured by the success-probability per unit cost ratio, P_i/E_i. As soon as a clause is satisfied, a classification conclusion can be drawn, and the algorithm can stop. Within a clause, however, the hardest-to-satisfy literal, as measured by the failure-probability per unit cost ratio r_{ij}, should be attempted first. Since all literals within a clause must be satisfied to satisfy the clause, there is no point in wasting resources on the easy ones at first: if it will not be possible to satisfy the whole clause, then this fact should be discovered as soon as possible. Therefore, the most difficult parts of the clause should be tried first.

These two heuristic guidelines lead to a computationally efficient procedure (Algorithm A) for identifying the next attribute to ask about. This procedure is much less myopic than recursive partitioning algorithms, since the selection of the next attribute to ask about has been heuristically optimized within and between clauses. However, the heuristic buys computational ease at the expense of global optimality in general. It ignores *informational spill-overs* among clauses in assigning priority scores -- e.g., situations in which even failure part way through one clause may be very valuable because of information that it yields that reduces the expected costs of resolving the remaining clauses. Ignoring these informational interdependencies makes Algorithm A tractable both for generating entire classification trees a priori (an "open-loop" inspection and classification strategy) and for generating questions adaptively as the values of an object's attributes are sequentially discovered (a "closed-loop" or on-line inspection and classification strategy.) But it leaves room for improved special-purpose algorithms.

An Example:

To understand how algorithm A works, consider the following problem. There are three attributes, a, b, and c, with prior probabilities of 0.1, 0.4, and 0.6 of being present, respectively. The cost of inspecting any attribute (to determine whether it is present) is 1. There are two classes. Objects are classified according to the following logical rule:

An object belongs to class C_1 if and only if at least two attributes are present; otherwise, it belongs to class C_2.

What is an optimal classification tree for this problem? Assume that the cost of a misclassification error is much greater than 1, so that it is always optimal to definitively determine an object's class. Algorithm A proceeds as follows. First, the logical rule must be translated into a canonical-form rule base, as follows:

1. C_1 if (a,b,*)
2. C_1 if (a,*,c)
3. C_1 if (*,b,c)
4. C_2 if (a',b',*)
5. C_2 if (a',*,c')
6. C_2 if (*,b',c').

Here, a means that attribute a is "present" (value = 1), a' means that attribute a is "absent" (value = 0), and * signifies the don't-care condition.

The first step is to sort attributes within clauses in order of decreasing r_{ij} ratio. This ratio is 0.9 for a, 0.6 for b, and 0.4 for c; and hence 0.1, 0.4, and 0.6 for a', b', and c', respectively. Thus the C_1 clauses are already correctly sorted, and the C_2 clauses should have their literals sorted in reverse order. The second step is to compute P_i/E_i for each of the six rules. For the clause (a,b,*), the expected cost is $E_1 = 1 + (0.1)(1) = 1.1$, while the overall success probability is $(0.1)(0.4) = 0.04$. The ratio is $P_1/E_1 = 0.04/1.1 = 0.036$. The corresponding ratios for the other rules are $P_2/E_2 = 0.06/1.1 = 0.054$, $P_3/E_3 = 0.24/1.4 = 0.17$, $P_4/E_4 = 0.54/1.6 = 0.34$, $P_5/E_5 = 0.36/1.4 = 0.26$, and $P_6/E_6 = 0.24/1.4 = 0.17$. Rule 4, namely "$C_2$ if (a',b',*)" ranks highest, and b' is the first-ranked attribute (recalling that the literals in C_2 clauses are sorted in the reverse of their original order.) Thus, *the first question generated should be whether b is absent* (or, equivalently, whether b is present.) (In a closed-loop implementation, calculation would stop at this point, with the first question being submitted to the user before any additional calculations are done.) If the answer is that b is <u>absent</u>, then rules 1 and 3 drop out. Iterating the steps of the algorithm again with the new information set, $I(1) = \{b'\}$, shows that rule 4 becomes the new winner, with a' being the left-most literal in it. Therefore, the next question on the b' branch will be whether a is present. If a is not present, then the classification tree terminates with a determination that the object belongs to class C_2; otherwise, it will ask whether c is also present, and will then assign the object to class C_1 if and only if the answer is "yes." If the answer to the first question is that b is <u>present</u>, then rules 4 and 6 can be eliminated from further consideration, and rules 1 and 3 must have their scores updated to $0.1/1 = 0.1$ and $0.6/1 = 0.6$, respectively. Rule 3 is now the highest-scoring rule, so the next question will be "Is c present?" The object will be classified as of class C_1 if the answer is "yes," and as class C_2 otherwise. This completes the derivation of the optimal classification tree via Algorithm A. That it is in fact optimal follows from the theorem presented above for k-out-of-n systems.

6. Application and Evaluation for Concept Classification: The Intersection-Sort Heuristic

An important special case of the classification problem in artificial intelligence is the *concept classification problem*. Here, each object is to be classified as either an instance or as not an instance of a given concept. In this case, with only two possible classes, say 0 and 1, Algorithm A can be further refined to take advantage of the symmetry between trying to prove that a given object belongs to class 0 and trying to disprove that it belongs to class 0. Algorithm A is asymmetric and "optimistic": it always seeks to prove that an object belongs to the class for which proving membership is currently judged to be easiest (based on the P_i/E_i ratio.) With only two classes, it is *a priori* equally sensible to try a "pessimistic" strategy in which one always seeks to disprove membership in the class for which disproving membership is judged to be easiest. This may be accomplished by sorting attributes within clauses in decreasing order of $s_{ij} = w_{ij}/c_j$ ratio [as opposed to the $r_{ij} = (1 - w_{ij})/c_j$ ratio in Algorithm A] in Step 1 and by choosing the rule with the greatest Q_i/E_i ratio in Step 3 (instead of the one with the greatest P_i/E_i ratio), where Q_i is defined as the product of the mismatch probabilities $(1 - w_{ij})$. This pessimistic version of the sorting heuristic, which is a symmetric dual to Algorithm A, we will call Algorithm B (the "backward" sort heuristic.) Algorithm B is also optimal for k-out-of-n classification problems.

Finally, it is possible to combine the features of Algorithms A and B, and hence to obtain the advantages of both the "optimistic" and the "pessimistic" heuristic perspectives, as follows. First, use Algorithm A to identify the rule with the greatest "optimistic" (P_i/E_i) score. Next, use Algorithm B to find the rule with the greatest "pessimistic" (Q_k/E_k) score. It is shown in [17] that *the right hand side clauses for these two different rules will have a nonempty intersection*. The next attribute to be inspected is then selected to be any attribute that belongs to the right-hand side clauses of both the winning "optimistic" rule and the winning "pessimistic" rule simultaneously. We refer to this as the *intersection-sort heuristic* for solving MEGCP.

At U S WEST Advanced Technologies, Algorithms A and B and the intersection-sort heuristic have been subjected to extensive experimental tests comparing them to each other and to a dynamic programming type algorithm that recursively constructs optimal classification trees for solving MEGCP exactly for arbitrary problems. They have been compared in terms of run time and quality of solution [quantified by the *relative error* $C(h)/C^* - 1$, where $C(h)$ is the expected cost of the classification tree built by the heuristic, h, being evaluated; and C^* is the expected cost of the optimal classification tree] on a variety of randomly generated binary concept classification tasks in which the two classes correspond to linearly separable subsets of an n-dimensional vector space. The principal empirical findings are as follows:

1. Both Algorithm A and Algorithm B as well as the intersection-sort heuristic have small average relative errors (less than 2%) over a wide range of classification task parameter combinations.

2. The intersection-sort heuristic substantially out-performs both of the other heuristics in terms of average relative error and largest observed relative error for all combinations of task parameter values tested. In several thousand randomly generated concept classification tasks, Algorithms A and B showed average relative errors of less than 2%, although the worst-case relative errors on a few isolated cases were as high as 113%. The intersection-sort heuristic did substantially better, with average relative errors of less than 0.05% (i.e., one twentieth of one percent) and worst-case errors on individual cases of less than 15%. In fact, the intersection-sort heuristic discovered the optimal classification tree in over 90% of the cases tested. This remarkably high performance was robust across task parameters such as the relative ranges of the p_j and c_j, which were varied systematically in an attempt to find and understand the worst-case performance characteristics of the heuristics.

3. None of the three heuristics uniformly dominates the other two. Despite the general superiority of the intersection-sort heuristic, in approximately 1% to 3% of the test cases observed (depending on the task parameters) it produced a solution having a greater relative error than that of one of the other two heuristics. (However, the difference was always less than 1% of the optimal cost.)

4. Algorithm A yields both lower average relative errors and lower maximum observed relative errors than does the pessimistic path-sorting heuristic when all probabilities in p are high (e.g., greater than 0.9), but this pattern is reversed at low probabilities (e.g., less than 0.1.) Since the two versions are symmetric duals of each other, this pattern is expected, as is the observation that they perform equally for probabilities close to 0.5. The intersection-sort heuristic outperforms both on average at all probability levels.

5. The performances of all three heuristics gradually decreased as n increased from n = 3 to n = 10. For n > 10, it became impractical to evaluate the true minimum expected cost solution, C^*, since the time complexity of the dynamic programming algorithm [which searches a state space that grows as $O(n!2^n)$] began to lead to solution times of greater than 10 hours for n = 10.

6. For problems with n = 10 that took approximately 10 hours to solve exactly with the dynamic programming algorithm (running in Common Lisp on a Compaq Deskpro 386/20), the time to compute the first attribute to be inspected was approximately 9 seconds for Algorithms A and B and approximately 19 seconds for the intersection-sort heuristic. The corresponding times to compute an entire ("open-loop") classification tree were 50 seconds and 90 seconds, respectively. Despite these orders-of-magnitude improvement in speed from using the heuristics, relative errors remained less than 5%.

The experimental designs used and the numerical results on which these findings are based are reported in greater detail in [17].

7. Summary and Conclusions

This paper has described several algorithms for using expert classification systems to classify objects efficiently when information about the objects is expensive to acquire. A goal for a pragmatic expert system in such cases is to establish the class of an object as quickly (or, more generally, as inexpensively) as possible. An intuitively appealing heuristic for accomplishing this goal is to direct the system's questioning by having it continually try to prove that the object belongs to the class for which proving membership is currently judged easiest. We have implemented this idea, and have discovered that it leads to joint questioning and decision-making strategies (represented as classification trees) that are usually close to optimal, at least over the range of problems tested. This is so even though the problem of discovering an exact optimal strategy is NP-hard. In the important case of classification with only two classes, we have found that augmenting the "easiest proof first" heuristic with a symmetrical "easiest disproof first" approach leads to a hybrid heuristic, called the intersection-sort heuristic, that performs substantially better on average than either approach alone. The intersection sort heuristic almost always gives optimal or near-optimal strategies, and yet takes only a tiny fraction of the amount of computational effort of the exact algorithm.

In addition to trying to establish class membership quickly, a pragmatic expert classification system that is concerned with minimizing the costs of classification should know when to stop asking questions and make a decision. We have formalized this idea using a statistical decision theory framework in which the known costs of continued information-gathering are balanced against the expected reduction in costs of misclassification from the additional information. A recursive partitioning algorithm of the type commonly used in A.I. to infer classification rules from data can be formulated for this purpose. However, the sorting algorithms introduced here require less computational effort and almost certainly produce better results. (The computational experiments to confirm this comparison empirically have not yet been performed.) Quitting before the class of the object has been determined with certainty requires drawing a *probabilistic conclusion*, denoted by q(t)

= $[q_1(t),...,q_m(t)]$, giving the posterior probabilities based on all attribute values observed so far that the object belongs to each of the m possible classes. An "optimal" terminal decision with respect to what is currently known can then be made by choosing the act in A that minimizes the expected loss with respect to q(t).

ACKNOWLEDGEMENT

An early version of this paper was presented at the 1988 European Knowledge Acquisition Workshop. I gratefully acknowledge the help of my colleagues Yuping Qiu, who helped to develop the intersection sort version of the sorting heuristic and who proved several interesting theoretical properties for it; and Warren Kuehner, who implemented the algorithms and expertly carried out the performance evaluations.

References

[1]   L. Breiman, J. Friedman, R. Olshen, and C. Stone, *Classification and Regression Trees*, Wadsworth, Belmont, CA, 1984.

[2]   K.A. Grajski et al, "Classification of EEG spatial patterns with a tree structured methodology: CART," *IEEE Transactions on Biomedical Engineering, Vol. EME-33*, 12, 1986, 1076-1086.

[3]   L. Gordon and R.A. Olshen, "Tree-structured survival analysis," *Cancer Treatment Reports, 69*, 10, 1985, 1065-1069.

[4]   M.R. Segal, "Recusive Partitioning Using Ranks," Tech. Report No. 15, Department of Statistics, Stanford University, August, 1985.

[5]   A. Ciampi et al, "Recursive partitioning algorithms: A versatile method for exploratory data analysis in biostatistics," in I.B. MacNeill and G. Umphrey (Eds.), *Biostatistics*, Reidel, Boston, MA, 1987.

[6]   J.R. Quinlan, "The effect of noise on concept learning," Chapter 6 in R.S. Michalski et al (Eds.), *Machine Learning: An Artificial Intelligence Approach*, Morgan Kaufmann, Los Altos, CA, 1987.

[7]   J.R. Quinlan, "Generating production rules from decision trees," *IJCAI '87, Vol. 1*, Morgan Kaufmann, Los Altos, CA, 1987.

[8]   J.C. Schlimmer and D. Fisher, "A case study of incremental concept induction," *AAAI-86 Proceedings, Vol. 1*, Morgan Kaufmann, Los Altos, CA, 1986, 496-501.

[9] C.J. Colbourn, *The Combinatorics of Network Reliability*, Oxford University Press, New York, 1987.

[10] D.H. Fisher, "Knowledge acquisition via incremental conceptual clustering," *Machine Learning, 2*, 2, 1987, 139-172.

[11] G.T. Duncan, "Optimal diagnostic questionnaires," *Operations Research, 23*, 1, 1975, 22-32.

[12] Y. Ben-Dov, "Optimal testing procedures for special structures of coherent systems," *Management Science, 27*, 12, 1981, 1410-1420.

[13] J. Halpern, "Fault-testing of a k-out-of-n system," *Operations Research, 22*, 1974, 1267-1271.

[14] D. Angluin and P. Laird, "Learning from noisy examples," *Machine Learning, 2*, 1988, 343-370.

[15] N. Littlestone, "Learning quickly when irrelevant attributes abound: A new linear-threshold algorithm," *Machine Learning, 2*, 1988, 235-318.

[16] L.G. Valiant, "A theory of the learnable," *Communications of the ACM, 27*, 1984, 134-1142.

[17] L.A. Cox, Qiu, Y., and Kuehner, W., "Heuristic least-cost computation of discrete classification functions with uncertain argument values," in F. Glover and H. Greenberg (Eds.), *Annals of Operations Research: Artificial Intelligence and Operations Research*, J.C.Baltzer Scientific Publishing Company, Basel, Switzerland, 1989.

VII. APPLICATIONS

The section presents two applied research papers, the first dealing with manufacturing systems and the second with wastewater treatment plants. The paper by Kusiak ("A Knowledge- and Optimization-Based Approach to Scheduling in Automated Manufacturing Systems") notes that the scheduling problem in an automated manufacturing environment involves a number of different manufacturing resources, such as machines, tools, fixtures, pallets, and material handling carriers. Due to its complexity, the scheduling problem cannot be effectively solved with the classical scheduling techniques. The tandem approach coupling a heuristic algorithm with a knowledge-based subsystem has lead to an effective tool for solving the scheduling problem. The knowledge-based scheduling system (KBSS) designed for an automated manufacturing environment is presented. Frames and production rules have been applied to represent declarative and procedural scheduling knowledge. The KBSS allows the consideration of basic and alternative process plans. Each process plan specifies manufacturing resources as well as their sequence. The knowledge-based system is illustrated with a numerical example. The system has been tested on a number of scheduling problems. Computational experience proves that the solutions generated are of good quality. The system presented can be implemented in any discrete manufacturing system. The tandem architecture underlying the KBSS applies to many other domains involving quantitative and qualitative elements.

Lai et al. ("An Integrated Management Information System for Wastewater Treatment Plants") note that the wastewater treatment plant operator's problem in collecting, storing, manipulating, and interpreting data is an information management problem. It is believed that the operator's performance in controlling the treatment process can be improved if he or she is equipped with a convenient and efficient method for extracting useful information from the data. This paper presents an approach to developing an integrated management information system to aid operation and control. The approach involves database management, an expert system, and statistical methods. The expert system provides the operator with timely advice based on the latest available process data as interpreted by a set of rules constructed with the help of experienced and skilled operators. Statistical methods are used not only because they are powerful aids for extracting information from data but also because they are useful in interpreting the fuzzy terms, such as 'high' and 'low', that describe the condition of the variables in the control rules. Using historical performance data from an activated sludge treatment plant, a preliminary evaluation of a prototype expert system has been conducted, and findings from the evaluation shows that the approach is feasible and efficient.

A KNOWLEDGE- AND OPTIMIZATION-BASED APPROACH TO SCHEDULING IN AUTOMATED MANUFACTURING SYSTEMS

Andrew Kusiak
Department of Industrial and Management Engineering
University of Iowa
Iowa City, Iowa 52242

ABSTRACT

In this paper a knowledge-based scheduling system (KBSS) designed for an automated manufacturing environment is presented. Frames and production rules have been applied to represent declarative and procedural scheduling knowledge. A heuristic algorithm is incorporated into the knowledge-based system. The KBSS allows to consider basic and alternative process plans. Each process plan specifies manufacturing resources such as: machines, tools, fixtures, material handling carriers as well as their sequence. The knowledge-based system is illustrated with a numerical example. The system has been tested on a number of scheduling problems. Computational experience proves that the solutions generated are of good quality.

1. INTRODUCTION

The scheduling problem in manufacturing systems has been of interest of many practitioners and researchers. It is even more important in automated manufacturing systems due to the following:

- considerable capital investment and therefore the need for high utilization of manufacturing resources such as machines, robots, etc,

- increased number of variables because of introduction of new resources, such as fixtures, pallets, automated material carriers, etc.

There are two basic approaches used for solving the scheduling problems in manufacturing systems:

- optimization, and
- knowledge-based approach.

The optimization approach for solving scheduling problems has been deeply explored in Baker (1974), French (1982), Bellman et al. (1982), and others.

Due to the high computational complexity of scheduling problems, optimization methods often fail to solve them. Since the optimization approach has not been fully successful in solving many scheduling problems, an expert system approach has been suggested. To date a number of prototype expert systems have been developed to solve the scheduling problems arising in manufacturing systems. Some of the recently developed expert systems are listed in Table 1.

Since both the optimization and knowledge-based approaches have certain advantages in solving scheduling problems, integration of the two approaches might be beneficial. The knowledge-based approach presented in this paper is based on the above premise. A heuristic algorithm is incorporated into the knowledge-based system. The algorithm includes a number of priority rules. The scheduling instances which cannot be handled by the algorithm are referred to the inference engine. The structure of the knowledge-based system and several sample production rules are presented in Section 3. A numerical example illustrating the algorithm and the knowledge-based system is presented in Section 4. Computational results for a number of sample problems solved are presented in Section 5. Conclusions are drawn in Section 6.

In many manufacturing systems, one associates with each part a basic process plan and one or more alternative process plans. A process plan specifies the operations belonging to the part, processing times of these operations, and the resources required such as machines, tools, pallet/fixtures, etc.

Before the heuristic algorithm will be presented, the following notation and definitions are introduced:

I set of all operations
K set of all parts

Table 1. Sample Scheduling Problems and Expert Systems

PROBLEM	KNOWLEDGE REPRESENTATION	CONTROL STRATEGY	REFERENCE
job shop scheduling	production rules	constraint based analysis	Erschler and Esquirol (1986)
	frames	constraint directed search	Fox (1983)
	production rules	heuristic	Shaw and Whinston (1986)
	production rules	meta-rules	Bensana et al. (1986)
	frames and production rules	heuristic	Grant (1985)
automated manufacturing system scheduling	frames	constraint directed search	Chiodini (1986)
	production rules	simulation	Bruno et al. (1986)
	production rules	forward chaining	Kerr and Ebsary (1988)
project scheduling	frames and a E-nets	heuristic	Morton and Smunt (1987)
robot scheduling	production rules	meta-rules	Thesen and Lei (1987)

L   set of all resource types
Q_l set of resources of type l, $l \in L$
d_k due date of part P_k, $k \in K$
f_i completion time of operation i, $i \in I$
rt_i remaining processing time of operation i, $i \in I$
t   current scheduling time
r_{lq} resource q of type l, $q \in Q_l$, $l \in L$

In particular, the following four types of resources are used:
- machine (l=1),

- tool (l=2),
- pallet/fixture (l=3), and
- material handling carrier (l=4)

Resource r_{lq} is *available* if it can be used without any delay, $q \in Q_l$, $l \in L$. The *status* sr_{lq} of such resource equals 1, otherwise $sr_{lq}=0$

A *process plan* $PP_k^{(v)}$ of a part P_k is a vector of triplets, each containing: operation number, processing time, and set of resources to process the operation. It is denoted as follows:

$$PP_k^{(v)} = [(a; t_a^{(v)}; R_a^{(v)}),...,(i; t_i^{(v)}; R_i^{(v)}),...,(b; t_b^{(v)}; R_b^{(v)})],$$

where: $R_i^{(v)} = (r_{lq}^{(v)}, r_{2q}^{(v)},...., r_{lq}^{(v)})$, $q \in Q_l$, $l \in L$

a,...,i,...,b denote operation number
v = 0 denotes the basic process plan,
v = 1,2,... denotes an alternative process plan

$t_i^{(v)}$ denotes processing time of operation i using process plan v.

For simplicity of further considerations two assumptions are made:

1. It is assumed that the number and type of operations in a basic process plan and the corresponding alternative process plan are identical. This assumption allows to switch processing from the basic process plan to an alternative process plan and reverse after completion of an operation considered.

2. Without loss of generality, it is assumed that $t_i^{(0)} \leq t_i^{(v)}$, $i \in I$ which holds in practice. A process plan $PP_k^{(v)}$ for part P_k and the corresponding operations is *available*, if each element in $PP_k^{(v)}$ has been specified.

Operation i is *schedulable* at time t, if:
    1) no other operation that belongs to the same part is being processed at time t,
    2) all operations preceding operation i have been completed before time t,

3) all resources required by the basic process plan to process operation i are available at time t.

Based on the above definitions, further notation is introduced:

operation status $\quad s_i = \{\begin{array}{l} 0, \text{ operation i is nonschedulable,} \\ 1, \text{ operation i is schedulable,} \\ 2, \text{ operation i is being processed,} \\ 3, \text{ operation i has been completed.} \\ 4, \text{ operation i satisfies the first} \\ \quad \text{two conditions in the definition of} \\ \quad \text{schedulability} \end{array}$

resource status $\quad sr_{lq} = \{\begin{array}{l} 1, \text{ resource } r_{lq} \text{ is available,} \\ 0, \text{ otherwise,} \end{array}$

$S_j \quad$ set of operations with $s_i = j$, $j = 0, 1, 2, 3, 4, i \in I$

$st_k \quad$ slack time of part P_k, $st_k = (d_k - t - \sum_{i \in S_0 \cup S_1} t_i^{(0)})$, $k \in K$

In the process of schedule generation, an operation might not be processed according to the basic process plan due to unavailability of the resources specified in the basic process plan. The scheduling heuristic presented in this section exits from step 3 and step 6, and enter the inference engine of the knowledge-based system.

Seven priority scheduling rules have been incorporated into the heuristic algorithm:

rule P1: selects an operation with the largest number of successive operations
rule P2: selects an operation belonging to a part with the minimum number of schedulable operations
rule P3: selects an operation with the largest number of immediate successive operations
rule P4: selects an operation belonging to a part with the largest number of unprocessed operations
rule P5: selects an operation with the shortest processing time
rule P6: selects an operation belonging to a part with the shortest slack time

rule P7: selects an operation arbitrarily.

Some of the terms used in the above priority rules are defined. Consider a part that has the structure shown in Figure 1.

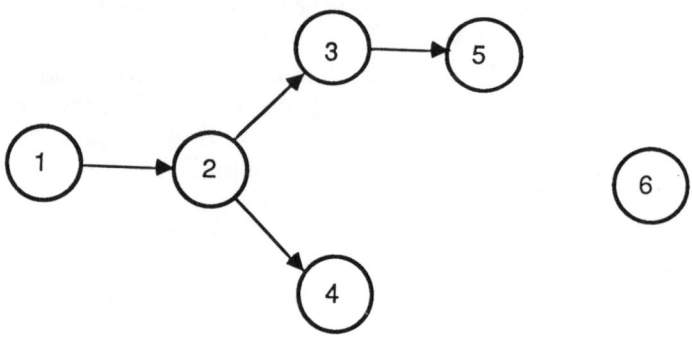

Figure 1. Sample part with 6 operations

As shown in Figure 1 all operations but 1 and 6 have precedence constraints. Assume that the part is being scheduled and operation 1 has been completed. In this case the number of unprocessed operations is five (operations 2,3,4,5 and 6). One also notes in Figure 1 that operation 2 is followed by three successive operations 3,4, and 5, while operation 6 has 0 successive operations. The operations 3 and 4 are the immediate successive operations of operation 2.

A slack time of a part is the difference between its due date and completion time of its final operation.

The priority rules P1 to P7 are used in step 2 of the algorithm. If more than one operation is selected by a rule, the next rules perform further selection. Different sequences of the priority rules have been tested for various problems and the results are reported later in this paper.

2. HEURISTIC ALGORITHM

Step 0. Set current time $t = 0$ and resource status $sr_{lq} = 1, q \in Q_l$ $l \in L$.

Step 1. Construct the following two sets:
- set S_0 of nonschedulable operations ($s_i = 0$)
- set S_1 of schedulable operations ($s_i = 1$).

Step 2. In the set of schedulable operations S_1, select an operation i^* based on the following priority rules:

    P1:   selects an operation with the largest number of successive operations

    P2:   selects an operation belonging to a part with the minimum number of schedulable operations

    P3:   selects an operation with the largest number of immediate successive operations

    P4:   selects an operation belonging to a part with the largest number of unprocessed operations

    P5:   selects an operation with the shortest processing time

    P6:   selects an operation belonging to a part with the shortest slack time

    P7:   selects an operation arbitrarily.

Step 3. Set:

- operation status $s_{i^*} = 2$ for operation i^* selected in step 2,

- operation status $s_i = 0$ for all the unprocessed operations of the part corresponding to operation i^*

  Delete operation i^* from S_1. If $S_1 \cup S_0 = \emptyset$, stop; otherwise, set:

- remaining processing time $rt_{i^*} = t_{i^*}^{(0)}$,

- resource status $sr_{lq} = 0$, for $r_{lq} \in R_{i^*}^{(0)}$, $q \in Q_1, l \in L$.

  Update sets of schedulable operations S_1 and nonschedulable operations S_0. If $S_1 \neq \emptyset$, go to step 2. If $S_1 = \emptyset$ and no resource is available, go to step 4; otherwise update set S_4, enter the inference engine, and return.

Step 4. Construct set of operations being processed S_2, and:

- calculate completion time $f_i = rt_i + t$, $i \in S_2$

- set current time $t = f_{\tilde{i}} = \min_{i \in S_2} \{f_i\}$,

- set operation status $s_{\tilde{i}} = 3$,

- delete operation i from S_2
- set resource status $sr_{lq} = 1$, $r_{lq} \in R_i^{(0)}$, $q \in Q_1$, $l \in L$
- set remaining time $rt_i = f_i - t$, $i \in S_2$

Update set of schedulable operations S_1 and nonschedulable operations S_0.

Step 5. If $S_1 \cup S_0 = \emptyset$, stop; otherwise, go to step 6.

Step 6. If set of schedulable operations $S_1 \neq \emptyset$, go to step 2. If $S_1 = \emptyset$ and no resource is available, go to step 4; otherwise update set S_4, enter the inference engine, and return.

The algorithm presented solves scheduling problems with due dates. If the due date is not imposed for a part P_k, the corresponding value of d_k is set an arbitrary large number.

The above algorithm is embedded into the knowledge-based system discussed in the previous section.

3. THE KNOWLEDGE-BASED SCHEDULING SYSTEM (KBSS)

Numerical algorithms have traditionally been used for solving scheduling problems. The approach presented in this paper involves not only algorithms but also declarative and procedural knowledge and an inference engine, all implemented as a knowledge-based system (KBSS). The KBSS performs the following two basic functions:

- selects an algorithm for the problem considered
- controls the schedule generation procedure of the algorithm selected.

The knowledge-based system is built using the tandem architecture proposed in Kusiak (1987) that involves:
- knowledge base
- algorithm base
- data base
- inference engine.

3.1. Knowledge Base

Knowledge in KBSS has been acquired from experts as well as the scheduling literature. Frames are used to represent the declarative knowledge related to the description of scheduling problems, parts and operations, and the schedules generated. Two sample frames representing a scheduling problem and the schedule generated are presented below.

Frame 1

(Problem_number (Problem_type (e.g. flow-shop))
  (Problem_features
    (Number_of_parts,
    Number_of_operations,
    Number_of_precedence_constraints)))

Frame 2

(Schedule_of_problem (Problem_number)
    (Generated_by (algorithm_number))
    (Idle_time (machine_1, machine_2, ..., machine_m))
    (Completion_time (part_1, part_2, ..., part_n))
    (Average_utilization_rate (machine, tool, pallet_fixture, material_handling_carrier))

The procedural knowledge of the knowledge-based system is in the form of production rules. To handle different problems the production rules are divided into following three classes:

- Class 1: selects an appropriate algorithm to solve the problem considered,
- Class 2: controls the procedure of selecting alternative process plans and modifying the sequence of the priority rules in the heuristic algorithm,
- Class 3: evaluates the schedules obtained and performs rescheduling.

Several sample production rules in each class are presented below:

Class 1

Rule 12: IF the scheduling problem involves less than 5 features
AND number of operations is less than 60
AND alternative process plans are not available

THEN use the mixed integer linear programming formulation
(see Kusiak , 1990)

Class 2

Rule 24: IF an alternative process plan is specified for an operation that is not schedulable due to unavailability of resources listed in the basic process plan

AND the required alternative resources for the operation are available

AND the sum of the waiting and processing time for the operation in the basic process plan is longer than one in the alternative process plan

THEN replace the basis process plan with the corresponding alternative process plan
AND add the corresponding operation to the set of schedulable operations

Rule 25: IF more than one operation has been added to the set of schedulable operations using production rule 24

THEN select one operation with the alternative processing time closest to the value of the corresponding basic processing time

Class 3

Rule 36: IF a part in a partial (or final) schedule generated by the heuristic

Scheduling Manufacturing Systems

                algorithm does not meet the required due date
    THEN schedule the part ensuring that the due date is satisfied
        AND reschedule other parts using the heuristic algorithm

3.2. Algorithm Base

In the algorithm base a number of algorithms discussed in the scheduling literature are stored. Additional algorithms can be easily incorporated into the base. The heuristic algorithm presented in the previous section is the most likely to be used for solving scheduling problems in large scale automated manufacturing systems.

3.3. Data Base

The data base contains parameters of the scheduling models and serves as a working space for the algorithms.

3.4. Inference Engine

The inference engine in the knowledge-based scheduling system (KBSS) controls the procedure of triggering rules in the knowledge base and the procedure of schedule generation of an algorithm. One of the greatest advantages of the tandem system architecture is the simplicity of the inference engine. The inference engine in KBSS employs a forward chaining control strategy. In a given class of rules it attempts to fire all the rules which are related to the context considered. If a rule is triggered, i.e. the conditions are true, then the actions of the rule triggered are carried out. Some rules stop the search of the inference engine and switch the control process to the algorithm.

The inference engine maintains a list of the rules which have been fired. This list is called "explain". The rules in "explain" are placed in the order that they were fired. The list forms a basis for building an explanation facility.

In the next section, the heuristic algorithm and sample production rules are illustrated with the following numerical example.

4. NUMERICAL EXAMPLE

Schedule twelve operations shown in Figure 2 on three machines.

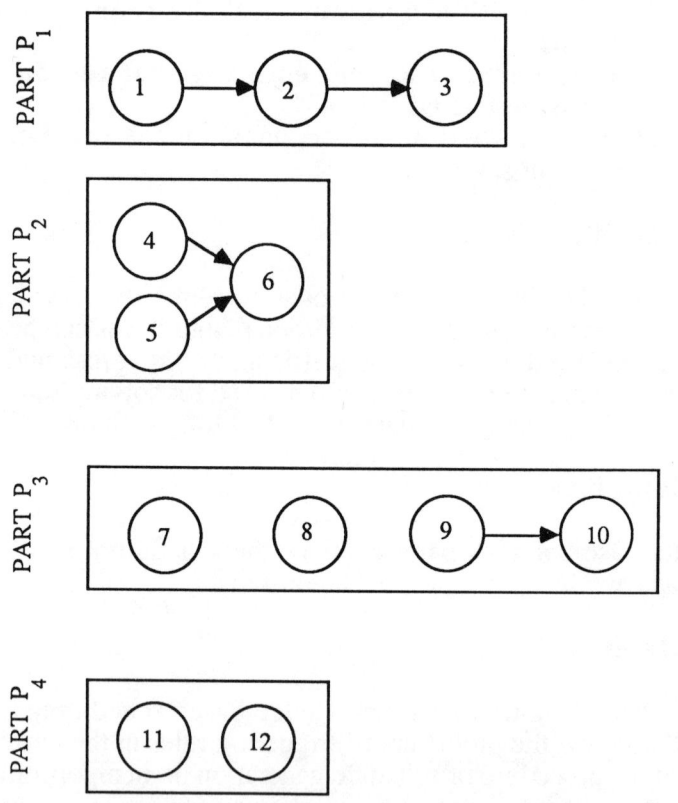

Figure 2. Parts with operations and precedence constraints

It is assumed that:
- three different tools are available to process the operations
- all other resources are unlimited, and
- due dates are not imposed.

The following notation is used for resources:
    r_{1q} denotes machine (resource type 1), q = 1,2,3,
    r_{2q} denotes tool (resource type 2), q = 1,2,3.

The machine and tool status are represented as follows:
    sr_{1q} denotes machine status, q = 1,2,3,
    sr_{2q} denotes tool status, q = 1,2,3, and
    t_i denotes the processing time $t_i^{(0)}$ of operation i in the basic process plan.

Scheduling Manufacturing Systems

The basic process plans of the four parts are shown below:

$PP_1^{(0)}$: [(1;4;2,2),(2;5;1,3),(3;2;3,2)],
$PP_2^{(0)}$: [(4;6;1,3),(5;3;2,2),(6;3;3,1)],
$PP_3^{(0)}$: [(7;3;3,1),(8;3;1,2),(9;6;3,1),(10;2;1,3)],
$PP_4^{(0)}$: [(11;4;3,2),(12;3;2,3)].

Note that for any triplet in the above process plans, the first element denotes operation number, the second denotes processing time and the third pair denotes the required machine number (resource type 1) and tool number (resource type 2).

The alternative process plans for the four parts are:

$PP_1^{(1)}$: [(1;6;3,1),(2;6;2,2),(3;4;1,1)];
$PP_1^{(2)}$: [(1;7;1,3),(2;7;1,2),(3;5;1,3)]

$PP_2^{(1)}$: [(4;6;2,2),(5;4;3,1),(6;5;1,2)];
$PP_2^{(2)}$: [(4;8;3,1),(5;8;1,3),(6;5;2,3)]

$PP_3^{(1)}$: [(7;4;3,2),(8;5;3,3),(9;7;2,1),(10;2;3,2)];
$PP_3^{(2)}$: [(7;4;2,2),(8;5;2,1),(9;9;1,3),(10;4;1,2)]

$PP_4^{(1)}$: [(11;4;1,3),(12;5;1,2)];
$PP_4^{(2)}$: [(11;4;3,1),(12;6;3,3)].

Solution Procedure

Step 0.   Set current time $t = 0$ and $sr_{lq} = 1$, $l = 1,2$, $q = 1,2,3$.

Step 1.   Construct the following two sets:
- $S_0 = \{2,3,6,10\}$,
- $S_1 = \{1,4,5,7,8,9,11,12\}$.

Step 2.   Using priority rule P1, operation 1 is selected.
Step 3.   Set:

- $s_1 = 2$,
- $s_2 = 0$, $s_3 = 0$.

Since $S_0 \cup S_1 = \{2,3,4,5,6,7,8,9,10,11,12\} \neq \emptyset$, set:
- $rt_1 = t_1 = 4$,
- $sr_{12} = 0$, $sr_{22} = 0$.

Set of schedulable operations $S_1 = \{4,7,9\} \neq \emptyset$.
Go to step 2.

Step 2. Using priority rules P1 and P2, operation 4 is selected.
Step 3. Set:
- $s_4 = 2$,
- $s_5 = 0$, $s_6 = 0$.

Since $S_0 \cup S_1 = \{2,3,5,6,7,8,9,10,11,12\} \neq \emptyset$, set:
- $rt_4 = t_4 = 6$,
- $sr_{11} = 0, sr_{23} = 0$.

Set of schedulable operations $S_1 = \{7,9\} \neq \emptyset$.
Go to step 2.

Step 2. Using priority rule P1, operation 9 is selected.
Step 3. Set:
- $s_9 = 2$,
- $s_7 = 0$, $s_8 = 0, s_{10} = 0$.

Since $S_0 \cup S_1 = \{2,3,5,6,7,8,10,11,12\} \neq \emptyset$, set:
- $rt_9 = t_9 = 6$,
- $sr_{13} = 0$, $sr_{21} = 0$.

Set of schedulable operations $S_1 = \emptyset$.
Since $S_1 = \emptyset$ and no resource is available, go to step 4.

Step 4. Construct $S_2 = \{1,4,9\}$, and
- calculate completion time $f_1 = 4$, $f_4 = 6$, $f_9 = 6$,
- set current time $t = f_1 = \min\{4,6,6\} = 4$,
- set $s_1 = 3$,
- delete operation 1 from S_2,
- set $sr_{12} = 1$, $sr_{22} = 1$,
- set remaining time $rt_4 = 6 - 4 = 2$, $rt_9 = 6 - 4 = 2$,

Set of schedulable operations $S_1 = \emptyset$ (there is no operation with the status equal to 1).

Step 5. Since $S_1 \cup S_0 = \{2,3,5,6,7,8,10,11,12\} \neq \emptyset$, go to step 6.

Scheduling Manufacturing Systems

Step 6. Since $S_1 = \emptyset$ and machine 2 and tool 2 are available, set $S_4 = \{2,11,12\}$ and enter the inference engine.

The inference engine activates production rule 24 in Class 2 to select an alternative process plan for operation 2.

Rule 24: SINCE the alternative process plan is specified for operation 2, in S_4.

           AND the alternative resources for operation 2 (r_{12} and r_{22}) are available

           AND the sum of the waiting and processing time for operation 2 in the basic process plan ($2 + 5 = 7$) is longer than one in alternative process plan (6)

           THEN operation 2 is moved from S_4 to S_1

           AND the basic process plan of operation 2 is replaced with the corresponding alternative process plan (2,6,2,2)

After the basic process plan has been replaced with the alternative process plan, the inference engine transfers the information to the heuristic algorithm and operation 2 is scheduled by the algorithm.

The Gantt chart of the final schedule obtained after 7 iterations is shown in Figure 3.

Note that in the above example, only machines and tools have been considered. The KBSS is general enough to consider other resources such as pallets/fixtures, material handling carriers, etc.

5. COMPUTATIONAL RESULTS

In this section, computational experience with the knowledge-based scheduling system KBSS is presented. The heuristic algorithm discussed is the most likely algorithm to be used while solving industrial scheduling problems in automated manufacturing systems. The results generated by the knowledge-based scheduling system can be improved by using other production rules. The degree of improvement depends upon the quality of the knowledge collected.

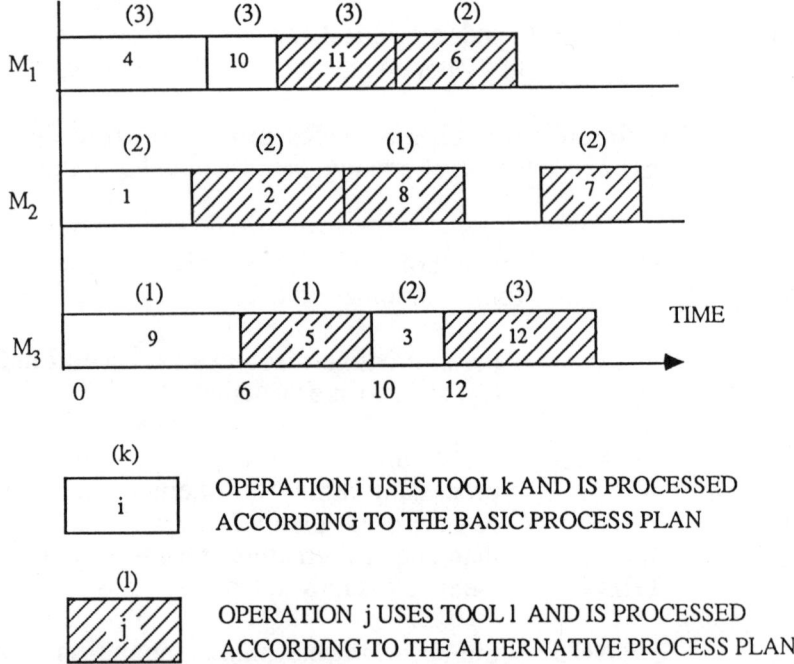

Figure 3. The final schedule

In order to evaluate the quality of solutions generated by the KBSS, sample problems have been solved. Three measures of performance were used:

- maximum flow time (F_{max})
$$F_{max} = \max_{j} \{ F_j \}, j = 1,...,N,$$

- average flow time ($\bar{F}$)
$$\bar{F} = \sum_{j=1}^{N} F_j / N$$

where: F_j is flow time for machine j, and
N is the number of machines.

- machine utilization (U_m)
$$U_m = \sum_{j=1}^{N} U_j / N,$$

where: $U_j = \sum_{i \in M(j)} t_i^{(v)} / F_j$

$t_i^{(v)}$ is the processing time of operation i using process plan v,

M(j) is the set of operations processed on machine j.

In order to evaluate the effect of the sequence of priority rules P1 to P6 on the solution quality, a number of scheduling problems have been solved for five different sequences of priority rules. Computational results for four sample problems are presented in Table 2. Problems in Table 3 include basic and alternative process plans where for a given operation, the processing time in an alternative process plan was assumed to be 10%, 20%, 40%, 60%, 80%, and 100% longer than the corresponding basic processing time.

To evaluate the impact of due dates on solution quality, two scheduling problems were analyzed subject various parameters (Table 3).

In all the problems presented in Tables 2,3, and 4 the number of resources, except machines, was assumed unlimited.

The computational results presented in Tables 2,3, and 4 allow one to draw the following conclusions:

1) schedules generated by the knowledge-based system are of good quality. Extensive computational experience showed that the results generated by KBSS are only a few percent worse than the optimal ones.

2) the results in Table 2 indicate that there is no sequence of priority rules that is dominant in terms of the solution quality. Most (frequently all) of the seven priority rules are involved in selection of an operation to be scheduled next and that is reason why their sequence does not significantly affect the solution quality.

3) using of alternative process plans increases scheduling flexibility and this in turn increases machine utilization rate and reduces scheduling makespan.

The results obtained indicate that CPU time increases with the increase of problem size and decreases with the increase of the number of precedence constraints. CPU time is also slightly affected by the sequence of priority rules.
It is obvious that as the problem size grows the computation time increases. Adding more precedence constraints to a scheduling problem, reduces the size of the space of feasible solutions and this reduces the CPU time.

4) the CPU time requirement of KBSS is modest. The KBSS has been designed for real time production scheduling in automated environment. Selecting a suitable scheduling horizon (problem size) allows to use the system in a real time mode.

Table 2. Computational Results for Scheduling Problems with Basic Process Plans

Problem 1 — 14 Machines, 20 Parts, 160 Operations, Minimum Maximum Flow Time: 24.0

Sequence of Priority Rules	F max	F̄	U m	CPU Time	F max	F̄	U m	CPU Time	F max	F̄	U m	CPU Time
P1-P2-P3-P4-P5-P6	25.5	24.3	0.99	9.59	27.0	24.5	0.98	8.44	26.5	25.5	0.94	7.19
P2-P1-P3-P4-P5-P6	25.5	24.3	0.99	9.37	26.5	24.7	0.97	7.92	26.0	24.8	0.97	7.47
P3-P2-P1-P4-P5-P6	25.5	24.3	0.99	9.43	26.5	24.5	0.98	8.36	26.0	24.8	0.97	7.79
P4-P2-P3-P1-P5-P6	25.5	24.2	0.99	9.44	26.5	24.4	0.99	8.46	27.0	25.4	0.95	7.81
P1-P3-P2-P4-P5-P6	25.5	24.3	0.99	9.51	27.0	24.5	0.98	8.56	26.5	25.5	0.94	7.27
P3-P1-P4-P2-P5-P6	25.5	24.2	0.99	9.78	26.0	24.4	0.98	8.31	26.5	25.5	0.94	7.49
P2-P3-P1-P4-P5-P6	25.5	24.3	0.99	9.29	26.5	24.7	0.97	8.03	26.0	24.8	0.97	7.76
Number of Precedence Constraints	0				48				140			

Problem 2 — 12 Machines, 18 Parts, 132 Operations, Minimum Maximum Flow Time: 27.0

Sequence of Priority Rules	F max	F̄	U m	CPU Time	F max	F̄	U m	CPU Time	F max	F̄	U m	CPU Time
P1-P2-P3-P4-P5-P6	28.5	27.3	0.99	6.63	29.0	27.3	0.99	5.95	29.5	27.7	0.98	5.08
P2-P1-P3-P4-P5-P6	28.5	27.3	0.99	6.52	30.0	27.5	0.98	6.00	32.5	28.5	0.95	5.06
P3-P2-P1-P4-P5-P6	28.5	27.3	0.99	6.56	29.0	27.3	0.99	5.99	32.5	28.5	0.95	5.34
P4-P2-P3-P1-P5-P6	30.0	27.5	0.98	6.47	30.0	27.5	0.98	6.16	29.5	27.8	0.97	5.41
P1-P3-P2-P4-P5-P6	28.5	27.3	0.99	6.54	29.0	27.3	0.99	6.05	29.5	27.7	0.98	5.10
P3-P1-P4-P2-P5-P6	30.0	27.5	0.98	6.75	28.5	27.2	0.99	6.21	28.0	27.5	0.98	5.08
P2-P3-P1-P4-P5-P6	28.5	27.3	0.99	6.47	30.0	27.5	0.98	6.05	32.5	28.5	0.95	5.29
Number of Precedence Constraints	0				36				114			

Table 2. (continued)

Problem 3 10 Machines 16 Parts 116 Operations	Sequence of Priority Rules	F_{max}	$\bar{F}$	U_m	CPU Time	F_{max}	$\bar{F}$	U_m	CPU Time	F_{max}	$\bar{F}$	U_m	CPU Time
	P1-P2-P3-P4-P5-P6	25.5	24.3	0.99	5.34	24.5	24.1	0.99	4.91	27.0	24.8	0.97	4.45
	P2-P1-P3-P4-P5-P6	25.5	24.3	0.99	5.27	25.0	24.1	0.99	4.90	28.0	25.3	0.95	4.40
	P3-P2-P1-P4-P5-P6	25.5	24.3	0.99	5.28	24.5	24.1	0.99	4.94	28.0	25.3	0.95	4.56
	P4-P2-P3-P1-P5-P6	25.0	24.2	0.99	5.35	25.0	24.1	0.99	4.83	28.0	24.9	0.96	4.59
Minimum Maximum Flow Time: 24.0	P1-P3-P2-P4-P5-P6	25.5	24.3	0.99	5.34	24.5	24.1	0.99	4.96	27.0	24.8	0.97	4.42
	P3-P1-P4-P2-P5-P6	25.0	24.2	0.99	5.48	25.0	24.2	0.99	4.99	27.5	25.0	0.96	4.46
	P2-P3-P1-P4-P5-P6	25.5	24.3	0.99	5.28	25.0	24.1	0.99	4.95	28.0	25.3	0.95	4.56
Number of Precedence Constraints		0				28				100			

Problem 4 12 Machines 18 Parts 94 Operations	Sequence of Priority Rules	F_{max}	$\bar{F}$	U_m	CPU Time	F_{max}	$\bar{F}$	U_m	CPU Time	F_{max}	$\bar{F}$	U_m	CPU Time
	P1-P2-P3-P4-P5-P6	21.0	19.4	0.98	3.52	21.0	19.4	0.98	2.98	21.5	19.7	0.96	2.74
	P2-P1-P3-P4-P5-P6	21.0	19.4	0.98	3.40	20.5	19.3	0.99	3.13	21.5	20.0	0.95	2.92
	P3-P2-P1-P4-P5-P6	21.0	19.4	0.98	3.43	21.5	19.6	0.97	3.01	21.5	20.0	0.95	2.92
	P4-P2-P3-P1-P5-P6	19.5	19.3	0.99	3.50	21.0	19.8	0.96	3.08	22.0	20.3	0.94	3.04
Minimum Maximum Flow Time: 19.0	P1-P3-P2-P4-P5-P6	21.0	19.4	0.98	3.47	21.0	19.4	0.98	3.06	21.5	19.7	0.97	2.76
	P3-P1-P4-P2-P5-P6	19.5	19.3	0.99	3.66	22.0	19.5	0.97	3.17	21.5	20.1	0.95	2.77
	P2-P3-P1-P4-P5-P6	21.0	19.4	0.98	3.44	20.5	19.3	0.99	3.13	21.5	20.0	0.95	3.03
Number of Precedence Constraints		0				24				76			

Table 3. Computational Results for Scheduling Problems with Basic and Alternative Process Plans

Problem Number	Number of Machines	Number of Parts	Number of Operations	Number of Preced. Constr.	Number of Alternative Process Plans for Each Part	KBSS Solution			
						F_{max}	$\bar{F}$	U_m	CPU Time
1	14	20	160	0	0	137.0	83.2	0.97	13.08
					1^1	109.0	84.1	0.98	12.91
					1^2	106.7	86.5	0.97	12.57
					1^3	111.5	86.2	0.98	12.63
					1^4	118.6	90.5	0.95	12.40
					1^5	132.5	89.9	0.95	13.33
					1^6	130.5	91.1	0.93	12.83
				48	0	137.0	88.5	0.89	12.56
					1^1	114.6	90.6	0.90	12.20
					1^2	114.1	91.2	0.91	12.07
					1^3	120.7	91.9	0.93	12.19
					1^4	121.5	92.9	0.93	12.44
					1^5	127.0	93.5	0.91	12.26
					1^6	128.5	94.6	0.88	11.96
				140	0	137.0	108.3	0.73	12.96
					1^1	109.8	98.5	0.83	12.22
					1^2	111.8	98.6	0.84	12.19
					1^3	116.5	100.6	0.84	12.52
					1^4	128.5	109.6	0.77	12.77
					1^5	126.5	108.4	0.78	11.77
					1^6	129.5	105.9	0.79	11.52

Table 3. (continued)

						0	263.5	106.2	0.95	15.31
						1^1	162.0	107.2	0.96	13.73
						1^2	178.0	109.4	0.95	14.36
					0	1^3	163.4	114.6	0.93	14.56
						1^4	174.1	114.1	0.95	14.27
						1^5	184.6	118.2	0.94	14.15
						1^6	208.5	115.5	0.94	14.18
						0	263.5	116.3	0.86	14.13
						1^1	180.5	113.3	0.90	13.64
						1^2	176.0	117.9	0.88	14.00
2	14	20	161	48	1^3	186.3	121.1	0.85	13.68
						1^4	176.5	122.5	0.87	13.74
						1^5	175.0	124.6	0.87	13.92
						1^6	220.0	121.1	0.86	13.78
						0	263.5	177.1	0.59	13.29
						1^1	153.5	127.4	0.82	12.88
						1^2	163.5	130.6	0.81	12.92
				140	1^3	160.0	135.0	0.80	13.17
						1^4	168.0	137.1	0.79	13.24
						1^5	174.0	143.7	0.77	13.01
						1^6	209.0	156.8	0.70	12.69

```
1    alternative processing time is 10 percent longer than basic processing time
2    alternative processing time is 20 percent longer than basic processing time
3    alternative processing time is 40 percent longer than basic processing time
4    alternative processing time is 60 percent longer than basic processing time
5    alternative processing time is 80 percent longer than basic processing time
6    alternative processing time is 100 percent longer than basic processing time
```

Scheduling Manufacturing Systems   475

Table 4. Computational Results for Scheduling Problems with Due Dates

Problem Number	Number of Machines	Number of Parts	Number of Operations	Number of Parts with Due Dates	Number of Preced. Constr.	Number of Alternative Process Plans for Each Part	KBSS Solution					
							F_{max}	$\bar{F}$	U_m	T_t	Number of Tardy Parts	CPU Time
1	14	20	160	5	0	0	137.0	83.0	0.97	64	3	11.10
						1^0	106.0	83.4	0.97	58	4	10.77
						1^1	107.3	84.3	0.97	66	4	11.34
						1^2	112.8	85.1	0.98	69	4	11.59
						1^3	115.5	86.3	0.98	57	3	11.69
						1^4	115.5	86.8	0.98	57	3	11.64
						1^5	119.0	87.3	0.98	57	3	11.38
					48	0	137.0	108.6	0.76	60	4	10.93
						1^0	105.0	93.8	0.87	45	3	10.24
						1^1	113.5	92.9	0.89	74	4	11.24
						1^2	119.6	98.1	0.86	55	5	11.27
						1^3	116.8	97.0	0.86	68	5	11.54
						1^4	120.0	96.5	0.88	99	5	11.30
						1^5	123.0	100.0	0.89	39	4	11.20
					140	0	137.0	122.4	0.66	88	4	11.39
						1^0	111.0	98.5	0.82	73	4	10.80
						1^1	113.4	106.0	0.78	99	5	12.07
						1^2	113.6	103.7	0.80	96	5	12.09
						1^3	119.1	103.1	0.84	93	5	11.89
						1^4	122.0	109.4	0.79	107	5	11.88
						1^5	121.6	107.3	0.81	106	5	11.99

Table 4. (continued)

2	14	20	161	7	0	0	263.5	109.8	0.90	79	6	13.64
						1^0	162.0	106.8	0.95	69	5	12.41
						1^1	165.0	106.2	0.97	79	5	13.73
						1^2	170.5	110.0	0.94	103	5	13.28
						1^3	174.5	113.1	0.94	95	6	12.73
						1^4	174.0	115.5	0.95	93	5	12.93
						1^5	176.5	119.8	0.94	70	6	12.56
					48	0	263.5	139.5	0.73	130	7	14.13
						1^0	170.0	113.3	0.89	143	7	12.55
						1^1	167.4	112.2	0.92	81	5	12.72
						1^2	172.0	119.5	0.86	100	6	12.68
						1^3	175.7	125.6	0.84	122	7	12.78
						1^4	181.5	125.7	0.85	94	7	12.78
						1^5	192.3	127.5	0.86	121	6	12.89
					140	0	263.5	184.8	0.53	154	7	12.61
						1^0	165.5	137.3	0.75	102	5	11.36
						1^1	164.5	137.9	0.76	136	6	12.53
						1^2	170.3	143.3	0.75	153	6	12.15
						1^3	178.0	149.3	0.72	106	6	12.24
						1^4	188.0	156.8	0.70	145	7	12.14
						1^5	187.9	158.5	0.69	172	7	11.99

0 alternative processing time is 0 percent longer than basic processing time
1 alternative processing time is 10 percent longer than basic processing time
2 alternative processing time is 20 percent longer than basic processing time
3 alternative processing time is 40 percent longer than basic processing time
4 alternative processing time is 60 percent longer than basic processing time
5 alternative processing time is 80 percent longer than basic processing time

6. CONCLUSIONS

In the paper the knowledge-based scheduling system (KBSS) suitable for automated manufacturing was designed. The KBSS contains a heuristic algorithm developed and a number of other existing algorithms for solving scheduling problems. While numerical calculations are performed by the algorithm selected, the KBSS performs the following functions:

- selects an appropriate algorithm (including the heuristic algorithm) the most suitable for the problem considered.

  Most of the scheduling algorithms available in the literature have a limited scope of application in manufacturing due to simplifying assumptions. The heuristic algorithm presented in Section 2 is the most likely to be used for solving scheduling problems in automated manufacturing systems.
- selects alternative process plans to schedule operations.

  Alternative process plans increase the scheduling flexibility.
- evaluates the generated schedules and performs rescheduling when required.

  Rescheduling is performed whenever a performance measure (eg. makespan) associated with the schedule obtained is not satisfactory.

The testing effort has been concentrated on the heuristic algorithm and production rules handling alternative process plans. First an attempt was made to identify the best possible sequence of priority rules. However, based on the computations performed, no conclusive results could be derived. For a selected sequence of priority rules, a number of sample problems with basic and alternative process plans typically resulted in solutions of better quality than problems with basic process plans only. This was especially visible in the case when machine breakdowns have occurred. From the comparison of the optimal schedules and those generated by the KBSS, the following two conclusions can be drawn:

1) the schedules generated by the KBSS are of the quality only slightly worse than the optimal solutions
2) incorporating alternative process plans increases scheduling flexibility and in most cases improves the solution quality.

3) using alternative process plans increases machine utilization rate and reduces makespan

The optimization approach is suitable for solving problems with high numerical computation requirement while knowledge-based systems are suitable for qualitative problems. The integration of the two into the knowledge-based scheduling system (KBSS) provides a powerful means for solving manufacturing scheduling problems.

REFERENCES

1. Baker, K. R. (1974) *Introduction to Sequencing and Scheduling*, John Wiley and Sons, New York, N.Y.

2. Bellman, R.E., Esogbue, A.O. and Nabeshima, I. (1982) *Mathematical Aspects of Scheduling and Applications*, Pergamon Press, Oxford, U.K.

3. Bensana, E., Correge, M., Bel, G. and Dubois, D. (1986) " An expert system approach in industrial job-shop scheduling", *Proceedings of 1986 IEEE International Conference on Robotics and Automation*, San Francisco, April 7-10, pp.1645-1650.

4. Bruno, G., Elia, A., and Laface P. (1986) "A rule-based system to schedule production", *IEEE Computer*, Vol.19, No.7, pp.32-40.

5. Chiodini, V. (1986) "An expert system for dynamic manufacturing rescheduling", *Symposium on Real Time Optimization in Automated Manufacturing Facilities*, National Bureau of Standards, Gaithersburg, MD, January.

6. Erschler, J. and Esquirol, P. (1986) " Decision-aid in job shop scheduling : A knowledge based approach", *Proceedings of the 1986 International Conference on Robotics and Automation*, San Francisco, April 7-10, pp.1651-1656.

7. Fox, M.S. (1983) "Constraint-directed research: A case study of job-shop scheduling", Ph.D. Thesis, Carnegie-Mellon University, Pittsburgh, PA.

8. French, S. (1982) *Sequencing and Scheduling: An Introduction to the Mathematics of the Job-shop*, John Wiley and Sons, New York, N.Y.

9. Grant, T.J. (1986) "Lessons for O.R. form A.I.: A scheduling case study", *Journal of Operational Research Society*, Vol.37, No.1, pp.41-57.

10. Kerr, R.M. and Ebsary, R.V. (1988) " Implementation of an expert system for production scheduling", *European Journal of Operational Research*, Vol.33, No.1, pp.17-29.

11. Kusiak, A. (1990) " *Intelligent Manufacturing Systems* ", Prentice Hall, Englewood Cliffs, N.J.

12. Morton, T.E. and Smunt, T.L. (1986) " A planning and scheduling system for flexible manufacturing", in A. Kusiak (Ed.) *Flexible Manufacturing Systems: Methods and Studies*, North-Holland, Amsterdam, pp.151-164.

13. Selen, W.J. and Hott, D.D. (1986) "A mixed-integer goal-programming formulation of the standard flow-shop scheduling problem", *Journal of Operational Research Society*, Vol.37, No.12, pp.1121-1128.

14. Shaw, M.J.P. and Whinston, A.B. (1986) " Application of artificial intelligence to planning and scheduling in flexible manufacturing", in A. Kusiak (Ed.), *Flexible Manufacturing systems: Methods and Studies*. North-Holland, Amsterdam, pp.223-242.

15. Thesen, A. and Lei, L. (1986) " An expert system for scheduling robot in a flexible electroplating system with dynamically changing work loads", in K.E. Stecke and S. Suri (Eds) *Flexible Manufacturing Systems: Operation Research Models and Applications*, Elsevier, New York, N.Y. pp 555-566.

AN INTEGRATED MANAGEMENT INFORMATION SYSTEM FOR WASTEWATER TREATMENT PLANTS

WENJE LAI, P. M. BERTHOUEX, and DON HINDLE

There are thousands of publicly owned wastewater treatment plants in the United States. They vary greatly in size, and consequently in staff numbers and skills, and in information system support to their managers. Many plants fail to meet effluent standards on frequent occasions (Schaefer et al. 1984; Ucbrin and Casper 1983). The wastewater treatment plant, in 1988, is under control to about the same extent of a blast furnace in 1968. The task of management is still largely an art (Berthouex, et al. 1985). There are many reasons why this is the case, but two are dominant. First, a wastewater treatment plant is, in production process terms, far more complicated than a blast furnace; and second, the management effort is dispersed over large numbers of small and variable sites, rather than concentrated in a few large sites with similar characteristics.

At the largest city plants, many aspects of the production system are electronically monitored and the data fed through analytical routines which provide various levels of diagnostic and prescriptive advice. Computerized databases are maintained, which provide many types of useful reports. In most plants, however, few variables are monitored electronically, little information is available in real time, and even batch processing of basic data is limited in scope. This study is primarily directed at these plants. However, it should be recognized that even the large plants rely to a great extent on the subjective judgments of their managers.

The management of wastewater treatment plants has improved, and will continue to do so, as a consequence of continual changes of many types: better education, improved equipment, enhanced understanding of chemical and biological processes, and so on. One critical aspect in which improvements are needed concerns the information available to managers to support their operational decisions.

This is the focus of interest here: the management information system, with emphasis on continuing operation of an existing plant. It is recognized that there are other important aspects, such as information required to support investment decisions with respect to new equipment. However, they cannot be addressed adequately unless there is a better basis for making routine decisions. While other types of data are required for investment decisions, they also require data of the types which are central to this study.

Three types of weaknesses: (1) data not available to the manager, (2) data not presented in ways which allow for their absorption, and (3) data inadequately analyzed in terms of derived statistics, may be the most important in some respects, in that the needs are generally recognized and simple technologies exist which would help eliminate them. They may not, however, be the most critical problems in terms of enhancement of management performance. A management information system includes not only the data but the managers themselves, who are the links between data and actions. The best data do not guarantee better decisions. Thus a critical component of a management information system is that which links the measures of the condition of the system to a set of decision options, and thereby to a production system performance measure. This gives rise to another weakness that is briefly enumerated below.

The relationships between operating characteristics of the plant and the decision options, and between the options and production system performance (however measured) are understood and able to be precisely predicted only to a limited extent. Specific causal relationships may be known. For example, precise formulations may have been established which relate the quantity of a chemical added to an influent and level of a pollutant in the effluent. However, this relationship will be affected by a large number of other factors, both controllable and uncontrollable, and a comprehensive model which takes account of all other factors may have much greater imprecision. Furthermore, action taken to control the level of one type of outflow constituent is likely to affect the levels of others; and therefore a set of highly complicated models must be used coterminously.

As has been the case in other complicated production systems, mathematical models become progressively more sophisticated over time. However, they must reach a threshold level of performance before they are able to be accepted as superior to more subjective approaches. There is an important human element in the process, which is a combination of reluctance to trust black boxes, to accept that traditional methods are flawed, and so on. The process of refinement and expansion of causal models will continue. There are, however, typically significant data measurement problems (Berthouex, et al. 1981; Hunter 1982). The models tend to be difficult to use, and the conjoint effects of large numbers of data errors and model approximations are such that predictive and explanatory powers are often disappointing. There is reason to believe that mathematical models of wastewater treatment plant operations are unlikely to become entirely accepted as replacements for experienced managers' judgments in the near future.

This study attempts to address all of the above issues. The data capture, manipulation, and display weaknesses are addressed by a microcomputer-based software package. The decision-making components will be addressed by development of an expert system (ES) module. The ES module eventually will incorporate control chart methods and other statistical methods which are well-known in other contexts, but not usually considered to be typical elements of expert systems.

DISCUSSION OF THE EXPERT SYSTEM APPROACH

An expert system (ES) is an analytical routine, which replicates the problem-solving process of a human expert. It usually takes the form of a computer program. The aim is to allow problems to be solved rapidly without relying on the presence of the human expert. The ES approach is applied to problems which are difficult to handle by formal mathematical modeling, and where it is therefore typically necessary to rely on human experts. These problem areas are characterized by data difficulties (such as incommensurability, intangibility, volume, and timeliness), as well as lack of knowledge regarding causal relationships.

In general it may be said that ES approaches are likely to be attractive where relationships are highly complicated, the system is dynamic, and many data are difficult to obtain or measure. This view is reflected in popular use of the term 'expert'. The implication is that some tasks are best handled by staff who have become more effective over time through acquisition of knowledge, that the knowledge is difficult to specify in precise terms, and that an element of individual judgment, taking account of non-explicit reasoning, contributes to success. Over time, progressively more tasks once regarded as 'expert' become understandable and hence replicable through performance of precise sets of operating rules. In this sense, they are essentially expedients and a more formal approach is preferable if feasible.

The central feature of the expert system is that it relies on heuristic rules suggested by human experts, rather than on detailed mathematical models of cause and effect. Human experts specify the rules, based on their knowledge and experience, which can then be incorporated in the computer program. By so doing, it should be possible to find solutions which are as good as those found by the human experts themselves.

The emphasis on knowledge rather than formal reasoning methods is thus the central component of the ES approach to problem-solving. The underlying assumption is that human experts are able to perform effectively, in spite of weaknesses in data and understanding of cause and effect at a detailed level, and, if the data available to them are also possessed by a computer model, all that is necessary is the ability to elucidate the human expert's analytical rules and to incorporate them into the computer program.

CONTROL RULES IN THE KNOWLEDGE BASE

The knowledge/expertise collected from the domain expert should be represented in a form that is implementable to a computer. Various schemes for representing knowledge for expert systems have been developed in the past years. A detailed overview and reference materials of knowledge representation can be found in Mylopoulos and Levesque (1984). In general, they are affected by the purpose of the expert system: for planning, diagnosis, prediction, control, and so on. In this application, the primary aim is control of the production system; that is, the adaptive management of the overall behavior of the wastewater treatment plant, and naturally the heuristic rules collected from the expert are represented in the form of IF-THEN statements(Waterman 1986).

For example, in controlling the wastewater treatment process, we may have a control rule, collected from the domain expert, such as "IF Mean Cell-Residence Time (MCRT) is low and concentration of NH3-N is high THEN decrease the sludge wasting rate". Here, since MCRT and NH3-N are determined as numerical values, numerical definitions of the qualitative or fuzzy terms "high" and "low" are needed before the rule can be used for inference. Unfortunately, operators from different treatment plants may agree with the logical relationship of the rule, but may not have a consensus on the numerical definition of the term "high" or "low". For instance, a 6 days MCRT may be low for plant A, but may be treated as normal in plant B.

There are arguments in favor of keeping qualitative or fuzzy terms in rule syntax. First, there is ease of elicitation of rules: it is generally easier to elicit the rule from the good operator or the expert, if he or she is required merely to categorize. Second, there is generalizability: The qualitative rule can be programmed into the ES, and be retained for use at all or most plants, whereas this would not be the case if numerical values were to be specified. Third, there is the related aspect of robustness for a single plant: for example, in one plant a low level of MCRT might produce the same kind of response irrespective of the season, while in others a value that is low in winter might be high for summer.

In the wastewater treatment plant context it is advantageous, therefore, to adopt a staged approach to rule specification. In the first stage, general rules are elicited from experts in largely qualitative terms. These rules may be intended to be generally applicable to a large set of plants. In the second stage, good operators or experts at individual plants would be asked to refine and extend the general rules in the light of their own experiences and local conditions. At this stage, all operating condition definitions should become numerical.

In handling of qualitative elements of rules, two approaches were frequently employed in this problem domain. They are direct logic controller and fuzzy logic controller, as referred by Magdol (1984). The approach of direct logic controller is straightforward, since qualitative terms are determined by some critical values set by the domain expert. For example, the condition of MCRT can be set as high if its value is above 6). The system developed by Vitasovic and Andrews (1987) and the system developed by Maeda (1984) are examples of rule-based systems that used direct logic controllers.

A fuzzy logic controller uses constructed membership function which relates numerical values and qualitative statements. Fuzzy logic has been discussed by Zadeh (1965, 1973, 1983) and will not be explained here. Some preliminary studies of a fuzzy logic controller for activated sludge system have been conducted by Tong, Beck and Latten (1980) and Johnston (1985), but no substantive results are available to date.

THE ANTICIPATED USE OF EXPERT SYSTEMS

An important goal has been to allow the expert system to incorporate new knowledge. It is generally recognized that it is not sufficient merely to allow the expert periodically to review ES performance and modify or extend the rules. Rather, it should also be possible for the expert to learn from the computer model's application, in a symbiotic way.

A related issue is that of incorporation of additional forms of information, such as objective measurements. In principle, the human expert's performance can be enhanced by provision of better data and analytical tools. For example, a skilled wastewater treatment plant operator could operate without an ES and his performance, while still being essentially judgment-based, could be enhanced by provision of more data, better ways of analyzing those data, graphical displays showing (say) seasonally adjusted trends, and so on. Such enhancement could be built into the expert system. It could also be used directly by the human expert, to influence understanding and the heuristics that would apply if the ES did not exist. As more data are incorporated into the man-machine system, opportunity progressively exists to increase the degree to which causal relationships are taken into account. Over time the gap between the expert system and the formal mathematical models of processes should be able to be eliminated.

In the operational context, a realistic goal is the development of expert systems as components in an integrated management information system. The ES would support the skilled plant manager, and he or she would contribute to its continual development. It would be particularly useful to the less experienced plant managers, as a practical learning aid. In both cases, the ES would provide reminders, and thus minimize omissions; and would also reduce time spent on some tasks. Given that ES rules could be of questionable precision initially, and especially in unusual conditions where the risks are greatest, it would be unwise to rely too much on its advice. Thus, other information system components need to be enhanced as backup facilities. Over time, the links between other components and the ES itself will be progressively strengthened. In short, the ES would be a useful addition to the plant manager's information support facilities. It cannot, however, function as a replacement for the plant manager given the current state of knowledge.

It is highly unlikely that, in the foreseeable future, there will be the need or desire to replace engineering professionals with less expensive and well-trained staff supported by a computer-driven expert system. Second, plants are so varied in characteristics that a prolonged period of experimentation at a large number of plants would be required before such an ES could be shown to be effective at all plants. Third, there are many plant tasks for which ES approaches are either infeasible or unsuitable, and thus the services of a professional engineer would still be required.

PROTOTYPE SYSTEM FOR WASTEWATER TREATMENT

A prototype statistics-based information management system was developed to improve everyday operation and control of the manually operated wastewater treatment process. The goal is to help the operator detect the problems in the process and to select appropriate control actions. An integration of a data base management system, a statistical analysis system, and a rule-based expert system provides a promising tool.

Several statistical methods have been investigated. These include discriminant functions for early detection of upsets, several kinds of control charts, and methods such as construction of external reference distributions and graphical methods. Some statistical methods have been incorporated into a user-friendly computer program that does data management and contains an expert system. The built-in statistical methods can be used to help define the fuzzy terms such as high and low for the conditions of the variables involved in the expert system's control rules. The system is written in dBase-III and Lotus-123 and has a user-friendly interface.

Information management includes data entry, data storage and retrieval, and data interpretation. Figure 1 shows the general structure of the integrated statistical, data base, and expert systems methods. The expert system also has utility functions to help the user add, delete, or modify the rules. A simulation environment that can be used to help the operator derive an appropriate definition of the fuzzy terms in the control rule is under development.

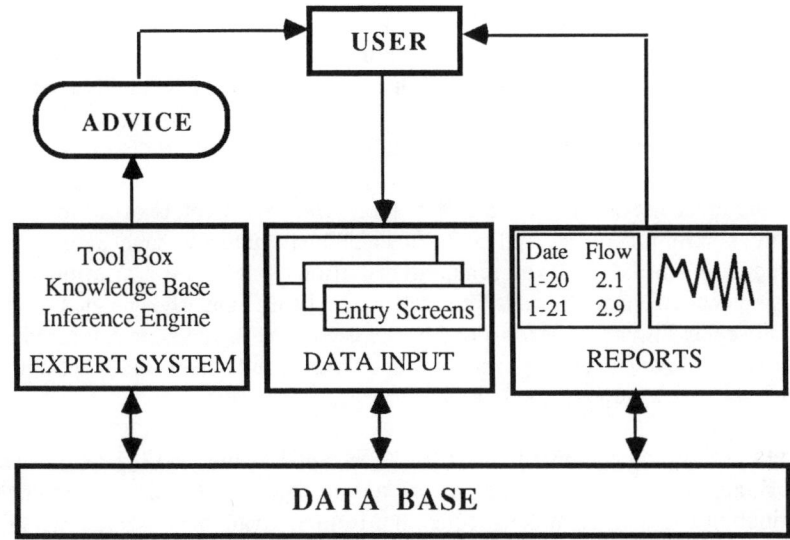

Figure 1. Structure of the Management Information System

Data storage and retrieval have been the usual focus of computer systems. Data analysis too often has been neglected, unfortunately, since this is the step that converts raw data into information. A valuable and frequently sufficient interpretative step is making suitable plots of the data. Our approach gives it considerable emphasis and importance.

Data Entry, Editing, and Retrieval

The data entry system facilitates entering, storing, and retrieving measured and computed variables. In addition to these functions, the data entry system provides the necessary links to the data analysis system and the expert advisory system. These two program systems are discussed separately in later sections. A more detail description of the data entry system can also be found in Berthouex et al. (1987, 1989) and Berthouex and Lai (1988).

The Data Analysis System

The data analysis system allows the operator to display data in either tabular or graphical form. The tabular form is intended for creating operating reports, for example to a state regulatory agency, or other summary or exception reports that can be used for evaluating the operator performance as well as the expert system's performance. In day to day operations a table of numbers is vastly inferior to graphical displays. The human eye and brain are intuitively skillful at recognizing patterns in graphical information.

The most useful displays are X-Y plots, time series plots of up to three variables simultaneously, and plots of calculated statistics such as moving averages, cusums, and reference distributions. Natural variation and measurement errors can be quite large relative to the true underlying signal so the task of seeing important patterns in the data is eased if the data are smoothed in some way. Moving average plots (7-day and 30-day) and exponentially weighted moving averages are very good for this purpose.

External reference distributions (Berthouex and Hunter 1983; Box, Hunter and Hunter 1978) can be prepared for any variable of interest. A reference distribution is a histogram constructed, not from all available observations, but from observations on an interval of operation selected to represent a condition of particular interest. Thus, it is actually a more powerful statistical tool than a simple histogram. A reference distribution might be prepared, for example, for all summertime daily effluent values from periods when the plant was in compliance with discharge permit requirements. Or, it could be constructed for weekly averages, for a particular day (e.g. Sunday), for k-day moving averages, or for differences between successive observations or between two variables. It also can be used for computed variables, which may have complex statistical properties.

The theoretical statistical properties of such quantities usually will not be known, and would be difficult to derive. The external reference distribution is constructed without making any assumptions about the properties of the data (normality, serial correlation, etc.). It, therefore, incorporates all these factors and provides a complete and honest picture of the statistic of interest. The simplicity and apparent empiricism of the method should not detract from its statistical validity and usefulness.

The Rule-Based Expert System

Figure 2 shows the three major components of the expert system which are: 1) a knowledge base that stores the domain knowledge, 2) an inference engine that holds the procedures for applying the knowledge, and 3) a "tool box" that contains utilities for the expert system. These interface with the data analysis system in order, for example, to build reference distributions for variables of interest, or to generate reports that are useful for evaluating the performance of the operator as well as the expert system performance.

Using Reference Distributions to Define Fuzzy Terms

One of the major facilities in the tool box is to interface with the data analysis system to construct reference distributions so as to define the high or low conditions for the variables of interest. Since reference distributions are constructed from the plant's own historical data, this method can reflect the particular characteristics of the plant and provide a more objective and suitable manner in defining the fuzzy levels for the variables involved in the control rules.

A detailed description of the procedures for building reference distributions was given in Berthouex and Lai (1988). In summary, to generate reference distributions, the tool box: 1) accesses the historical data from the database; 2) displays time-series plots of the data; 3) identifies the periods when the operation is under control based on the quality of the effluent data; and finally 4) generates reference distributions for variables of interest based on good periods of data.

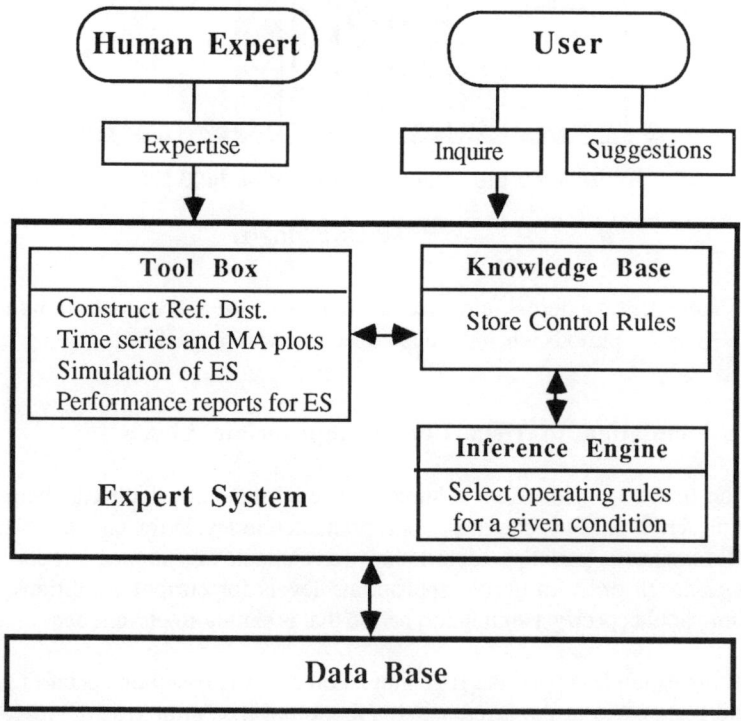

Figure 2. Structure of the expert system.

Figure 3 shows a reference distribution of the daily MLSS data. The median, upper, and lower 5% levels are automatically produced by the tool box. The median represents the target condition of the variable, based on past conditions, while the 5% upper or lower levels indicate that conditions are extraordinary and they might be used as the levels for defining the condition is "high" or "low".

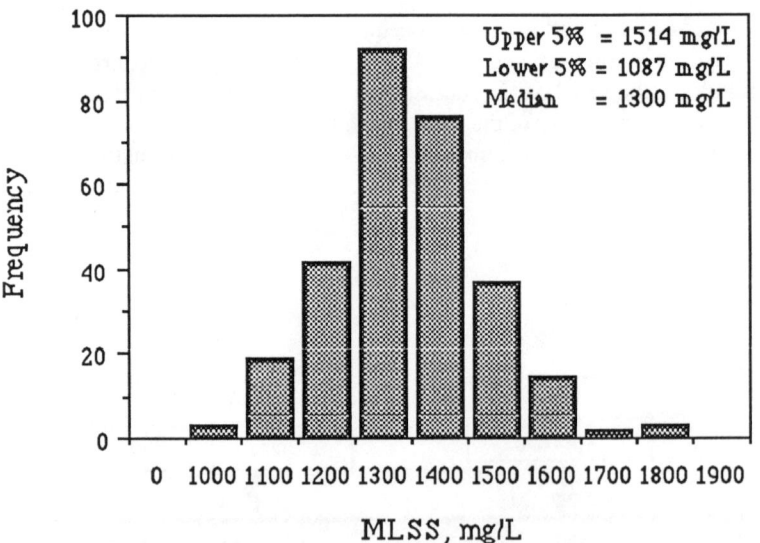

Figure 3. An external reference distribution of MLSS concentration for periods when the plant operation was normal.

Using Simulation to Help Derive Appropriate Levels

The tool box also offers a simulation environment which can help the operator derive possibly the most appropriate action levels for the variables of interest for specific periods. Table 1 shows an example of a summary report of a simulation. In order to derive appropriate levels for current conditions, the operator should specify a simulation period that is similar to current conditions.

In the example report, the first simulation run was based on specifying the top and bottom 5% of the reference distribution of the variable as the high and low conditions. With this definition, the ES missed giving a warning or suggesting a control action 15 times, and once it suggested taking action when it should not have. A second calibration run was done using the upper 10% and lower 5% critical levels for the variable MLSS. The definition of the critical levels for the other variables were unchanged. Additional settings were explored until in the fourth calibration run, the definitions of the critical MLSS concentrations were set at the lower 10% and the upper 20% levels.

The performance index shown in the last column of the Table 1 was calculated by a simple utility function: $P_j = W1\, N1_j + W2\, N2_j$, where P_j denotes the performance index based on run j, W1 is the relative weight for miss, W2 is the weight for a false alarm, and $N1_j$ and $N2_j$ are, respectively, the number of misses and false alarms for run j.

The performance index serves as reference guide for the operator. Basically, a lower index represents consistently better operating decisions. Hence, in this example, assuming that $W1 = W2 = 0.5$, run 3 would be the choice and the upper 15% and lower 5% critical levels would be used as the high and low conditions for the variable MLSS.

Table 1. Example of a Summary Report for a Sample Simulation

Run	Misses	False-alarms	Index
1	15	1	8
2	7	1	4
3	2	2	2
4	1	7	4

Using Performance Reports to Improve Control Rules

Occasionally, the operator may request the expert system to generate a performance report which lists the control action actually taken by the operator and the actions suggested by the ES. The tool box is designed to allow the operator to obtain a performance report for any variables and periods that are of interest.

This utility can help the operator evaluate his own past performance as well as the ES's performance, since it displays both the actual and suggested actions and, if desired, other important data, including those that were are not available at the time when the operator was making the daily decision. Based on this information, the operator can be more objective in judging the quality of his decisions and the expert system's decisions. The information provided should be useful for continually improving and refining the control rules.

Using an Exception Reports to Resolve Conflicts

As additional, and perhaps more complex, rules are collected and stored in the system, it is possible that in some cases the expert system may suggest actions that are contradictory. For example, it might suggest, based on one triggered control rule, that the operator increase the level of variable A while simultaneously invoking another rule that suggests decreasing the level of A. To help the operator resolve this type of conflict, the tool box can generate a statistical report that gives the number of times in the past that this problem occurred, the dates when it occurred, and the action that was actually taken. The operator can use this type of report to investigate the causes and the results of the conflicts, and hopefully,to resolve the problem by modifying the existing rules or adding new hierarchical rules to the ES knowledge base.

Evolution of the Expert System

An expert system should always be regarded as unfinished. Experimentation and experience should gradually increase knowledge about the system and, gradually, this knowledge should be incorporated in an up-dated the rule base. The strategy for development of action rules consists of their progressive refinement through an interactive process between the expert system module and competent plant operators. The development process might proceed as illustrated in Figure 4. The curve depicts the true (but unknown) relationship between two monitoring variables, $MV(1)$ and $MV(2)$, and a control action X. At an initial stage experts might be able to do no better than to approximate it by the rule
   IF $MV(1) > 10$ and $MV(2) > 2500$ THEN do X.
By continuing interaction, and by evaluation of performance, it will become evident that improvements are possible and the step-wise approximations in the expert system will more closely approach the true response curve.

TESTING THE PERFORMANCE OF THE EXPERT SYSTEM

Having developed a rule base, the problem of evaluating the expert system arises. Some issues involved in testing an expert system are discussed and a very preliminary test result is given to illustrate an approach that may be helpful.

Two kinds of evaluations are needed. The first is to check that the computer system reproduces the advice of the domain expert. This is straightforward and while essential is not sufficient. It is based on the assumption that the 'expert' who helped build and calibrate the system (set the state limits) is in fact highly skilled and that his knowledge and judgment have been captured in the rule base.

Even if this is true, we ultimately want the expert system to do more. We want it to improve the operation of the treatment plant. The second kind of test seeks to establish whether this is accomplished. Lacking identical parallel plants, one controlled by the expert and one by the computer system, this is quite difficult to arrange. Various approaches to this kind of evaluation have been considered but they will not be discussed here. An evaluation of the first kind, the knowledge capture/calibration problem, is presented.

The purpose of the evaluation is to examine the accuracy of the expert system, where the degree of accuracy is defined as the percent of cases where the expert system advice agrees with the control action actually taken by the plant operators. The evaluation is retrospective, the expert system was used to analyze

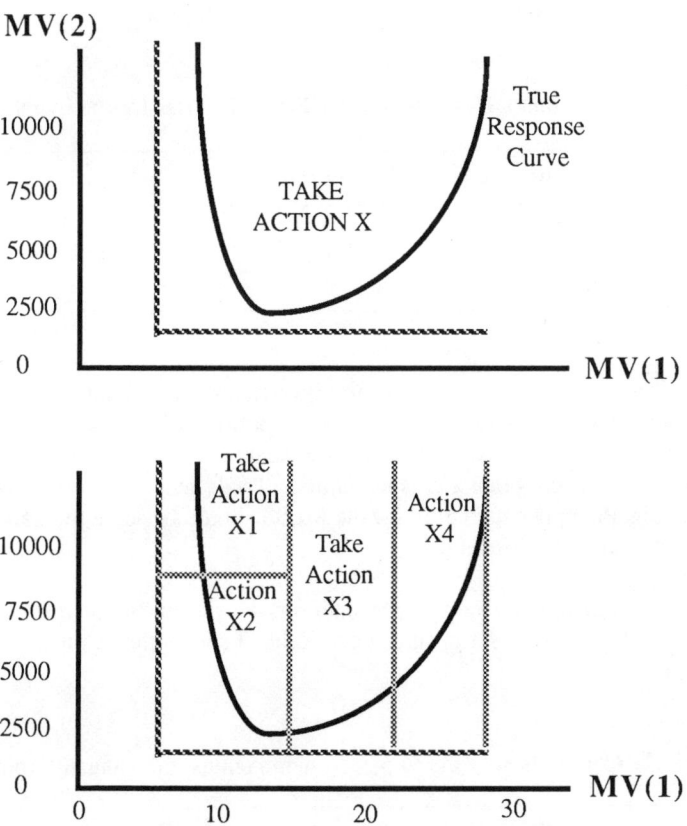

Figure 4. Progressive improvement of the rule base should be possible as the relationship between monitoring variables is discovered by experimentation.

a series of historical data and the advice offered was compared with the record of control actions taken at the plant. A key working assumption is that the plant was operated by good operators (experts). This assumption was accepted after interviewing the operators and evaluating the general quality of their performance.

A sequence of one hundred daily data records were used for the evaluation. These records were divided into two sets of fifty records each, called the estimation set and the test set. As shown in Table 2, in the estimation set there were control actions taken in 16 of the the 50 cases, and 34 cases with no action taken. On the other hand, in the test set, there were control actions taken in 24 of the 50 cases, leaving 26 with no action taken. We are interested in how often the expert system recommended action when action actually had been taken, and how often it agreed with the expert that no action was needed.

Table 2. A Preliminary Evaluation of the Problem Cases for the Evaluation

Decision	Estimation	Test
Action	16	24
No Action	34	26
Total	50	50

In the first stage of the evaluation process, reference distributions for each of the variables were constructed based on the periods of data when the plant operation was stable and the system was run against the estimation set. This is an opportunity to calibrate the state limits. Trials using the estimation set indicated that the upper state limit for the MLSS should be set at the upper 20% point on the reference distribution.

The second stage of evaluation was to run the system on the test set using the state limits that were set in stage one. Table 3 shows the evaluation results for the test set.

Table 3. Comparison of the Expert System Against the Human Experts

Decision	Human Expert	Expert System	Agreement
action	24	22	92 %
no action	26	21	81 %
total	50	43	86 %

In 43 of the 50 cases (86%), the suggestions given by the system were identical to the human expert. Of the remaining seven cases, two were due to the system's inability to suggest any actions where in reality the expert did take control actions. In subset of the cases in which the expert actually did take control actions, there were only 2 out of the 24 cases in which the system missed suggesting the appropriate control action to the user. A similar analysis was applied to the 26 problem cases in which there no action actually was taken. In five of these cases, the system suggested taking action.

SUMMARY

This paper describes work toward development of an integrated system to improve everyday operation and control of the wastewater treatment process. The goal is to help the treatment plant operator make a timely detection of operating problems and to select appropriate control actions.

Statistical methods, with emphasis on plots of various forms external reference distributions, have been integrated with program modules for data management and an expert system. The built-in statistical methods can be used to help define the fuzzy terms such as high and low for the conditions of the variables involved in the expert system control rules. The system is written in dBase-III and Lotus-123 and has a user-friendly communication interface.

ACKNOWLEDGEMENT

Although the research described in this article has been funded in part by the United States Environmental Protection Agency under assistance agreement CR–812655–01–0 to The University of Wisconsin-Madison, it has not been subjected to the Agency's peer and administrative review and therefore may not necessarily reflect the views of the agency, and no official endorsement should be inferred.

REFERENCES

Berthouex, P. M., W.G.Hunter and Lars Pallesen. 1981. Wastewater Treatment: A Review of Statistical Applications, *Environmentrics 81, Selected Papers,* SIAM, Philadelphia, 77-99.

Berthouex, P. M. and W. G. Hunter. 1983. How to construct reference distributions to evaluate treatment plant effluent. *Journal WPCF*, 55, 1418-1424.

Berthouex, P. M., Hunter, W. G., and Fan, R. 1985. Characterization of

Treatment Plant Upsets. *Instr. and Control of Water and Wastewater Treatment and Transport Systems*. Pergamon Press, London

Berthouex, P.M., W. Lai and A. Darjatmoko. 1987. A Statistics-Based Information and Expert System for Plant Control and Improvement. In *Proceedings of 5th National Conference on Microcomputers in Civil Engineering*. ASCE. Orlando. Fl., pp.146-150.

Berthouex, P.M.and W. Lai. 1988. A Statistically-Aided Expert System for Wastewater Treatment Process Control. In *Proceedings of the 1988 ASME Computers in Engineering Conference* ASME. San Francisco. Ca 199-203.

Berthouex, P.M., Wenje Lai, and A. Darjatmoko 1989. Statistics-Based Approach to Wastewater Treatment Plant Operations, *Jour. Envir. Engr. Div., ASCE*, 115, 650-671.

Box, G.E.P., W.G.Hunter and J. Stuart Hunter. 1978 *Statistics for Experimenters*, John Wiley and Sons, New York.

Hunter, W.G. 1982. *Environmental Statistics Encyclopedia of Statistical Sciences*, Vol. 2, Kotz and Johnson, eds., John Wiley and Sons.

Johnston, D.M. 1985. Diagnosis of Wastewater Treatment Processes, In *Proceedings, Computer Applications in Water Resources*, ASCE, 601-606.

Maeda, K., 1985. *An Intelligent Decision Support System for Activated Sludge Wastewater Treatment Process*, In Instr. and Control of Water and Wastewater Treatment and Transport Systems, Drake, R.A.R., Ed., Pergamon Press, New York, 629-632.

Magdol, J. 1984. Heuristic Control of a Wastewater Treatment Process, M.S. Thesis, Univ. of Wisconsin-Madison

Mylopulos, J. and J.H. Levesque. 1984. An Overview of Knowledge Representation, In *On Conceptual Modelling*, Brodie, M., J. Mylopulos and J. Schmidt, Ed., Springer-Verlag, New York, 3-17.

Schaeffer, D.J. et al. 1984. Municipal Compliance - Another View, *Journal WPCF*, 56, 8, 924-927.

Tong, M.R., B. M. Beck and A. Latten. 1980. Fuzzy Control of the Activated Sludge Wastewater Treatment Process, *Automatica*, 16, 695-701.

Ucbrin, C.G. and R.L. Caspe. 1983. Operational Characteristics of Federally Funded Wastewater Treatment Plants, *Journal WPCF*, 55, 935- 940.

Waterman, D.A. 1986. *A Guide to Expert Systems*, Addison-Wesley, Reading, MA.

Vitasovic, Z. and Andrews, J.F., 1987. *A Rule-Based Control System for the Activated Sludge Process*, In Systems Analysis in Water Quality Management, Beck, M.B., Ed. Pergamon Press, New York, 423-432.

Zadeh, L.A. 1965. Fuzzy sets, *Inf. & Control*, 8, 338-353.

Zadeh, L.A. 1973. Outline of a New Approach to Analysis of Complex Systems and Decision Process, *IEEE Trans. Sys, Man Cybernet.*, 28-44.

Zadeh, L.A. 1983. Commonsense Knowledge Representation Based on Fuzzy Logic, *IEEE Computer, Oct.*, 61-65.

About the Authors

A. Terry Bahill received a B.S. in Electrical Engineering from the University of Arizona, Tucson, in 1967, an M.S. in Electrical Engineering from San Jose State University, in 1970, and a Ph.D. in Electrical Engineering and Computer Science from the University of California, Berkeley, in 1975.
He served as a Lieutenant in the U.S. Navy, and as an Assistant and Associate Professor in the Departments of Electrical and Biomedical Engineering at Carnegie Mellon University, and Neurology at the University of Pittsburgh. Since 1984 he has been a Professor of Systems and Industrial Engineering at the University of Arizona in Tucson. His research interests include systems engineering theory, modeling physiological systems, head and eye coordination of baseball players, and expert systems. He has published over 100 papers and has lectured in a dozen countries. Dr. Bahill is a member of the following IEEE Societies: Systems, Man, and Cybernetics; Engineering in Medicine and Biology; Computer; Automatic Controls; and Professional Communications. For the Systems, Man, and Cybernetics Society he has served three terms as vice president, six years as associate editor, as Program Chairman for the 1985 conference in Tucson, and as Co-chairman for the 1988 conference in Beijing and Shenyang, China. He is an associate editor for the Computer Society's magazine, *IEEE Expert*. He is a member of Tau Beta Pi, Sigma Xi, Psi Chi, and is a Registered Professional Engineer.

Judith Barlow is a Visiting Research Fellow in the Computer Science Department at the University of Liverpool, Commission of the European Communities Senior Research Assistant, and Faculty Associate in Advanced Systems Research at US WEST Advanced Technologies, Englewood, Colorado. She has authored published articles in the fields of multicriteria optimization, expert systems, hypertext and information retrieval.

Colin E. Bell is Professor and Chairperson in the Department of Management Sciences at the University of Iowa. His research areas include optimal operation of congestion systems and knowledge-based approaches to planning and scheduling problems. In 1984-85, Dr. Bell was a visiting research scholar in the Department of Artificial Intelligence and the Artificial Intelligence Applications Institute at Edinburgh University. He has published widely, particularly in *Management Science* and *Operations Research*. He is an associate editor of *Naval Research Logistics*.

P.M. Berthouex is Professor of Civil and Environmental Engineering, University of Wisconsin-Madison. His work is mainly in water pollution control and statistical methods for environmental engineering.

Raj Bhatnagar is currently a graduate student in the Computer Science Department of the University of Maryland, College Park, and is involved in research relating to reasoning with uncertain knowledge.

Donald E. Brown received the B.S. degree from the U.S. Military Academy, West Point, the M.S. and M.Eng. degrees from the University of California, Berkeley, and the Ph.D. degree from the University of Michigan, Ann Arbor. He is an Assistant Professor of Systems Engineering and Associate Director of the Institute of Parallel Computation at the University of Virginia. He also serves as an advisor to the Foreign Science and Technology Center. His research interests are in data analysis, the integration of artificial intelligence and operations research techniques for decision aiding.

Tony Cox is Technical Director of Modeling at US WEST Advanced Technologies. He leads an applied research team of operations research and computer science specialists who provide analytic modeling support to US WEST business and engineering decision makers in the areas of work flow simulation, resource allocation, network design and performance evaluation, and microeconomic and marketing modeling applications. The group also conducts original research in advanced combinatorial optimization algorithms, automated modeling, knowledge representation, and queuing networks. Dr. Cox holds an A.B. from Harvard University (1978) in Mathematical Economics and received his S.M. in Operations Research (1985) and his Ph.D. in Risk Analysis (1986) from MIT's Department of Electrical Engineering and Computer Science. His current research is focused on integrating and extending artificial intelligence and operations research techniques for representing, reasoning about, and managing uncertainties in applied decision-making and risk analysis.

James R. Evans is Professor and Head of the Department of Quantitative Analysis and Information Systems in the College of Business Administration at the University of Cincinnati. Professor Evans' research interests lie in the areas of combinatorial optimization and mathematical programming, as well as applications of operations research, operations management, and artificial intelligence. Dr. Evans is the author or co-author of several books and over sixty papers in the professional literature.

Peter H. Farquhar is Associate Professor of Industrial Administration and Director of the Center for Product Research at Carnegie Mellon University. He received his Ph.D. in Operations Research from Cornell University and

ABOUT THE AUTHORS

has been on the faculties of Northwestern University, Harvard University, and the University of California at Davis. Dr. Farquhar was also a member of the technical staff at the RAND Corporation. Dr. Farquhar is known for his research work in decision analysis, multiattribute utility theory, new product development, consumer behavior, and marketing strategy. His current research focuses on adaptation in temporal decision making and on the design of heuristic evaluation functions for machine learning. He is the Area Editor for Decision Analysis, Bargaining, and Negotiation for *Operations Research* and an Associate Editor for the *Journal of Forecasting* and *Information and Decision Technologies*.

Fred Glover is the US West Chair in System Science and Research Director of the Center for Applied Artificial Intelligence at the University of Colorado, Boulder. He has authored or co-authored more than one hundred ninety published articles in the fields of mathematical optimization, computer science and artificial intelligence. Professor Glover is the first US West Distinguished Fellow, and is also an honorary fellow and reward recipient of several national societies and research organizations, including the American Association for the Advancement of Science, Alpha Iota Delta, NATO, IBM, the Energy Research Institute, the Decision Sciences Institute, the Institute of Management Sciences College of Practice and Miller Institute for Basic Research in Science. He has served on the boards of directors of several companies and consults widely for industry and government.

Othar Hansson is a Ph.D. candidate in the Department of Electrical Engineering and Computer Science at the University of California, Berkely. In 1986, he earned his A.B. in English Literature and Computer Science from Columbia College, where he was a John Jay Scholar. In 1988, he received his M.S. in Computer Science from the University of California, Los Angeles, where he was a University Fellow and a Rand Corporation Fellow.

Michael Hilliard, a Phi Beta Kappa graduate of Furman University with a B.S. in Mathematics, received his M.S. and Ph.D. in Operations Research from Cornell in 1983, specializing in Optimization. As a Research Staff member at the Oak Ridge National Laboratory, he has pursued an interest in combining techniques from operations research and artificial intelligence to solve practical problems such as scheduling. For the past four years he has concentrated on techniques from the machine learning literature and particularly genetic algorithms. Currently he is working with a team of researchers to develop a scheduling system for the Military Airlift Command.

Don Hindle has a management consulting firm, Hindle and Associates, Canberra, Australia. He has worked in many countries, most recently Brazil, Indonesia, Egypt, and Portugal, mainly on health systems planning and management.

Gerhard Holt is an associate in the Financial Analytics and Structural Transactions Group at Bear, Stearns Inc. He earned his A.B. from Columbia College in 1985, and his M.S. and M.B.A. from Columbia University in the following year. Prior to joining Bear, Stearns, he worked for Paribas as a quantitative analyst in capital markets. His interests include decision making and risk evaluation in real world, financial and consumer domains.

Musar J. Jafar was born in Tyre, Lebanon, on July 3, 1953. He received a B.S. in Mathematics from the Hygazian College Beirut, Lebanon, in 1977, an M.S. in Mathematics from the American University of Beirut, in 1982, and M.S. in Systems Engineering from the University of Arizona, Tucson, in 1985, and a Ph.D. in Systems and Industrial Engineering from the University of Arizona, in 1989.
His research interests include systems engineering theory, probabilistic modeling and artificial intelligence applications in the area of man-machine systems such as expert systems. He has taught mathematics, statistics and computer science at Pima Community College and the University of Arizona. He is currently a knowledge engineer with BICS (Tucson, Arizona).

Laveen Kanal is a Professor of Computer Science at the University of Maryland, and Managing Director of LNK Corporation. His research interests include artificial intelligence, pattern recognition, computer vision, and artificial neural networks.

Andrew Kusiak is a Professor and Chairman of the Department of Industrial and Management Engineering at the University of Iowa. He holds the Ph.D. degree in Operations Research (Polish Academy of Science), M.Sc. degree in Mechanical Engineering, and B.Sc. degree in Precision Engineering (both from the Warsaw Technical University). Andrew Kusiak is interested in applications of artificial intelligence and optimization in modern manufacturing systems. He is the editor-in-chief of the *Journal of Intelligent Systems*.

Wenje Lai has a Ph.D. in Industrial Engineering from the University of Wisconsin-Madison and works on data-base management and expert systems for AT&T Bell Laboratories, Napierville, Illinois.

ABOUT THE AUTHORS

Hau L. Lee is an associate professor in the Department of Industrial Engineering and Engineering Management of Stanford University. He received his M.Sc. in Operations Research from the London School of Economics, and his Ph.D. in Operations Research from the Wharton School of the University of Pennsylvania. His research interests include production planning, inventory control, quality assurance and health care management. Prior to joining Stanford, Dr. Lee was on the faculty of the Department of Decision Sciences at the Wharton School, University of Pennsylvania. He is currently a Member of AIIE, TIMS, ORSA, and the Institute of Statisticians.

Gunar Liepins received his A.B. in Mathematics from the University of California, Berkeley in 1969, and his Ph.D. in Mathematics from Dartmouth College in 1974. He taught for two years as a Visiting Lecturer in mathematics at Texas Tech University before returning to Stanford University, where he earned an M.S. in Engineering Economic Systems in 1977. Since that time he has been employed at Oak Ridge National Laboratory where he has pursued research in data quality control, computer security, optimization, and machine learning. He serves concurrently as an Adjunct Associate Professor in Computer Science at the University of Tennessee and teaches courses in artificial intelligence, neural networks, genetic algorithms, and machine learning. He has published more than 50 technical papers and journal articles in statistics, operations research, computer security, and artificial intelligence.

Stuart H. Mann is Professor of Operations Research in the School of Hotel, Restaurant and Institutional Management at the Pennsylvania State University. He has a Ph.D. and an M.S. in Operations Research from Case Western Reserve University, and a B.S. in Mathematics from the University of Illinois. His research interests are in the application of decision-making methods to problems in the hospitality and service industries. He is an active consultant in the strategic management of retails operations.

Andrew Mayer is a Ph.D. candidate in the Computer Science Division of University of California at Berkeley, and a member of the AI research staff at the NASA Ames Research Center. He received A.B. and M.S. degrees in Computer Science from Columbia and Berkeley respectively. Currently, he is studying the application of probabilistic inference and decision-theoretic control to develop new resource-bounded search techniques for AI, and exploring the compilation of such techniques as a general method for designing efficient algorithms.

Daniel E. O'Leary is on the faculty of the University of Southern California in the Graduate School of Business. He received his Ph.D. from Case Western Reserve University and his masters degree from the University of Michigan. Professor O'Leary's research areas include validation and verification of expert systems, methodologies of expert systems and representation of uncertainty in expert systems. Professor O'Leary is the editor of the *Expert Systems Review*. He has published a number of papers in the area of expert systems and artificial intelligence. Those papers have appeared in journals such as *European Journal of Operational Research, Decision Sciences, Annals of Operations Research, Interfaces, International Journal of Man-Machine Studies, International Journal of Expert Systems: Research and Applications* and *IEEE Expert*. Professor O'Leary also has published a number of papers in edited collections and has presented papers at conferences such as IJCAI, Conference for Knowledge Acquisition for Knowledge-based Systems.

Mark R. Palmer earned a B.S. in Mathematical Science from Memphis State University in 1982 and an M.S. in Computer Science from the University of Tennessee in 1987. He has been employed since 1984 as a Research Assistant and then Research Associate with the University of Tennessee. His research interests include design techniques for large scale software projects, non-deterministic optimization techniques and machine learning.

Panos Pardalos received his Ph.D. degree in Computer Science from the University of Minnesota in 1985, and his B.S. degree from Athens University, Greece in 1978. He is the author (and co-author) of over forty publications in the area of mathematical programming, graph algorithms, and parallel computation. He coauthored (with J.B. Rosen) the book Global Optimization: Algorithms and Applications: Springer-Verlag, Lecture Notes in Computer Science 268 (1987) and (with A. Phillips and J.B. Rosen) the book Parallel Computing in Mathematical Programming: SIAM Publications (1989). He is also co-editor (with J.B. Rosen) of a special issue of *Annals of Operations Research* on "Computational Methods in Global Optimization." Since 1985, Dr. Pardalos is an Assistant Professor of Computer Science and Operations Research at the Pennsylvania State University. He is a member of ORSA, SIAM, AMS, ACM and Mathematical Programming Society.

Elizabeth Paté-Cornell is an Associate Professor in the Department of Industrial Engineering and Engineering Management at Stanford University. Her field of expertise includes risk analysis, decision analysis, and engineering economy, with emphasis on engineering risk management problems of the public sector. She obtained her B.S. in Mathematics and Physics in 1968 in Marseilles, France, an Engineer Degree in Computer Science and Applied Mathematics in 1971 in Grenoble, France, a Master's Degree in Operations

Research in 1972 at Stanford University, and a Ph.D. in Engineering-Economic Systems in 1978, also at Stanford University. Professor Paté-Cornell taught at MIT from 1978 to 1981 in the Department of Civil Engineering. Her research in recent years has focused on the analysis and the optimization of warning systems with application to a large variety of problems. Her current work involves the extension of risk analysis methods to include organizational factors in the assessment of engineered systems reliability. This work was recently applied to the management of offshore platforms and of the thermal protection system of the space shuttle.

Roger Alan Pick is an Assistant Professor of Information Systems at the University of Cincinnati. He earned a Ph.D. in Management Science with a major in Systems Analysis and Computer Science from Purdue University in 1984. He also holds an M.S. in Mathematics from Purdue and a B.S. in Mathematics from Oklahoma University. His papers have appeared in Communications of the ACM, Journal of MIS, and a number of conference proceedings. He was born in Ithaca, New York in 1955.

Stephen D. Post, president and principal scientist of Applied Information Technology, specializes in creating computer systems that solve client problems through a combination of operations research, artificial intelligence, decision theory, and computer science. During his 16 years as a computer scientist he has developed systems for medical, manufacturing, and C3I applications, expert system tools and applications, and natural language processing tools and applications. He was most recently director of artificial intelligence applications at Planning Research Corporation. He received a B.S. in Secondary Education, Mathematics, from the University of Maryland in 1972, an M.S. in Systems Management from the University of Southern California in 1984, and a Ph.D. in Information Technology and Engineering from George Mason University in 1989.

J. Richardson received his B.S. in Mathematics from the University of Alabama, Tuscaloosa in 1986. From 1986 until 1988 he worked as a research assistant at Oak Ridge National Laboratory. Currently, he is studying for his Master's degree in Computer Science at the University of Tennessee.

David A. Schum has since 1985 been member of the faculty of the Department of Operations Research and Applied Statistics in the School of Information Technology and Engineering at George Mason University. Prior to 1985 he held the rank of professor in the Departments of Psychology and of Mathematical Sciences at Rice University. His major research interests concern both formal and behavioral issues in the study of inferential reasoning.

Margaret M. Sklar is an Assistant Professor of Information Systems at Northern Michigan University and a Ph.D. candidate at the University of Cincinnati. She earned an M.A. in Computer Science from the University of Detroit and a B.A. in Mathematics from Rosemont College.

Katia P. Sycara is a Research Scientist in the School of Computer Science at Carnegie Mellon University. She is also the Director of the Laboratory for Integrated Manufacturing Systems Architecture. She received her B.S. in Applied Mathematics from Brown University, M.S.s in Applied Mathematics and Electrical Engineering from the University of Wisconsin, and Ph.D. in Computer Science from the Georgia Institute of Technology. She has served as Head of the Computing Section at the Center of Planning and Economic Research, a government research institute, in Athens, Greece. While there, she participated in the development of econometric models of the Greek Economy and acted as Liaison for Regional Planning with the European Economics Community. Her research interests include knowledge-based systems, case-based reasoning, distributed systems, negotiation models for multi-agent planning, and constraint-directed reasoning. She is particularly interested in integrating Operations Research and Artificial Intelligence methods in addressing complex problems. She is a member of the American Association of Artificial Intelligence, ACM, the IEEE Computer Society and the Institute for Management Science.

Evangelos Triantaphyllou is a computer consultant and a research member in the Center for Academic Computing at the Pennsylvania State University while he is working on his dissertation for the Dual Ph.D. in Operations Research and Industrial and Management Systems Engineering at the same university. He received the Diploma in Architectural Engineering from the National Technical University of Athens, Greece, in 1983. He received a Dual M.S. degree in Operations Research and Man-Environment Relations and an M.S. degree in Computer Science from the Pennsylvania State University in 1984 and 1988, respectively. His research interests are expert systems, decision support systems, fuzzy sets, and the application of optimization techniques to logical problems.

G. Brian Vesprani is currently a Systems Analyst for the Procter & Gamble Company. He received a B.S. in Computer Science from the University of Cincinnati in June of 1986 and has completed course work for an M.S. in Quantitative Analysis/Information Systems from the University of Cincinnati.

Chelsea C. White received his Ph.D. from the University of Michigan in 1974 in the Computer, Information, and Control Engineering Program. He initially was a faculty member at Southern Methodist University. Since 1975, he has been a member of the University of Virginia faculty, where he is currently Professor and Chairman of Systems Engineering. He has authored numerous publications on the control of finite stochastic systems and knowledge-based decision support systems with application to health care, strategic planning, and military decision-making. He is co-author (with A.P. Sage) of the second edition of Optimum Systems Control (Prentice-Hall, 1977). He is an associate editor of the following journals: *Operations Research*, the North-Holland journal *Information and Decision Technologies* (formerly *Large Scale Systems: Theory and Applications*), the *American Journal of Mathematical and Management Sciences*, and the IEEE *Transactions on Systems, Man, and Cybernetics*. He is Chairman of the IEEE Systems, Man, and Cybernetics Technical Committee on Optimization, a Fellow of the IEEE, and a full member of ORSA. His research interests include stochastic optimization and the integration of formal reasoning techniques and concepts in artificial intelligence for problem-solving with application to health care, military decisionmaking, and strategic planning.

Fatemeh Zahedi is Associate Professor at the Department of Management Sciences, University of Massachusetts/Boston. She has received her doctoral degree in Decision Sciences from Indiana University. Her papers have been published in a number of journals including *Operations Research, Decision Sciences, Computers and Operations Research, Interfaces, Journal of Operational Research, Mathematical Modelling, MIS Quarterly, Review of Economics and Statistics*, and *Socio-Economic Planning Science*. Her main research interest is in combining quantitative and qualitative information in modeling decision problems. She has served as the vice president and president of the TIMS Boston Chapter.

Index

A* algorithm, 13
adjudication, 249
applications, 449-495
arbitration, 249
artificial neural networks, 173
automated manufacturing, 451
belief structure, 267
branch-and-bound, 12, 13
Case-Based Reasoning (CBR), 251, 253, 256
Case Knowledge Base, 253
causality, 88, 101, 102
causal structures, 84, 102, 125
circular reasoning detection, 411
classification entropy, 438
classification tree, 432
classifier system, 61, 427
combining probabilities, 398
complexity, 427
complexity management, 387
complexity measures, 417
complexity, measurement of, 389, 415
computational complexity, 432
concurrent verbalization, 286
conflict resolution, 68, 218
context focusing, 286
cyclomatic number, 420
decision analysis, 167-214, 215-248
decision support, 81, 215-248
decision support systems, 141-165, 167-214
decision theory, 91
decomposition principle, 414
default reasoning, 87, 171
default rules, 169
Dempster's rule, 117
Dempster-Shafer, 118, 191
diagnostic tasks, 106
dispute mediation, 249

distributive justice, 266
EMYCIN, 398
equity theory, 266
evidential reasoning, 116-120
expert classification systems, 427-446
failure diagnosis, 81, 141
frames, 283
fuzzy logic controller, 484
fuzzy sets, 169, 197
fuzzy set theory, 173, 191
fuzzy systems, 117
fuzzy, 109, 123
generalized classification problem (GCP), 430
genetic algorithms, 9, 10, 29-58, 61
group decision support, 218
heuristic search, 7-78,
Human Rationality Assumption, 201
imprecise reasoning, 167-214
incidence matrix, 398
incomplete evidence, 105-139
inclusion-exclusion algorithm, 435
inconclusive evidence, 105-139
inference path, 409
information-seeking strategy, 427
inheritance networks, 284
integer programming, 169, 171-195, 326
integrated management information system, 481-496
intersection-sort heuristic, 443
knowledge acquisition, 285
knowledge engineer, 281
knowledge-based scheduling system (KBSS), 453
knowledge representation, 281
knowledge representation schema, 279-316
Least Exception Logic (LEL), 172, 173, 174
linear programming, 169, 171
logic, 171-195
machine learning, 10, 59-78
Markov model, 15, 16

Markov modeling, 9, 24
mathematical programming, 171, 169, 197-214, 387
mathematical programming and AI, 275-367
mediation, 249
mediation/negotiation, 249, 270
metaknowledge-base, 322
meta-OR, 317
meta-rule, 322
meta-rules for large-scale systems, 329
meta-rules for network design, 330
meta-rules for scheduling, 330
meta-rules for screening and assignment, 330
minimum expected-cost generalized classification problem (MEGCP), 430
multiattribute utlity theory, 9, 14, 21, 222-236, 249
multicriteria, 217, 237
negotiation, 249
nonmonotonicity, 193
non-monotonic logic, 171, 190
object classification, 427
optimal inference strategies, 390
organizational schema, 298
parallel architectures, 414
pattern recognition systems, 427
performance and complexity management of expert systems, 369-447
PERSUADER, 249, 269
preference analysis, 251, 261
primary effect, 406
probabilistic reasoning, 106
probabilistic risk analysis, 141-165
probabilistic search, 13
production rules, 284
project selection, 243
qualitative knowledge, 317
recency effect, 406
recursive partitioning, 427, 437
representation of ignorance, 402
retrospective verbalization, 287
risk analysis, 219, 237

search, 7-78
semantic LP domain knowledge, 282
semantic LP formulation knowledge, 282
semantic world knowledge, 282
set covering problems, 33
similarity class, 257
simulated annealing, 68
situation assessment, 251
Situation Assessment Packets (SAPs), 254
software engineering, 415
sorting heuristics, 427
structural analyses, 110-116
structured modeling, 321
supervised learning, 427
Syntactic Error Checker, 373
syntactic LP knowledge, 282
TABU search, 9
traveling salesman problems, 9, 29-58
truth maintenance, 193
uncertainty management, 79-166
utility, 11-28, 87, 91, 217, 219
utility theory, 217, 219-236
validation, 373, 390
Validator, 373
verification, 373
wastewater treatment, 481-496

DEC 5 1990